TRAFFIC FLOW AND TRANSPORTATION

The Fifth International Symposium on the Theory of Traffic Flow and Transportation

Under the auspices of
  The College of Engineering and the Institute of Transportation and Traffic Engineering, University of California, Berkeley
  with the collaboration of the
  Transportation Science Section, Operations Research Society of America

Sponsors
  National Science Foundation, Mathematical and Physical Sciences Division
  Automotive Safety Foundation
  Ford Motor Company
  General Motors Corporation
  International Business Machine Corporation
  Operations Research Society of America
  The Port of New York Authority

Co-Chairmen
  Adolf D. May and Gordon F. Newell

Advisory Committee

Leslie Edie	Alan Miller
Denos Gazis	Renfrey Potts
Alan Goldman	Tsuna Sasaki
Robert Herman	Reuben Smeed
Wilhelm Leutzbach	John Tanner

TRAFFIC FLOW
AND
TRANSPORTATION

PROCEEDINGS OF THE
FIFTH INTERNATIONAL SYMPOSIUM
ON THE THEORY OF
TRAFFIC FLOW AND TRANSPORTATION

Held at Berkeley, California, June 16–18, 1971

Edited by GORDON F. NEWELL

Professor of Transportation Engineering
University of California, Berkeley, California

AMERICAN ELSEVIER
PUBLISHING COMPANY, INC.

New York London Amsterdam

AMERICAN ELSEVIER PUBLISHING COMPANY, INC.
52 Vanderbilt Avenue, New York, N.Y. 10017

ELSEVIER PUBLISHING COMPANY
335 Jan Van Galenstraat, P.O. Box 211
Amsterdam, The Netherlands

International Standard Book Number 0-444-00128-X

Library of Congress Card Number 72-85806

Previous Symposia and Proceedings

I General Motors Research Laboratory, Detroit, Michigan, 1959: *Theory of Traffic Flow*, Robert Herman, Editor, Elsevier Publishing Company, Amsterdam, The Netherlands, 1961

II Road Research Laboratory, London, England, 1963: *Proceedings of the Second International Symposium on the Theory of Road Traffic Flow*, London 1963, Joyce Almond, Editor, Organisation for Economic Co-Operation and Development (OECD), Paris, 1965

III Transportation Science Section ORSA, New York, 1965: *Vehicular Traffic Science*, L. C. Edie, R. Herman, and R. Rothery, Editors, American Elsevier Publishing Company, Inc., New York, 1967

IV Technical University, Karlsruhe, Germany, 1968: *Beiträge zur Theorie des Verkehrsflusses*, Wilhelm Leutzbach and Paul Baron, Editors, Strassenbau und Strassenverkehrstechnik Heft 86, 1969

Contributors

RICHARD E. ALLSOP — Research Group in Traffic Studies, University College London, London, England

MARTIN J. BECKMANN — Technische Hochschule, Munich, Germany and Brown University, Providence, Rhode Island

HEINZ BEILNER — University of Stuttgart, Stuttgart, Germany

JAMES G. BENDER — Department of Electrical Engineering, The Ohio State University, Columbus, Ohio

BERNARD F. BYRNE — Towne School of Civil and Mechanical Engineering, University of Pennsylvania, Philadelphia, Pennsylvania
Present address: De Leuw Cather and Co., San Francisco, California

J. M. CLARK — Transport Assessment Group, Cranfield Institute of Technology, Cranfield, Bedford, England
Present address: Centre for Transport Studies, Cranfield Institute of Technology, Cranfield, Bedford, England

M. COHEN — Institut de Recherche des Transports, Arcueil, France

JOHN N. DARROCH — The School of Mathematical Sciences, The Flinders University of South Australia, Bedford Park, South Australia

JOHN W. DICKEY — Department of Civil Engineering, Virginia Polytechnic Institute and State University, Blacksburg, Virginia
Present address: Urban and Regional Planning Program

P. H. FARGIER — Institut de Recherche des Transports, Arcueil, France

ROBERT E. FENTON — Department of Electrical Engineering, The Ohio State University, Columbus, Ohio

NATHAN GARTNER — Technion-Israel Institute of Technology, Haifa, Israel
Present address: Department of Civil Engineering, Massachusetts Institute of Technology, Cambridge, Massachusetts

THOMAS F. GOLOB Transportation Research Department, General Motors Research Laboratories, Warren, Michigan

EZRA HAUER Department of Civil Engineering, University of Toronto, Toronto, Ontario, Canada

ROBERT HERMAN General Motors Research Laboratories, Warren, Michigan

TETSUZO HOSHINO Tokyo Branch Office, Japan Highway Public Corporation, Tokyo, Japan
Present address: Sapporo Branch Office, Japan Highway Public Corporation, Sapporo, Japan

FRIEDRICH JACOBS University of Stuttgart, Stuttgart, Germany

I. JEEVANANTHAM University of California, Berkeley, California

MASAKI KOSHI Institute of Industrial Science, University of Tokyo, Tokyo, Japan

TENNY N. LAM General Motors Research Laboratories, Warren, Michigan

ALLAN MARCUS Department of Statistics, The Johns Hopkins University, Baltimore, Maryland

ALAN J. MILLER Transport Section, University of Melbourne, Melbourne, Australia
Present address: Division of Mathematical Statistics, C.S.I.R.O., Sydney, Australia

C. B. G. MITCHELL Transport Research Assessment Group, Transport and Road Research Laboratory, Department of the Environment, Crowthorne, Berkshire, England
Present address: Royal Aircraft Establishment, Farnborough, Hampshire, England

MAURICE NETTER Institut de Recherche des Transports, Arcueil, France
Present address: 28 Avenue du Panorama, 92 Bourg La Reine, France

K. J. NOBLE Applied Mathematics Department, University of Adelaide, Adelaide, South Australia

IWAO OKUTANI   Department of Transportation Engineering,
                       Kyoto University, Kyoto, Japan
                       Present address: Department of Civil Engineer-
                       ing, Shinshu University, Nagano, Japan

R. H. OLDFIELD   Transport and Road Research Laboratory,
                       Department of the Environment, Crowthorne,
                       Berkshire, England

C. PEARCE   Mathematics Department, University of
                       Adelaide, Adelaide, South Australia

R. B. POTTS   Applied Mathematics Department, University
                       of Adelaide, Adelaide, Australia

RICHARD W. ROTHERY   General Motors Research Laboratories, Warren,
                       Michigan

TSUNA SASAKI   Department of Transportation Engineering,
                       Kyoto University, Sakyo-ku, Kyoto, Japan

M. W. SIDDIQEE   Stanford Research Institute, Menlo Park, California

PIPPA SIMPSON   Mathematics Department, University of
                       Adelaide, Adelaide, South Australia
                       Present address: South Australian Institute of
                       Technology, Levels Campus, Pooraka, Adelaide,
                       South Australia

J. C. TANNER   Transport and Road Research Laboratory,
                       Department of the Environment, Crowthorne,
                       Berkshire, England

RODNEY J. VAUGHAN   Operational Research Group, University of
                       Sussex, Sussex, England
                       Present address: Research Group in Traffic
                       Studies, University College London, London,
                       England

W. VENABLES   Statistics Department, University of Adelaide,
                       Adelaide, South Australia

V. VIDAKOVIC   Department of Public Works, Urban Research
                       Division, Amsterdam, The Netherlands

VUKAN R. VUCHIC   Towne School of Civil and Mechanical Engineer-
                       ing, University of Pennsylvania, Philadelphia,
                       Pennsylvania

JOHN G. WARDROP — Research Group in Traffic Studies, University College London, London, England

F. V. WEBSTER — Transport and Road Research Laboratory, Department of the Environment, Crowthorne, Berkshire, England

C. C. WRIGHT — Enfield College of Technology, Queensway, Enfield, Middlesex, England
Present address: Research Group in Traffic Studies, University College London, London, England

SAM YAGAR — Department of Civil Engineering, University of Waterloo, Waterloo, Ontario, Canada

Contents

Preface

This volume is the fifth in a series of international symposia on the theory of traffic flow and transportation. The series began in 1959 when Dr. Robert Herman organized a meeting in Detroit among some of the physicists, mathematicians, statisticians, economists, and engineers from around the world who had, in a variety of ways, tried to make some scientific contributions to problems of transportation. Many of the people who participated in this first symposium as amateur enthusiasts of an infant subject have since become mature contributors to an active and growing subject which has since been labeled transportation science.

These symposia have been perpetuated by a succession of ad hoc committees which have been free to conduct each symposium in any way they wish. No society has been organized to define the subject matter or create rules to impede its natural evolution. The series survives because of the enthusiasm of the participants who anxiously welcome the opportunity each three years to renew old acquaintances, meet new contributors, and exchange ideas which will influence the direction of progress in research for at least the next three years.

The thirty papers presented at this symposium were selected from a total of about ninety papers submitted. The distribution of papers among subjects is, however, representative of all ninety papers and seems to be a faithful indication of the present interests of the participants. The title of this fifth symposium was modified by the addition of "and Transportation" because previous trends had already shown that the interests of the group had broadened during twelve years and that the symposium could no longer be confined to what had previously been labeled "the Theory of Traffic Flow."

Activity in highway traffic theory already started to decline after the second symposium. Activity in·traffic signal systems has shifted from small networks to large networks during four previous symposia and appears now also to be declining. Interest in "transportation planning" (generation, distribution, assignment, modal split, etc.) has increased considerably during the last few years. This is reflected more by the improved caliber of the work than by quantity. Perhaps this field has also reached its peak of activity at this symposium where it represents about half of the papers. The appearance of several papers dealing with public transportation systems is possibly an indication of the next wave of interest.

Although it is difficult to measure the growth of activity in any field, a rough count of the number of papers published each year during the 1950s indicated that the activity in transportation science was doubling about every four or five years. This exponential growth appears to continue also through the 1960s. The size of these proceedings does not show this growth because the last few symposia have intentionally been maintained at about the same size. The task of selecting papers and participants, however, becomes more difficult each time.

Preface

Many people, too numerous to list, helped make this symposium possible. Professor Adolf May as co-chairman handled all correspondence and arrangements with participants and contributors. Several dedicated scientists not even listed on the advisory committee helped referee papers, some refereed as many as fifteen papers within one month. The success of the symposium, however, rests primarily with those who showed their support and interest by contributing good papers (many of which were not presented), and with the organizations listed on the previous page which helped finance the symposium.

GORDON F. NEWELL

The Starting Characteristics of Automobile Platoons

Robert Herman, Tenny Lam, and Richard W. Rothery
General Motors Research Laboratories, Warren, Michigan

Abstract

The results of a series of experiments carried out to determine the starting characteristics of automobile platoons are reported. In particular, the space–time trajectories of the lead and last vehicles have been examined in detail in order to determine the macroscopic properties of vehicular platoons. The effects of such factors as initial intervehicle spacing, speed, starting delay, and acceleration behavior are presented. In addition, calculations have been made for the flow of the platoon as a function of position along the roadway as well as for the speed of propagation of the starting wave as the platoon accelerates up to speed.

Introduction

In a recent study,[1] the authors examined the transient characteristics of bus platoons. The motivation of that earlier work was to shed some light on the dynamics of a bus system operating much like that of a train or subway where platoons of buses would start and stop at stations along an exclusive right of way. From these studies it was found that the cyclic dynamics of bus platoons starting at one station and stopping at another possessed repeatable features which could be described relatively simply. The present paper in which we focus our attention on the transient characteristics of automobile platoons is an outgrowth of this earlier work. There is little doubt that the dynamics of a platoon of vehicles discharging from a queue at a signalized intersection play an important role in saturation flow and intersection capacity, and therefore are an important element of the traffic complex in urban areas.

Several studies have been reported on different facets of the dynamics of automobile platoons. For example, Foster[2] examined the starting phase of queues of vehicles immediately downstream from a signalized intersection. From trajectories averaged over a number of cycles, an estimate of the speed of propagation of the starting wave was made. Other investigators[3-6] have also studied various characteristics, such as flow, speed, and dispersion, of platoons discharging from signalized intersections. In addition, Herman and Rothery[7] have studied the dynamics of the transition of a vehicular platoon going from one steady state to another.

In the present work we are concerned mainly with the starting dynamics of a queue composed of similar automobiles on a test track facility. In particular,

References p. 17

1

our efforts are directed towards studying the initial starting delay, the starting wave as it propagates back through the platoon, a description of the dynamics of starting as well as the transit time of platoons moving past an observation point as a function of distance down the roadway.

In the next section we discuss the techniques that were used in carrying out these experiments.

Experimental Details and Techniques

A platoon of six automobiles was used throughout the studies. These vehicles were standard, full-size production model 1970 Chevrolets with the exception of the lead vehicle which was a standard full-size Oldsmobile. This latter vehicle housed all of the recording equipment.

The roadway used was the four lane divided test track facility located at the General Motors Technical Center. The test track is approximately 1¼ miles long including turn arounds at each end. Only the straight section was used for each test run and the platoon was confined to one lane in each direction. This facility thus provided a straight, wide and level road section, with no entrances, exits or cross roads, 0.7 of a mile long.

The lead vehicle was controlled by the same driver in every test run. He was provided with a monitor that displayed the acceleration of his vehicle and was instructed to accelerate his vehicle at a nearly constant rate to a predetermined speed. The five other vehicles were also driven by employees of the General Motors Research Laboratories. They could not, in any way, be considered professional drivers and were not selected because of any particular attribute. Their position in the platoon was changed each of the five days that were required to complete the total number of test runs. The only instruction given to these drivers was to "follow the vehicle ahead in a 'normal' and 'safe' manner."

Ninety-eight test runs in all were recorded which provided approximately 11 repeated trials for each combination of the two control parameters. These control parameters were the maximum attained speed of the lead vehicle, u_0, and the mean acceleration, $\bar{a}$, used by the lead vehicle to attain this speed. Three levels were used for each variable. Table I gives the matrix of the experimental conditions investigated and the number of repetitions of each condition. The chronological order of the runs for the various combinations of speed and acceleration was sequenced by a quasirandom scheme.

The acceleration control variable requires special comment. The initial objective was to have three levels for the mean acceleration of the lead vehicle with the specific values of 4, 8, and 12 ft/sec². Operationally, this was not achieved. First of all, the lead vehicle's performance was below that of a standard vehicle due to the abnormal loading caused by the electronic equipment that was housed in it. This reduction in performance was particularly detrimental to the runs calling for the highest acceleration. The resulting average acceleration of the lead vehicle for this high acceleration case was ~9.18,

TABLE I

The Control Variables Used in the Experiments:
Mean Acceleration, $\bar{a}$, and Final Cruising Speed,
u_0, of the Lead Car. The entry in the Table is
the number of runs for each condition.

u_0		30 (mph)	40 (mph)	50 (mph)
	Low	11	11	11
$\bar{a}$	Medium	9	12	12
	High	11	10	11

~8.52, and ~7.73 ft/sec² for the three speed levels of 30, 40, and 50 miles/hour, respectively.

Throughout each test run magnetic tape recordings were made of the motion of the lead vehicle and the last vehicle of the six-vehicle platoon. These recordings consisted of the time series of events for every foot of forward motion of each instrumented vehicle. This information was obtained by using fifth wheels. The positional information from the last vehicle was transmitted via a telemetering link to the lead vehicle and was recorded on magnetic tape simultaneously with similar information obtained from the lead vehicle as well as a 3000 cycles/sec synchronizing clock signal.

The information recorded during the experiments was later reduced to a digital format by accumulating the time taken by each vehicle to travel a distance of 14 ft. In order to measure starting times, the time at which the first forward foot of motion was completed was also obtained for the lead and last vehicles.

The data was processed by an IBM 360 Model 65 computer to obtain trajectory information interpolated at equal ¼ sec intervals. Speed, platoon length and relative speed histories were also obtained in both numerical and graphical form (see Fig. 1 of Ref. 1).

A number of analyses were performed on the basis of this data, the results of which are discussed in the next section.

Results

Transient Characteristics of Starting

The transient characteristics of a vehicular platoon from the time the platoon starts to move to the time it reaches a steady state can be separated into two phases: the acceleration phase defined from the time the lead car starts to move to the time the last car reaches the cruising speed of the lead car; and the relaxation phase defined from the end of the acceleration phase to the time the platoon reaches a steady state. These two phases of the starting transient are

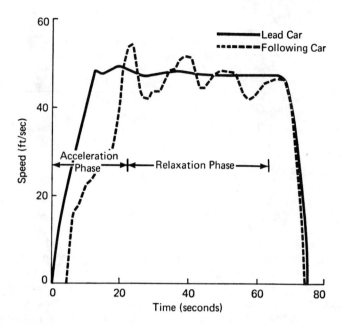

Fig. 1. Speed histories of the lead and last cars for a typical test run. Low acceleration and 30 mph maximum speed were used by the lead car in this case. The acceleration phase and the relaxation phase of the starting transient are indicated.

illustrated in Fig. 1 which is a typical test run in which the lead car used low acceleration to attain a 30 mph cruising speed. The dynamical characteristics of the platoon during the acceleration phase is discussed in this section and the relaxation phase is discussed later.

The acceleration phase of the platoon can be described by the initial platoon length, the starting delay, the acceleration of the lead car, and the acceleration of the sixth car. Such parameters from each of the experimental runs have been analyzed in terms of their averages and variances and with respect to the acceleration and the cruising speed of the lead car. Given these parameters an overall description can be obtained for the platoon during the acceleration phase.

The initial platoon length was the spacing between the same points, e.g., rear bumper to rear bumper, on the lead and last cars at the beginning of the test run. Since no instruction was given to the drivers regarding how they should space their cars, this initial platoon length can be interpreted as the sum of the normal spacings used by the five following drivers when they approach an intersection at low speed ($\approx$20 mph) and then stop. The initial platoon length in the 98 test runs has an average of 129.62 ft and a standard deviation of 5.97 ft. This gives an average spacing between pairs of vehicles of 25.92 ft with a standard deviation of 2.57 ft. In order to establish its generality, this result has been compared with the initial spacings of a four-car platoon in a preliminary experiment with 15 test runs that involved a different set of drivers. The average spacing from these two different sets of experiments show no statistically significant difference.

The averages and standard deviations of the starting delays for the nine experimental conditions are shown in Table II. The results of an analysis of variance on the starting delays give an indication that the starting delay depends on the acceleration of the lead car (at the 0.5% significance level) but not on the final cruising speed (at the 10% significance level). Since at the time the sixth car started to move the lead car was still accelerating, even for the 30 mph cases, the starting delay in these experiments is thus expected not to depend on the cruising speed of the lead car. However, the result may be different if the cruising speed of the lead car were sufficiently low or the platoon size were large. After the starting delays for different lead-car cruising-speed cases have been combined,

TABLE II

The Averages and Standard Deviations (in parentheses)
of the Starting Delays (in sec) for Each Experimental
Condition $(u_0, \bar{a})$

u_0		30 (mph)	40 (mph)	50 (mph)	All speeds
		(sec)	(sec)	(sec)	(sec)
$\bar{a}$	Low	5.35 (1.08)	4.77 (0.65)	4.74 (0.55)	4.95 (0.82)
	Medium	4.62 (1.22)	4.25 (0.70)	4.44 (0.96)	4.44 (0.95)
	High	4.29 (0.68)	3.98 (0.79)	3.97 (1.00)	4.08 (0.82)

the results of Behrens–Fisher tests show the difference of the mean starting delays between low and medium acceleration cases to be significant at the 2.5% level; between medium and high acceleration cases to be significant at the 10% level; and between low and high acceleration cases to be significant at the 0.5% level.

The mean acceleration as a function of speed has been calculated as the ratio of a given speed and the time that it took the vehicle to reach that speed from a standing start. The mean acceleration averaged over runs of the same controlled conditions is shown in Fig. 2 for the lead and the last car and for the nine conditions of lead car acceleration and cruising speed. Also shown together with the averages are the 99% confidence limits on the averages.

The acceleration of the lead car is one of the control variables of the experiment. In the low acceleration cases, the lead car was able to maintain consistently a mean acceleration of about 4 ft/sec^2, as intended, with only small variations from run to run. Although the mean acceleration was a little higher than the instructed value when the vehicle started to move, the variation from run to run

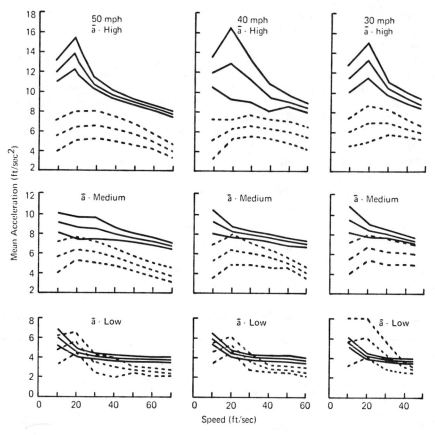

Fig. 2. Averages of mean accelerations calculated from a standing start to various intermediate speeds versus intermediate speed. Also shown are the 99% confidence limits of the averages. The solid curves are for the lead car and the dashed curves are for the last car. Each graph shows the results for one of the nine experimental conditions.

was small and the mean acceleration intended was very quickly reached and maintained. In the high acceleration cases, due to the affect of the equipment on the instrumented lead car, the intended accelerations of 8 and 12 ft/sec^2 could not be maintained when the vehicle reached higher speeds. However, as can be seen in Fig. 2, two distinctly different acceleration characteristics were produced by the lead car for the medium and high acceleration cases. Although the variations in the acceleration characteristics were large from run to run at low speeds, the mean acceleration became quite consistent by the time the lead car reached higher speeds. In other words, the mean acceleration of the lead car from a standing start to the specified cruising speed was replicated well from run to run for each of the acceleration conditions.

The mean acceleration of the last car has, in general, larger variation between runs of the same condition than the lead car. The variations from run to run were, however, drastically reduced at speeds of 40 ft/sec or higher implying that most of the variation occurred at speeds lower than 40 ft/sec. Such variations at low speeds, which will be discussed later, were averaged out by the time the last car reached higher speeds.

Comparing the mean accelerations of the last car for runs with the same lead-car acceleration but different cruising speeds, it can be seen that the mean accelerations of the last car as a function of speed are similar in that portion of the curves where they overlap in speed. This result is expected because the last car driver did not have information of the final cruising speed of the lead car. He adopted the cruising speed of the lead car only when this information was propagated to him through the car immediately in front of him. With the averages of three different samples agreeing, this result further demonstrates the consistency (in the sense of averages) of the mean acceleration characteristics of the last car in runs with the same lead car acceleration condition. Furthermore, the mean acceleration of the last car for medium and high lead-car acceleration cases is only slightly different despite quite different mean acceleration characteristics for the lead car in the two cases.

In the high and medium lead car acceleration cases, the mean acceleration of the last car was always below that of the lead car; whereas in the low acceleration cases, the mean acceleration of the last car at speeds between 10 and 20 ft/sec exceeded that of the lead car in most of the test runs. Mean accelerations were not calculated for speeds below 10 ft/sec because the first data point available was the travel time of the first 14 ft made by the vehicle. The average speed for the first 14 ft traveled was already greater than 10 ft/sec. Where the mean acceleration of the last car exceeded that of the lead car it was quickly corrected by the last-car driver to a level below that of the lead car by the time the last car had reached approximately 30 ft/sec.

Despite the consistent average acceleration behavior of the last car as illustrated in Fig. 2, the detailed speed history of the last car was quite different from one test run to another. Such variations are attributable primarily to the adjustments made in the acceleration of the last car during the early part of the acceleration phase. For most runs where it appeared that the driver of the last car was using too high an initial acceleration, the last car showed adjustments in its acceleration at speeds below 40 ft/sec. Since there was no information recorded for the dynamics of the four intermediate cars of the platoon, the observed behavior of the last car could be a response to disturbances created in the platoon and propagated to it, as well as a result of adjustments to the last car driver's individual action. Such acceleration adjustment may vary from a sharply defined discontinuity in the speed history to a slow continuous reduction in acceleration. Figure 3 shows the speed histories and the phase plots of the speed of the last car versus the platoon length for two runs to illustrate the two extreme types of acceleration adjustments made by the last car.

The acceleration adjustments made by the last vehicle of a vehicular platoon

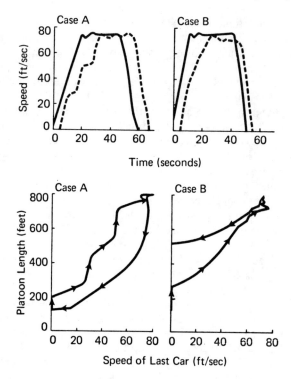

Fig. 3. Speed histories of the lead and last cars and the corresponding plot of the speed of the last car versus platoon length are given for two test runs. In both cases the final speed of the lead car was 50 mph. The lead car used low acceleration in case (A) and high acceleration in case (B) to reach the final speed. In case (B), the last car did not join the platoon at the end of the test run, but instead stopped at a pre-designated position for fifth wheel calibration.

during starting was first studied with a bus platoon. The result of that study[1] showed the sixth bus of the platoon making acceleration adjustments when the platoon reached a certain critical state expressed as a transition region on a plot of the speed of the last bus versus platoon length. Whenever such an acceleration adjustment occurred, it was a well defined one-step correction. Because of the low acceleration capability of the buses, the lead bus, in those experiments, used maximum performance to reach the final cruising speed and the last bus used an acceleration almost identical to that of the lead bus in its initial phase of acceleration. The dynamics of the bus platoon thus had very high repeatability from run to run. Furthermore, since at most one distinct acceleration adjustment was necessary to allow the last bus to reach the final cruising speed of the lead bus, the occurrence of such acceleration adjustments was also highly predictable.

The transient characteristics of the automobile platoon during the acceleration phase of starting are far less predictable. Although the lead car could replicate

its programmed accelerations quite well, the acceleration capability of the automobiles was much higher than the buses, thus giving the following cars flexibility
in accelerating into the same near-steady-state platoon configuration at the time
the platoon reached the specified cruising speed. The reproducibility of the
dynamics of an automobile platoon during the acceleration phase of the starting
transient was therefore reduced. However, as indicated in Fig. 2, most of the
variations occurred at low speeds over which the automobiles have the highest
acceleration capability.

The adjustments made by the last car driver on the acceleration of his vehicle
also showed higher variability than those observed with the bus platoon. Unlike
the adjustments observed for the bus platoon, there was often more than one
distinguishable acceleration adjustment made by the last car before it reached
the final cruising speed of the lead car; or there could even be a continuous
adjustment in the acceleration. Nevertheless, it was possible to obtain a few
general qualitative results regarding the acceleration adjustments made by the
last car. Table III shows, for each of the experimental conditions, the number
of runs where a discontinuity in the speed history of the last car was observed
during the acceleration phase. Also shown in the same table are the number of
test runs and the average number of distinct acceleration reductions for those
runs where such acceleration adjustments were made. Acceleration adjustments
occurring after the lead car had reached its final cruising speed were not included,
for it was not clear whether such adjustments were related to features in the
acceleration phase or they were responses to the disturbance created by the lead
car when it discontinued its acceleration. It can be seen in Table III that whenever the lead car used low acceleration the last car made at least one acceleration
reduction prior to reaching the final cruising speed. Such adjustments were
absent in runs with high lead-car accelerations. This phenomenon is undoubtedly
related to the last car using accelerations higher than that of the lead car in low
acceleration cases, as discussed before.

TABLE III

The Number of Runs in Which Acceleration Adjustments Occurred Expressed as a
Ratio of the Total Number of Runs for Each of the Experimental Conditions
and in the Parentheses the Average Number of Corrections Per Run for
Those Cases Where Such Adjustments Took Place.

u_0		30 (mph)	40 (mph)	50 (mph)
	Low	10/11 (1.5)	11/11 (1.5)	11/11 (1.8)
$\bar{a}$	Medium	0/9 (–)	3/12 (1.0)	10/12 (1.0)
	High	0/11 (–)	0/10 (–)	0/11 (–)

10

Herman, Lam, and Rothery

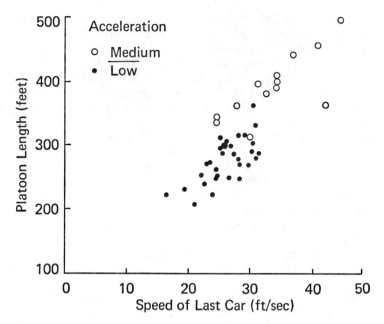

Fig. 4. Platoon length versus speed of the last car at the time the last car initiated its first acceleration adjustment.

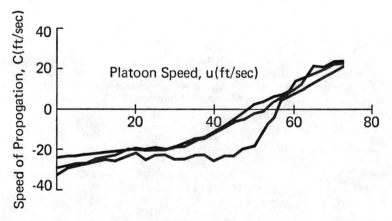

Fig. 5. The speed of propagation, C, of the starting wave versus platoon speed, u, for three individual test runs where the lead car used high acceleration to reach the speed u_0 of 50 mph.

The transition points, at which acceleration adjustments of the last car were made, in terms of the speed of the last car and the platoon length are plotted in Fig. 4 for the first acceleration corrections. It can be seen that practically all these first transitions took place at last-car speeds of less than 30 ft/sec for low lead-car acceleration cases and less than 40 ft/sec for medium lead-car acceleration cases. The locations of the transition points on this graph indicate that the acceleration corrections in runs with low and in runs with medium lead-car accelerations occurred at different platoon configurations. If the difference in vehicle lengths were taken into consideration, the results shown in Fig. 4 for the six car platoon are similar to the transition region obtained for the six bus platoon (see Fig. 8 of Ref. 1).

Speed of Propagation of the Starting Wave

In this section we are concerned with estimating the speed of propagation of the starting wave. In particular, we have calculated from all of the test runs for the case u_0 = 50 mph the quantity C which is the speed of the transition bringing the sixth vehicle up to a speed u. The calculations have been made using the roadway as a frame of reference and therefore C corresponds to the 'standard' definition of the speed of propagation of a transition as defined, e.g., by Lighthill and Whitham.[5] If hydrodynamical theories are applicable for this case, the numerical values for C would be equivalent to the slope of a flow versus concentration diagram, i.e., dq/dk (see Discussion on pp. 14–15 of Ref. 5).

Our estimates for C have been made for each test run at increments of 2.5 ft/sec. Each estimate is made by noting the elapsed time, Δt, between the lead and last vehicle reaching a speed, u; the distance between the positions of the vehicles on the roadway, Δx, where these events occur; and taking the ratio $\Delta x/\Delta t$.

The results for several test runs of the high acceleration case are shown in Fig. 5. It should also be mentioned that the initial starting wave, i.e., C at $u \simeq 0$ is also given in Fig. 5. This latter estimate was made by noting the elapsed time between the first forward foot of travel of the lead and last vehicle and the initial platoon length, L_0. We have termed this starting interval as the 'starting delay' and denote it by T_0.

The cases shown in Fig. 5 were selected as being representative of the types of results that were obtained. The results derived by averaging over all runs for each acceleration condition are shown in Fig. 6. Superimposed on this figure are the results of Foster[2] who calculated C for the starting wave of platoons discharging from a signalized intersection. Foster's experimental technique was to estimate the trajectory of each vehicle from arrival times at six positions spaced at 50 ft intervals downstream from the intersection and to estimate the slope (i.e., speed) of the average trajectory of each vehicle in the platoon. From this information he calculated C as a function of u. In particular the solid line in Fig. 6 is the result of a least squares fit of the equation

$$C = u - \bar{\lambda} \tag{1}$$

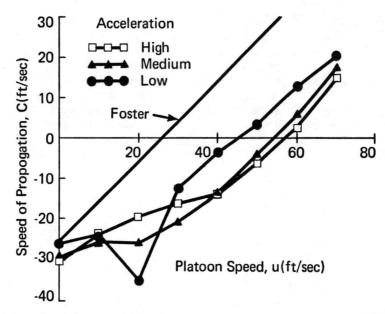

Fig. 6. Speed of propagation, *C*, of the starting wave, versus platoon speed, *u*. Each curve is derived by averaging over all runs of the same lead car acceleration. The maximum speed for the lead car in these cases was 50 mph. The result obtained by Foster is also shown.

to his data. The value of $\bar{\lambda}$ was estimated by Foster to be 25.7 ft/sec. Equation (1) simply states the algebraic relationship between the speed of the traffic stream, *u*, and the speed of propagation of a disturbance with respect to the roadway, *C*, and the speed with respect to the moving traffic stream, $\bar{\lambda}$. While Eq. (1) is valid for small fluctuations, it is not appropriate for our case where we have large perturbations. Here we have a substantial change in platoon length during the propagation time Δt. If complete information were available on each vehicle trajectory one could trace the propagation of a disturbance more accurately and possibly circumvent this problem.

One of the most interesting features of Fig. 5 showing *C* versus *u* is that in a large number of cases the curve for a given test run initially begins with a plateau or at least as a slowly changing function of speed up to a speed of about 30 ft/sec. A constant value of *C* would imply that the lead and last vehicle trajectories were the same, apart from a translation in space and time to take into account the initial delay time and the initial platoon length. Above the platoon speed of 30 ft/sec, the data fitted to Eq. (1) approximates the relationship between *C* and the platoon speed *u*. A "least-squares" fit of Eq. (1) to the data shown in Fig. 6 above 30 ft/sec provides an estimate of $\bar{\lambda}$. These estimates of $\bar{\lambda}$ for the low, medium, and high acceleration cases are 46.7, 53.5, and 55.6 ft/sec, respectively.

Flow Characteristics of the Platoon

The dynamics of the automobile platoon in the relaxation phase of the starting transient have been studied through observations of the transit time of the platoon passing fixed points along the roadway. Observations of this type have been studied previously in Refs. 1, 3-5. The averaged transit time as a function of distance downstream from the initial stopped position of the last car is plotted in Fig. 7 for each of the nine experimental conditions. It can be seen that after both the lead and last cars had reached the final cruising speed, usually within 1000 ft of the starting position, the platoon spread out. Despite very

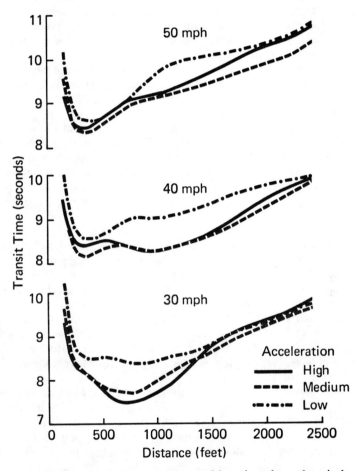

Fig. 7. The transit time of the platoon at varous positions along the roadway is shown versus downstream distance from the initial stopped position of the last car for each of the experimental conditions investigated. Each curve represents the average of all the runs with the same lead car acceleration and speed u_0.

different transit time characteristics immediately downstream from the starting position for runs of different control conditions, the transit times at 2400 ft are almost identical with runs of the same cruising speed. In the early part of the relaxation process, the different configurations reached at the end of the acceleration phase due to different sets of initial conditions were corrected to a near-steady-state condition with moderate adjustments; and then, final adjustments were made by the following cars of the platoon to achieve a steady state with primarily small perturbations of approximately ±5 ft/sec around the cruising speed of the lead car. Since in all the runs the platoon was moving to a steady state by increasing platoon length the average speed of the last car in the relaxation phase was necessarily less than that of the lead car. However, with the small perturbations the difference between the time averaged speeds of the lead and the last car was very small, resulting in a slow drift towards higher transit times as shown in Fig. 7.

At the time the platoon reached the end of the predetermined 0.6 mile roadway section used for these experiments, the data indicates for each run that the platoon was continuing to expand. This suggests that steady states had not yet been achieved in these experiments.

An alternative vantage point for viewing the starting dynamics of vehicular platoons is the flow-concentration diagram. Using the transit time of the platoon at each 50 ft interval on the roadway, the flow, q, vehicles/hour, as measured by an observer at each of these positions is readily calculated. The corresponding concentration, k, vehicles/mile, for each position was estimated by taking the arithmetic average of the length of the platoon at the beginning and end of the transit time.*

Figure 8 illustrates three examples of such calculations. The stopping maneuver has not been included in these graphs. The curves labeled A, B, and C in Fig. 8 are for the following test runs: high acceleration and u_0 = 50 mph; high acceleration and u_0 = 30 mph; and low acceleration and u_0 = 50 mph, respectively. Since each point plotted is for 50 ft increments along the road, these plots graphically display the flow-concentration history for a test run as the platoon moves along the roadway. The case given by curve A illustrates that a quasi steady state is reached. For a considerable period of time the lead and last vehicles are traveling at about the lead car control speed, u_0. However as can be seen from this curve, the platoon slowly expands its length as the q, k point moves along the constant speed line.

The case shown in curve B of Fig. 8 is very typical of the test runs for high acceleration to the low speed of 30 mph. In every case of this group, the q, k graphs indicated an "accordian" type motion of the platoon as reflected by the 'loop'. This motion is mainly caused by the overshoot in speed of the following vehicles. In this case the platoon approaches the steady state from above the

---

*The reader is cautioned to interpret these calculations of concentration with care since the platoon length during the initial starting phase changes considerably during the transit time of the platoon.

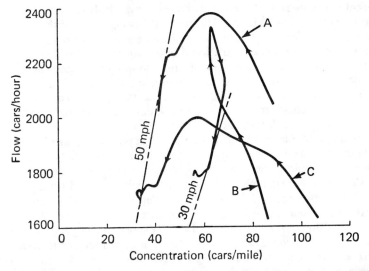

Fig. 8. Platoon flow versus average concentration measured at positions along the roadway for three test runs. Each run shown is a typical case among runs of the same experimental condition. The broken lines are constant speed lines of speed u_0. Curves (A), (B) and (C) are test runs for the following conditions: high acceleration and u_0 = 50 mph; high acceleration and u_0 = 30 mph; and low acceleration and u_0 = 50 mph, respectively.

operating speed, u_0. It also expands slowly with a perturbation near the end. The case given by curve C of Fig. 8 is for the same test run whose lead and last car speed histories are exhibited as Case A in Fig. 3. Here the operating speed is just reached before the stopping maneuver of the lead vehicle is executed. A careful comparison of case A in Fig. 3 and curve C in Fig. 8 is instructive.

Discussion

This paper has been devoted to a study of the starting characteristics of a platoon composed of six full-size automobiles. The approach taken has been primarily an experimental one and the measurement of the vehicle trajectories of the lead and last vehicles provides a data base from which a number of interesting dynamical features have been determined. Specifically, the characteristics of the starting phase, including the initial acceleration and relaxation phases, have been described in detail. It is noted that this starting phase cannot be so simply described as was the case with bus platoons reported in an earlier work by the authors.[1] It would appear that drivers using automobiles which have a considerably wider range in vehicle performance, make a number of 'corrections' as they accelerate up to speed. This is in contradistinction to the case with bus platoons where it appears that usually only one 'correction' is employed.

The speed at which the starting wave propagates along the roadway has also been examined in detail. Reasonably consistent results have been obtained for platoon speeds above 30 ft/sec. The most interesting aspect of these results, however, is that for low platoon speeds the data did not fit Eq. (1) which relates the speed of the platoon with the speed of propagation of disturbances. Indeed, it would appear that in any specific test run the speed of propagation of a disturbance is approximately a constant determined by the initial platoon length and the starting delay. Because of the high accuracy of the experimental information and the consistency of this phenomenon, this result sheds possible doubt on the validity of earlier related work.

The study of the transit time of the platoon at points downstream should be useful for the planning of pre-signals and signal funnels of the types described by von Stein[8] and for the location and synchronization of progressive traffic signal systems. For example, Fig. 7 shows that with standard American vehicles and 30 mph cruising speed for the queue leader, there existed a minimum for the transit time at approximately 700 ft downstream measured from the initial stopped position of the last car, or ~130 ft (the platoon length on starting) less if the lead car position is used as the distance origin. If the queue leader has a low acceleration capability, there is a very flat plateau from 400 to 1000 ft in the neighborhood of the transit time minimum for the 30 mph runs. Similar results regarding the location of minimum transit time can also be observed in Fig. 7 for other cases. Since the reciprocal of the transit time is the flow rate of cars in the platoon, the maximum flow that can be established at different locations downstream from a traffic signal can also be estimated from the transit times shown in Fig. 7. This, however, would require assumptions regarding the headway times of the 7th, 8th, 9th, etc., vehicles.

One of the primary limitations of the analysis presented here is that all of the calculations have been based on the trajectory information of the lead and last vehicles. This information on the dynamics of the platoon provides limited knowledge of what is occurring during the starting transient. Most of the results obtained in the present study bring out rather sharply the need to obtain the details of what is happening within the platoon so that driver-to-driver coupling can be better understood.

On the other hand, the information reported in this paper, limited as it is, does demonstrate a number of features of platoon behavior in the starting transient that car-following models to date cannot take into account. This latter point is of particular importance with respect to simulation calculations of urban networks where car-following models are frequently used in simulating the dynamics of vehicular platoons.

Acknowledgments

The authors acknowledge with pleasure Mr. George Gorday's assistance with the instrumentation, data handling and for driving the lead vehicle. We also thank Dr. Leonard Evans who developed, with one of the authors, the calibration procedure used in the numerical analysis. Lastly, we thank Mrs. Susan Klemmer for her help with some of the computer programming.

References

1. R. Herman, T. Lam, and R. Rothery, Further studies on single-lane bus flow: Transient characteristics, *Trans. Sci.* **4**, 187–216 (1970).
2. J. Foster, An invesitgation of the hydrodynamic model for traffic flow with particular reference to the effect of various speed-density relationships, *Australian Road Res. Board Proc.* **1**, Pt. 1, 229–257 (1962).
3. B. J. Lewis, Platoon movement of traffic from isolated signalized intersection, *Highway Res. Board Bull.* 178, 1–11, (1958).
4. C. C. Wright, Some characteristics of traffic leaving a signalized intersection, *Trans. Sci.* **4**, 331–346 (1970).
5. Z. A. Nemeth and R. L. Vecellio, Investigation of the dynamics of platoon dispersion, *Highway Res. Board Record* 334, 23–33 (1970).
6. R. Herman, R. B. Potts, and R. W. Rothery, Behavior of traffic leaving a signalized intersection, *Traffic Eng. Control* **5**, 529–533 (1964).
7. R. Herman and R. Rothery, Propagation of disturbances in vehicular platoons, in *Vehicular Traffic Science* (L. C. Edie et al. Eds.), pp. 14–25, American Elsevier Publishing Co., Inc., New York, 1967.
8. W. von Stein, Traffic flow with pre-signals and the signal funnel, in *Theory of Traffic Flow* (R. Herman, Ed.), pp. 28–56, Elsevier Publishing Co., Amsterdam, 1961.

Some Properties of the Fundamental Relations of Traffic Flow

C. C. Wright

Enfield College of Technology, Queensway, Enfield, Middlesex, England

Abstract

The fundamental relations of traffic flow are examined in relation to the way in which their constituent variables, speed, flow, and concentration, are defined. It is shown that even if instantaneous point values of the variables obey a unique functional relationship, their values averaged over time or distance cannot, in general, do so.

In the case of traffic in which the arrivals of vehicles are partly random, it is shown that the slope of the relationship between average speed and flow varies with the size of the space-time interval over which the speed and flow are defined or measured, if the latter are serially correlated. The amount of variation is calculated for a simple autoregressive flow model.

Some experimental evidence is given to support the theory, and its practical implications are briefly discussed.

Introduction

The behavior of road traffic systems in response to changes in their traffic loading has often been expressed as a set of "fundamental relations" between the variables speed, volume, and density (v, q, and k, respectively), whose precise statistical definitions vary with the circumstances of their application. When speaking of the macroscopic behavior of traffic as distinct from that of individual vehicles these variables are meaningful only when expressed as average values either over a length of road, a period of time, a combination of both, or a group of vehicles. The term "sampling domain" will be used here to denote a distance-time interval in which an observation of concurrent average values of q, k, and v is made.

The relationship between any pair of these variables defines the characteristics of a model since by definition $q = kv$. The relationship between q and k has been called the "Fundamental Diagram of Traffic" or the traffic equation of state, for which many theoretical forms have been suggested. Most have been fitted to data consisting of pairs of average values of q and k, or other closely related variables, observed over sampling domains of arbitrary size at selected sites. For example, Greenberg[1] has fitted a theoretical fluid continuum model to (among other data) reciprocal mean distance headways and space mean speeds measured by Huber[2] over 5-min sampling intervals at a temporary bridge on the Merritt Parkway, Connecticut. Herman et al.,[3] Newell,[4] and Edie[5] have derived steady-state equations relating q, k, and v from hypotheses concerning the car-following behavior of individual vehicles, and have fitted them to average values observed over various sampling intervals at tunnels in New York.

In contrast, theoretical fundamental relationships for freely flowing traffic have been proposed, but not experimentally verified, by Newell[6] and Prigogine et al.[7,8], based on the time-independent homogeneous solutions of equations simulating the dynamics of the traffic stream as a process of molecular inter-actions between gas particles.

Many purely empirical studies have been carried out in attempts to find agreed formulas for the fundamental relationships associated with roads of different kinds, although published information on the efficiency, lack of bias, or even theoretical validity of the sampling methods commonly in use is limited. Gafarian et al.[9] in a recent empirical investigation computed flows and concen-trations from observations of vehicles on a highway in Virginia by three different methods, using average values over five selected periods of approximately constant flow as a yardstick for comparison. The methods were, firstly to divide the data into $\frac{1}{50}$ th of an hour time slices and plot space mean speed, flow rate and density for each time slice, secondly to classify vehicles by virtual concentration (inverse space headway) and plot space mean speed, flow rate and average concentration for each concentration class, and thirdly to classify vehicles by speed and to plot the same quantities for each speed class. The first and third methods gave results which did not appear to be inconsistent with the constant flow data, although the distributions of points in each case were rather different.

Coburn[10] compared the relationships between average speed and flow rate for 3 and 15-min sampling intervals on the A5 near St. Albans, England, and found no significant difference.

Rothrock and Keefer,[11] however, plotted the average travel time per vehicle along a congested urban street in Charleston, West Virginia, against flow rate for each of a number of sampling intervals ranging from 6 to 60 min and found small differences in the distributions of data points. These results also appeared in the Highway Research Board's *Highway Capacity Manual*,[12] where the inference was drawn that the effect of flow on space mean speed was probably more pronounced for short sampling intervals than for long ones; certainly the extent of curve delineated by the short-sampling interval observations was greater. However, a strict interpretation of that statement would be that the rate of change of speed with flow was numerically greater in the short sampling interval results—this did not appear to be the case.

Previous work by the author[13] showed that the regression coefficients between values of time mean speed and flow rate at a point 2000 ft downstream from a traffic signal on one lane of a two-lane road in London varied with size of sampling interval. The signal cycle time was approximately 1 min, and vehicle speeds and arrival times were sampled over a continuous period of 140 min. Average speeds of all vehicles in various flow rate classes are given in Fig. 1 for sampling intervals of 6 sec, 30 sec, and 5 min as an illustration of the general trend, which was clearly such that the effect of flow rate on speed was more

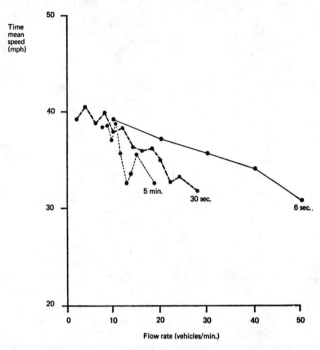

Fig. 1. Time mean speed–flow rate relationships, Lea Valley Road, London, 17.42–20.00 hr, 26/6/68. The points represent overall average speeds for time slices classified by flow rate. Curves are given for three values of sampling interval.

pronounced for *longer* sampling intervals. A qualitative theoretical argument was given to support this, and it was also shown that a sampling effect, arising purely from the operation of calculating time mean speeds, could account for some of the observed variation. More recent results obtained on an open rural road are discussed below.

It is the purpose of this paper to examine the behavior of samples of traffic stream variables in relation to the underlying fundamental relations which they are intended to estimate, and in particular to show that the estimates may be biased in the sense that they may vary with size of sampling domain. Such samples are also subject to a scatter effect if they are drawn from a traffic population which conforms to a nonlinear fundamental relationship, and to an attenuation effect whose consequences can be predicted in the linear case.

Fundamental Relations–the Nonlinear Case

Traffic engineers are well aware that independent random fluctuations in measured values of q, k, and v tend to obscure any underlying relationships between them. What is perhaps not clear is that some of the scatter of results

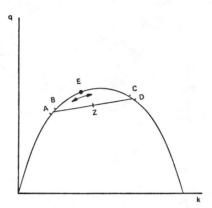

Fig. 2. Fluid continuum equation of state.

typically found in curvilinear relationships is entirely a consequence of the
measuring process itself, in which average values of the variables are sampled
over domains of arbitrary size. As an illustration, consider an ideal traffic
stream for which instantaneous values of flow rate and density at all points are
uniquely related by a nonlinear equation of state of the familiar form shown in
Fig. 2. This is essentially a deterministic fluid flow model, which would be most
likely to apply to tunnel traffic and bottleneck situations where the behavior
of individual vehicles is totally controlled by those ahead. The point E in Fig. 2,
representing the instantaneous "state" of traffic at a given location, will move
along the curve as demand conditions vary with time at that point; for instance,
it might lie within the arc AB for t seconds and move into arc CD for a further
t seconds on the arrival of a dense group of vehicles. During the combined
period of $2t$ seconds the average values of q and k would then be represented
approximately by the point Z, the midpoint of the chord joining the arcs AB
and CD. In theory Z may lie almost anywhere between the curve and the k
axis if there are no constraints on the state of the traffic stream in each of the
subintervals t. That is to say, average values of q and k measured over finite
sampling domains are distributed over the area of the diagram in a way which
is controlled by variations in demand conditions, and they do not therefore
conform to a functional relationship, even though their respective instantaneous
values are deterministically related. Furthermore, any attempt to deduce from
those values alone the original equation of state, e.g., by grouping them into
density classes and computing the mean flow rate for each class, will result in
an estimated curve lying between the true equation of state and the k axis. (An
alternative sampling procedure, which approaches the ideal of a sampling domain
of very small size and which has been used in particular by Edie et al.,[14] is to
sample vehicles individually, computing for each a "virtual concentration"
value equal to the inverse of its estimated space headway.)
   Stochastic models for freely flowing traffic do not possess a deterministic

equation of state for obvious reasons; Prigogine et al.[7,8] have, however, computed stationary time independent q/k relationships for their model. They express the relation between average values of q and k measured over long periods of time under steady flow conditions. To verify such a model one would therefore use sampling domains as *large* as possible. We may again imagine an experiment on an idealised traffic stream: suppose the long-term steady-state condition is satisfied and that average values of q and k are measured at a point on a lane in n successive time slices of length t, and that local statistical fluctuations in q and k cause the measured values to be distributed about some mean curve $M_t(q,k)$. These values may then be averaged over pairs of successive time slices of total length $2t$. By the same process of reasoning which was applied to the fluid continuum equation of state in Fig. 2, it can be seen that the resulting curve $M_{2t}(q,k)$ for the paired intervals lies between $M_t(q,k)$ and the k axis. The pairing procedure may be repeated until all or nearly all of the original data are compiled into a single average flow rate and concentration value for the whole experiment—these values would be highly likely to lie close to the theoretical steady state curve. Unfortunately, in real traffic, if the steady state exists at all it does so only for relatively short periods of time and is difficult to detect. The best one can do is to use the largest sampling interval consistent with local fluctuations in traffic conditions.

It can be seen from the foregoing that the measurement of stream variables in both congested and freely flowing traffic over finite sampling domains may give a biased or misleading estimate of a nonlinear fundamental relation. However, traffic whose instantaneous values of flow rate and density conform at any point to a unique linear equation of state has the property that observed average values of q and k for any size of sampling domain also obey the same equation;

(1) exactly, if the averages are calculated as simple averages with respect to the units of the sampling domain, or

(2) approximately, if they are calculated in any other way (which is the more usual practice).

But this is not necessarily true of traffic in which the variables q, k and v are subject to independent random fluctuations, as will be shown in the next section.

Speed-Flow Relationships for Freely Flowing Traffic

Whereas the traffic equation of state or q/k relationship holds a central position in flow theory, in practical terms it is probably true that speed-flow relationships are more important. For instance, the economic assessment of a road scheme requires an estimate of the average journey time per vehicle under projected demand conditions. Such predictions are made from speed-flow or travel time-flow relationships measured on existing roads of a similar type, and may vary with the sampling procedure used, and, in particular, with the size of the sampling domain. Of the theoretical arguments for this proposition referred to in a previous paper,[13] the first concerns an effect associated with the operation of computing time mean speeds; there is probably an equivalent effect for space mean speeds but it has not been possible to express it in quantitative terms. The second is a regression effect which was examined qualitatively from the point of view of its occurrence in bunched traffic streams; in fact it is a quite general phenomenon described by Yule and Kendall[15] as an attenuation effect occurring in correlated data which is expressed in "modifiable units," i.e., values averaged over domains of arbitrary size on a common sample space. The relationship between wheat and potato yields in Britain was given by Yule and Kendall as an example; the sample space common to both variables is geographical area and the division of area into sampling domains is clearly arbitrary in the mathematical sense of the word. The coefficient of correlation between average yields per acre for sample domains consisting of groups of equal number of counties was shown to increase with group size, and a theoretical argument was given to account for this. Using a similar technique it can be shown that the regression coefficient b in the regression equation, $v = a + bq$, for average speeds and flow rates also varies with size of sampling domain—in this case road length or sampling interval depending on the experimental procedure used. Suppose that v and q may be written

$$q = E + e$$
$$v = F + f, \tag{1}$$

where e and f are random variables independent of E, F and each other, E, F are correlated, and v and q are average speed and flow rate with respect to the sampling domains $\mathcal{D}$ over which they are measured. Without loss of generality we may stipulate that $q, v, e, f, E,$ and F are measured about their mean values. Then:

$$\text{var}(q) = \text{var}(E) + \text{var}(e)$$
$$\text{var}(v) = \text{var}(F) + \text{var}(f)$$
$$\text{covar}(q,v) = \text{covar}(E,F) \tag{2}$$

Denoting the linear regression coefficients between E and F by a' and b' such that $F = a' + b' E$, then

$$b' = \frac{\text{covar}(E,F)}{\text{var}(E)} \tag{3}$$

and

$$b = \frac{\text{covar}(q,v)}{\text{var}(q)} = \frac{\text{covar}(E,F)}{\text{var}(E) + \text{var}(e)}$$
$$= b' \left(\frac{1}{1 + \dfrac{\text{var}(e)}{\text{var}(E)}} \right). \tag{4}$$

Equation (4) expresses the fact that the regression coefficient b between q and v is attenuated by the presence of the random element e. If the sampling domain

$\mathcal{D}$ is now increased in size, i.e., q and v are averaged over longer periods of time or greater lengths of road, the ratio var $(e)/$var (E) is likely to decrease; this is a common sampling phenomenon which, to paraphrase Yule and Kendall, can be more simply described as a decrease in sampling fluctuations relative to systematic fluctuations with increasing sample size.

While Yule and Kendall did not elaborate on the meaning of the term "systematic," it will be taken here to denote the event that the E are positively serially correlated. The equivalent notion that continuity between data in samples is responsible for variation in regression coefficients between alternative sampling schemes has been suggested by Duncan et al.[16] in a discussion on parallel problems in the analysis of areal data; that it is a necessary and sufficient condition may be demonstrated in the case of the following statistical model.

Let v_r and q_r denote the rth values of average speed and flow rate sampled in succession over N contiguous and equal time slices each of length t. It is convenient to introduce an intermediate dummy variable w_r with zero mean such that

$$q_r = A_1 + B_1 w_r + e_r$$
$$v_r = A_2 + B_2 w_r + f_r, \tag{5}$$

where the e_r and f_r are independent random variables uncorrelated with w_r and each other and have zero means; A_1, A_2, B_1, and B_2 are constants. The mean values of any subset of q_r or v_r are taken as simple arithmetic means, and in particular

$$\bar{q} = \frac{1}{N} \sum_{r=1}^{N} q_r \simeq A_1.$$

Similarly $\bar{v} \simeq A_2$. Let the sequence of variables w_r constitute an autoregressive stationary time series generated by

$$w_r = C w_{r-1} + g_r \qquad (0 < C < 1), \tag{6}$$

where the g_r are independent random variables uncorrelated with e_r, f_r, q_r, v_r, and w_{r-s} $(s = 1, 2 \ldots)$, and have zero mean, and C is a constant. Equation (6) implies that

$$w_r = C^s w_{r-s} + \sum_{h=0}^{s-1} C^h g_{r-h} \qquad (s = 1, 2 \ldots). \tag{7}$$

Denoting by b_t the gradient regression coefficient of v_r on q_r we have

$$b_t = \frac{\sum_{r=1}^{N} (v_r - \bar{v})(q_r - \bar{q})}{\sum_{r=1}^{N} (q_r - \bar{q})^2}. \tag{8}$$

Substituting Eq. (5), discarding terms whose expectations are zero, and writing var (w) and var (e) for

$$\frac{1}{N}\sum_{r=1}^{N} w_r^2 \quad \text{and} \quad \frac{1}{N}\sum_{r=1}^{N} e_r^2,$$

respectively, we obtain

$$b_t \simeq \frac{B_1 B_2 \; \text{var}\,(w)}{B_1^2 \; \text{var}\,(w) + \text{var}\,(e)} = \frac{B_2}{B_1} \left\{ \frac{1}{1 + \left(\dfrac{\text{var}\,(e)}{B_1^2 \; \text{var}\,(w)}\right)} \right\} \qquad \text{[c.f. Eq. (4)]}.$$

(9)

Now consider the regression between arithmetic mean values of q and v taken over p successive groups of n observations each $(pn = N)$. Denote the ith values of q and v in the jth group by $q_{i,j}$ and $v_{i,j}$, and denote the required gradient regression coefficient by b_{nt}, since we are in effect sampling over time slices of length nt.

Now

$$b_{nt} = \frac{\sum\limits_{j=1}^{p}\left\{\left[\dfrac{1}{n}\sum\limits_{i=1}^{n}(v_{i,j}) - \bar{v}\right]\left[\dfrac{1}{n}\sum\limits_{i=1}^{n}(q_{i,j}) - \bar{q}\right]\right\}}{\sum\limits_{j=1}^{p}\left[\dfrac{1}{n}\sum\limits_{i=1}^{n}(q_{i,j}) - \bar{q}\right]^2}$$

$$= \frac{\sum\limits_{j=1}^{p}\sum\limits_{i=1}^{n}[(v_{i,j} - \bar{v})(q_{i,j} - \bar{q})] + \sum\limits_{j=1}^{p}\sum\limits_{i \neq m}[(v_{i,j} - \bar{v})(q_{m,j} - \bar{q})]}{\sum\limits_{j=1}^{p}\sum\limits_{i=1}^{n}(q_{i,j} - \bar{q})^2 + \sum\limits_{j=1}^{p}\sum\limits_{i \neq m}[(q_{i,j} - \bar{q})(q_{m,j} - \bar{q})]}.$$

(10)

Irrespective of the model under consideration, if the variables v and q are serially correlated the second terms in both numerator and denominator of (10) may be nonzero, and, by comparison with (8), b_t and b_{nt} will not in general be equal even though they are computed from the same raw data. An obvious exception [see Eq. (11) below] in this case occurs when var $(e) = 0$; the existence of an independent random component in q is, therefore, a necessary condition to the argument although by itself it does not guarantee the effect predicted by Yule and Kendall.

After substituting Eq. (5) the second numerator term becomes

$$\sum_{j=1}^{p}\sum_{i \neq m}[B_2 w_{i,j} + f_{i,j}]\,[B_1 w_{m,j} + e_{m,j}].$$

After discarding terms with zero expectation this is approximately

$$B_1 B_2 \sum_{j=1}^{p} \sum_{i \neq m} w_{i,j} \, w_{m,j}$$

$$= 2 B_1 B_2 \sum_{j=1}^{p} \sum_{i=2}^{n} \sum_{m=1}^{i-1} \left(C^{i-m} w_{m,j} + \sum_{h=0}^{i-m-1} C^h g_{i-m-h,j} \right) w_{m,j}$$

on inserting Eq. (7). Again discarding terms with zero expectation and executing the second summation we obtain approximately

$$\frac{2 B_1 B_2 C}{(1 - C)} \sum_{j=1}^{p} \sum_{m=1}^{n-1} (1 - C^{n-m}) \, w_{m,j}^2 \, .$$

This expression is a weighted sum of squares whose expectation, since the $w_{m,j}$ constitute a nonoscillating stationary series, is

$$\frac{2 B_1 B_2 C p \, \text{var}\,(w)}{(1 - C)} \left[(n - 1) - C \left(\frac{1 - C^{n-1}}{1 - C} \right) \right] = B_1 B_2 \, \text{var}\,(w)\, Z, \text{ say.}$$

By similar reasoning it can be shown that the second term of the denominator of Eq. (10) $\simeq B_1^2 \, \text{var}\,(w)\, Z$, so that after some manipulation,

$$b_{nt} \simeq \frac{B_2}{B_1} \cdot \left[\frac{1 + \dfrac{Z}{N}}{1 + \left(\dfrac{\text{var}\,(e)}{B_1^2 \, \text{var}\,(w)} \right) + \dfrac{Z}{N}} \right] \cdot \tag{11}$$

Clearly, $|b_{nt}| < |b_t|$ if (subject to sampling fluctuations which this crude argument has ignored) $C < 0$. Conversely, $|b_{nt}| > |b_t|$ if $0 < C < 1$, which is the case we might reasonably expect in road traffic. In addition, as $n \rightarrow \infty$ we have that $Z \rightarrow 2CN/(1 - C)$ and $b_{nt} \rightarrow \text{constant} \times B_2/B_1$. Therefore, as larger and larger sampling intervals are chosen the regression coefficient b_{nt} approaches a finite limit.

The model seems to be consistent with the available data (some recent results are discussed below), although there are equally feasible alternatives; for instance, one might allow for different types of serial correlation between the v and between the q. Furthermore, it is not yet known with any certainty what proportion of the total observed variation in the regression coefficient may be ascribed to serial correlation effects as distinct from the previously noted averaging effect, which seems to act on b in the same direction and with a similar order of magnitude.

A series of experiments were recently conducted by the author to establish whether the earlier results[13] referred to in the Introduction were also typical of freely flowing traffic on a rural road, in that numerical values of the regression coefficient between time mean speed and flow rate increased with increasing

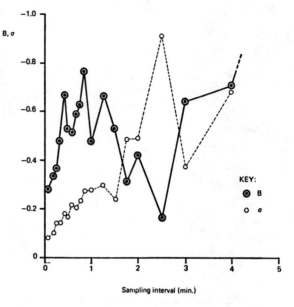

Fig. 3. Time mean speed–flow rate regression coefficients *B* and their estimated standard errors *σ*. A11, Six Mile Bottom 11.40–12.40 hr, 5/8/70, northbound lane.

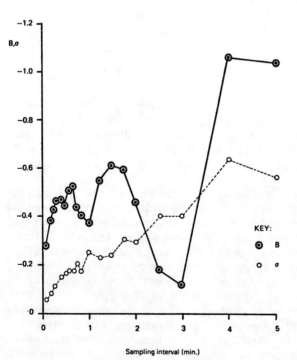

Fig. 4. Time mean speed–flow regression coefficients *B* and their estimated standard errors *σ*. A11, Six Mile Bottom 15.25–16.45 hr, 5/8/70, southbound lane.

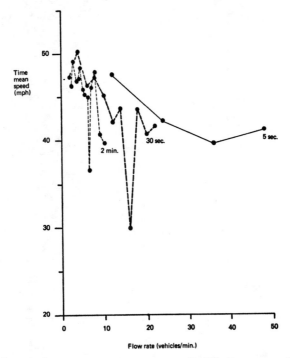

Fig. 5. Time mean speed-flow rate relationships, A11, Six Mile Bottom, northbound lane, 11.40–12.40 hr, 5/8/70. The points represent overall average speeds for time slices classified by flow rate. Curves are given for three values of sampling interval.

values of sampling interval of up to several minutes in duration. Two sites were chosen on the A11, a two-lane two-way road in Cambridgeshire carrying relatively light traffic. Four pneumatic tube detectors were placed at 100-ft intervals in the northbound lane at the first site and in the southbound lane at the second, and arrival times of successive vehicles at each tube were recorded on an Esterline Angus multiple pen recorder for continuous periods ranging from 60 to 80 min at different times of the day at each site. The arrival times were then coded on punched Hollerith cards and processed by computer to give time mean speeds and average flow rates in equal consecutive time slices ranging from 5 sec to 20 min in length. For each sampling interval value, linear regressions of the form $v = A + Bq$ were performed; values of B obtained in the first two experiments are plotted against sampling interval in Figs. 3 and 4, together with their estimated standard errors. For sampling intervals up to about $\frac{3}{4}$ min B apparently rises steadily, being subject to sampling fluctuations of considerable size elsewhere. In Fig. 5 individual time slices have been classified into flow rate groups and the overall time mean speed for each group plotted, for three values of sampling interval. Figure 6 shows the scatter of q and v values for a sampling interval of 30 sec. The results in Figs. 3-6 are fairly

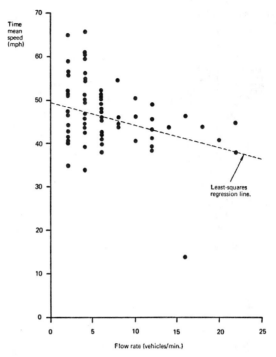

Fig. 6. Time mean speeds and flow rates in successive 30-sec time slices. A11 Six Mile Bottom, northbound lane, 11.40–12.40 hr, 5/8/70.

typical of all five experiments carried out. Work now continues on the problem of improving the accuracy of estimates of B for sampling intervals greater than $\frac{3}{4}$ min at other sites.

Conclusions

The fundamental diagram or equation of state of traffic as derived from fluid flow theory expresses the mechanism by which instantaneous point values of flow rate and density change with changing demand conditions; average values of these quantities sampled over finite space-time domains do not necessarily provide accurate estimates of those relations unless either:

(a) The traffic stream is in a steady state (i.e., flow rate and density do not change markedly during each time slice).

(b) Average flow rate and density are defined as simple mean values with respect to the sample space and the equation of state is linear over the range of instantaneous point values of q and k occurring during the sampling process.

In the absence of more positive evidence that a given sampling domain is appropriate for a given experimental situation, the best criterion would seem to be that it should not mask (and, therefore, subsequently misrepresent through the averaging process), significant variations in the local dynamics of the traffic stream. Measurements at a point in congested traffic of the stop-and-start variety may well call for a sampling interval of less than 30 sec if reasonably accurate results are required.

In freely flowing traffic, relations between the measured flow parameters may be affected by serial correlation and the estimation of an underlying steady-state relationship requires sampling intervals of the largest possible size.

The practical implications of the phenomena discussed are probably most serious where speed-flow relationships are concerned. Speed-flow or travel time-flow relationships are measured on existing roads and used in the evaluation of new road schemes; the objective here is not to verify a theoretical concept but to obtain the best possible estimate of average speed or travel time at given projected flow rates during selected periods of the day. Since speed-flow relations may vary with sampling domain, what are the criteria for an appropriate choice? Unless it can be shown that a smaller sampling interval gives a sufficiently close approximation, one may be forced to choose a value equal to the time period (e.g., peak hour) for which the estimate is required, and to accept the experimental consequences.

Furthermore, any measure of road capacity should be related to the period for which that capacity is expected to be maintained. The maximum average flow rate observed at a given site over a given size of sampling interval cannot be accepted as more than a relative quantity, a point which has been made elsewhere by several authors from different viewpoints.

To summarize, an attempt has been made to examine the validity and consequences of measuring and describing traffic flow through flow, speed, and density values averaged over space or time intervals of finite size. Some of the effects described are common to other stochastic processes similarly characterized by relationships between two or more variables.

Acknowledgments

The author would like to thank Mr. J. G. Wardrop, Dr. R. E. Allsop, and Professor R. J. Smeed of the Research Group in Traffic Studies, University College, London, and Messrs. T. Hyde and A. Dockerty of Enfield College of Technology, Enfield, Middlesex for their help, encouragement and advice.

References

1. H. Greenberg, An analysis of traffic flow, *J. Opns. Res.* **7**, 79-85 (1959).
2. M. J. Huber, Effect of temporary bridge on parkway performance, *Highway Res. Board Bull.* No. 167, 63-74 (1957).
3. D. C. Gazis, R. Herman, and R. B. Potts, Car following theory of steady-state traffic flow, *J. Opns. Res.* **7**, 499-505 (1959).
4. G. F. Newell, Non-linear effects in the dynamics of car following, *J. Opns. Res.* **9**, 209-228 (1961).

5. L. C. Edie, Car-following and steady-state theory for non-congested traffic, *J. Opns. Res.* **9**, 66-76 (1961).
6. G. F. Newell, Mathematical models for freely flowing highway traffic, *J. Opns. Res.* **3**, 176-186 (1955).
7. I. Prigogine, A Boltzmann-like approach to the statistical theory of traffic flow, in *Theory of Traffic Flow* (R. Herman, Ed.), Elsevier Publishing Co., Amsterdam, 1961.
8. R. L. Anderson, R. Herman, and I. Prigogine, On the statistical distribution function theory of traffic flow, *J. Opns. Res.* **10**, 180-195 (1962).
9. A. V. Gafarian, R. L. Lawrence, P. K. Munjal, and J. Pahl, *An Experimental Validation of Various Methods for Obtaining Relationships between Traffic Flow, Concentration and Speed on Multi-lane Highways*, System Development Corporation and UCLA Institute of Transportation and Traffic Engineering, Contractors; US Bureau of Public Roads Contract No. FH-11-6623, 1970 (unpublished).
10. T. M. Coburn, Speed and Flow on a Two-lane Section of A5 near St. Albans, Herts, *Road Research Laboratory Research Note* No. RN/2427/TMC, 1955 (unpublished).
11. C. A. Rothrock and L. A. Keefer, Measurement of urban traffic congestion, *Highway Res. Board Bull.* No. 156, 1-13 (1957).
12. *Highway Capacity Manual*, Highway Research Board, 67 (1966).
13. C. C. Wright, Some characteristics of traffic leaving a signalized intersection, *Trans. Sci.* **4**, No. 4 (1970).
14. L. C. Edie, R. S. Foote, R. Herman, and R. Rothery, Analysis of single lane traffic flow, *Traffic Eng.* **33**, 21-27 (1963).
15. G. U. Yule and M. G. Kendall, *An Introduction to the Theory of Statistics*, 14th ed., pp. 310-315, Charles Griffin, London, 1968.
16. O. D. Duncan, R. P. Cuzzort, and B. Duncan, *Statistical Geography*, The Free Press, Glencoe, 1961.

On Vehicle Longitudinal Dynamics*

James G. Bender and Robert E. Fenton
Department of Electrical Engineering, The Ohio State University, Columbus, Ohio

Abstract

The longitudinal dynamics of a highway vehicle are frequently represented by a first-order, constant-coefficient, differential equation with two fixed parameters. It is well known that this representation is not valid in practice, for these parameters vary erratically with time both within and across vehicles; however in some studies, particularly those dealing with the determination of macroscopic traffic characteristics such as flow versus speed, the parameter variations—both within and across vehicles—are effectively averaged out and will cause little or no effect on the final results. The opposite is true when individual vehicle control is considered; here, the selected model must realistically represent the vehicle dynamics or the derived results will be of limited practical value. In the research reported, the quantitative limitations of the commonly used first-order model were experimentally determined for steady-state driving conditions. Subsequently, a more realistic model was developed and experimentally validated.

Model Development

A block-diagram representation of vehicle longitudinal dynamics is shown in Fig. 1. Note that these dynamics have been divided into two parts with one part corresponding to vehicle mass (m) and the effective force acting on the vehicle, and the second part associated with the carburetor-engine-drivetrain combination. The former part is described by

$$F + f_d = mpV + F_f(V) \qquad (p \equiv d/dt), \qquad (1)$$

where

F = net force applied to vehicle by drivetrain,
f_d = disturbance input from external environment,
$F_f(V)$ = effective friction force at vehicle speed V.

(Wheel braking was not included in this model as subsequent experimental work involved only small accelerations where braking, when required, was obtained from engine action and environmental effects.) In the situation where a vehicle is traveling at a nearly constant speed V_0, it is convenient to linearize

*This study was sponsored by the Ohio Department of Highways and the Federal Highway Administration. The opinions, findings, and conclusions expressed in this publication are those of the authors and not necessarily those of the State of Ohio or the Federal Highway Administration.

References p. 46

33

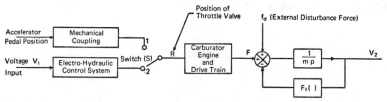

Fig. 1. Block diagram representation of vehicle longitudinal dynamics.

Eq. (1). To this end, let

$$V = V_0 + v \tag{2}$$

and

$$F = F_0 + f, \tag{3}$$

where

   v = variational component of vehicle speed,
F_0 = net average force acting on vehicle,
   f = variational component of force applied to vehicle.

If it were further assumed that $F_f(V)$ were a suitable differentiable function of V, then it could be adequately represented by the first two terms of a Taylor series expansion for small values of v; thus

$$F_f(V) \simeq F_f(V_0) + f_{fo}v \tag{4}$$

where

$$f_{fo} = \frac{dF_f}{dV}\bigg]_{V=V_0}.$$

On substituting Eqs. (2), (3), and (4) into (1) and cancelling the constant terms, there results

$$f_d + f = mpv + f_{fo}v. \tag{5}$$

This result has been used by other investigators who determined the control force f (in the absence of external disturbances) which would optimize certain performance measures.[1-5] A primary difficulty with this approach is that the derived control force is located at an inaccessible point in the system and can only be controlled indirectly; for example, by control of the carburetor throttle valve in an internal combustion engine as shown in Fig. 1 or by control of the electric power flow in an electric car. A second difficulty is that the assumed characteristics of $F_f(V)$ may not be realistic in so far as a practical driving situation is concerned. An additional difficulty is that a disturbance force acts as a system input signal which adversely affects the controlled variable (v). Such a disturbance is invariably present in practice–even on a nearly ideal road. It should be noted that similar difficulties would probably be present regardless of the type of engine-drivetrain combination used.

The second part of vehicle longitudinal dynamics are associated with the carburetor, engine, and drivetrain–at least for a conventional vehicle. These dynamics are nonlinear exhibiting such effects as deadzone, backlash, saturation and hysteresis–especially under large-signal conditions. However, for steady-state driving, it seems appropriate to represent these dynamics by a pure gain G_0. The corresponding relation between a small motion (r) of the throttle valve about an average displacement of R_0 and the corresponding vehicle speed variation is, from Eq. (5)

$$v = \frac{G_0 r + f_d}{mp + f_{fo}}.$$ (6)

Note from Fig. 1 that the position of the throttle valve is shown as being controlled in two ways–the conventional approach using the accelerator pedal and via an electrohydraulic control system. The latter approach was used in this study, so that a voltage signal V_i could be used to control vehicle speed. If V_{io} is the voltage corresponding to the average throttle position R_0 (and hence the average speed V_0), then

$$V_i = V_{io} + v_i,$$

where v_i is a variational input voltage which will cause speed variations about V_0. Since the dynamics of the electrohydraulic control system were characterized by a dominant natural frequency of some 32 rad/sec, and thus were negligible at the much lower frequencies involved in vehicle steady-state control, the system model can be expressed in the form

$$v = \frac{G_0 G_1/f_{fo}}{(m/f_{fo})p + 1} v_i + \frac{1/f_{fo}}{(m/f_{fo})p + 1} f_d.$$ (7)

Here the first term is vehicle response due to v_i and the second that due to f_d.

In some cases it might be convenient to use the following simplified representation for vehicle longitudinal dynamics:

$$v = \frac{C_1}{C_2 p + 1} v_i$$ (8)

and then empirically estimate suitable values for (C_1, C_2). (However, considerable variability might be expected in run-to-run estimates of (C_1, C_2) because of variations of the system parameters $(G_0, G_1, \text{ and } f_{fo})$, the intrinsic nonlinearity of the system, and/or small disturbance forces.) Henceforth, C_1 and C_2 will be referred to as the vehicle gain and time constant, respectively, and the model defined by Eq. (8) as the uncompensated vehicle dynamics.

These vehicle dynamics were subsequently modified by the use of control compensation in order to reduce the adverse effects of substantial variations in

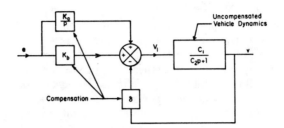

Fig. 2. Modification of vehicle longitudinal dynamics.

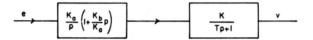

Fig. 3. Reduced block diagram.

C_1 and C_2. Such variations were partially overcome by using internal feedback with a gain of δ as shown in Fig. 2. The remaining blocks in this figure correspond to system compensation. This block diagram can be simplified to the one shown in Fig. 3 where

$$T = \frac{C_2}{1 + C_1 \delta}$$

and

$$K = \frac{C_1}{1 + C_1 \delta}.$$

If the system compensation were properly chosen (i.e., the pole at $-(1 + C_1\delta)/C_2$ were cancelled by the zero of the integral compensator), the resulting function would be

$$\frac{v}{e} = \frac{(1/\tau)}{p},$$

where τ equals $1/K_a K$, K_a is a constant in the compensation network, and e is a voltage input signal as shown in Fig. 3.

The resulting block diagram of the overall compensated system is shown in Fig. 4, from which the overall system function can be found as

$$\frac{v}{v_L} = \frac{1}{1 + \tau p} \tag{9}$$

which will hereafter be referred to as the compensated system dynamics. Here, v_L is an incremental change of command velocity. [It is worth noting that if v_L corresponded to the velocity of a lead car (expressed in terms of a voltage), then Eq. (9) would correspond to a relative velocity controller in which the

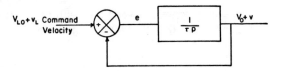

Fig. 4. Compensated vehicle dynamics.

command acceleration of the controlled following car would be proportional to the relative velocity between it and the nearest lead car.][6]

Since Eq. (9) is a theoretical ideal, it will be replaced for discussion purposes by a more practical model which allows for possible *dc* variations in gain; thus

$$\frac{v}{v_L} = \frac{K_c}{1 + \tau p}. \tag{10}$$

If, of course, the term K_c is deterministic and is equal to unity, then Eqs. (9) and (10) would be identical.

Experimental Procedure

The validity of the specified system models—Eqs. (8) and (10)—was intensively investigated under quasi-steady-state conditions. A 1965 Plymouth sedan was instrumented with both electrohydraulic systems for "voltage" control of the throttle valve and brake-line pressure and an analog computer for implementing the required compensation networks and for use in data collection.[7]

The two parameters (C_1 and C_2) associated with the uncompensated vehicle dynamics were obtained via the following procedure:

The test vehicle was first set at a constant speed and then a step change in the throttle input signal was superimposed on a preset throttle position and the vehicle's velocity–time history was recorded. This sequence of events is portrayed in Fig. 5.

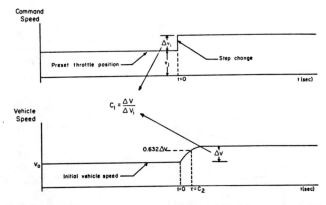

Fig. 5. Vehicle response to a step change in throttle input signal (V_i).

If the command change in speed were Δv_i and the resulting steady-state change in controlled car speed were Δv, then it follows from Eq. (8) that

$$C_1 = \frac{\Delta v}{\Delta v_i}.$$

Further, since the time constant is defined as that time required for the response to reach 0.632 of its final value, C_2 can be determined graphically as shown in Fig. 5. (It is well to note that the time lag which occurred before the vehicle responded to the step input was neglected here. Typically, this was some 250–400 msec and thus was small compared to the measured values of C_2 of some 6 to 40 sec.)

Data were collected at average speeds of 58.6, 73.2, and 88 ft/sec (40, 50, and 60 mph) and the corresponding magnitude of the command step change was ±5.0 ft/sec. Some 1000 estimates of (C_1, C_2) were made with a nearly equal number of data points being collected for each speed.

The parameters K_c and τ associated with the compensated vehicle dynamics were obtained in precisely the same way and under the same conditions; however, as a subsequent examination of the collected data will show, it was not necessary to collect as many data points as for the uncompensated case. The compensation circuits were adjusted in advance of any testing so that $K_c = 1$ and $\tau = 4$ sec. It is important to note that τ could have been set anywhere in the

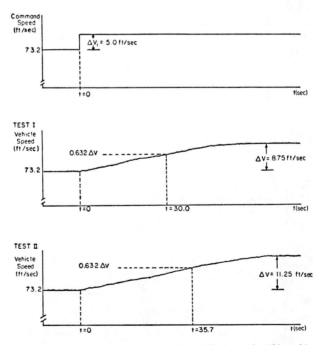

Fig. 6. Step response–uncompensated vehicle dynamics (50 mph).

interval 1-20 sec with essentially the same results as presented in the next section.

The test site was chosen so that there would be minimal effects from the environment on the collected data. These were collected over a two-month period on a nearly ideal road which was an essentially straight and approximately level section of typical interstate highway. Testing was only conducted on days when the concrete road surface was dry and the wind speed was low.

Experimental Results—Uncompensated Dynamics

The velocity-time histories obtained from two typical tests conducted at an average speed of 73.2 ft/sec are shown in Fig. 6. These data were both obtained in a 7-min period on the same section of roadway under virtually identical environmental conditions; however, note the differences in (C_1, C_2) which were

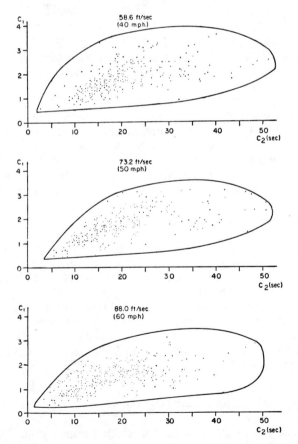

Fig. 7. Experimental data points—uncompensated vehicle dynamics (C_1 vs C_2).

obtained in each case. Such differences were present in all of the collected data as can **be** seen from Fig. 7 where the array of experimentally obtained values of (C_1, C_2) for each average test speed is presented. Note that the ranges of variation of C_1 and C_2 are enclosed by a solid line in each case.

The mean values of C_1 and C_2 at each average speed, together with the corresponding standard deviations are listed in Table I. Note both the similarity of

TABLE I

Statistics of Uncompensated Vehicle Parameters

Vehicle speed	Mean values		Standard deviations	
ft/sec	C_1	C_2 (sec)	C_1	C_2 (sec)
58.6 (40 mph)	1.940	22.075	0.672	10.654
73.2 (50 mph)	1.784	21.314	0.614	9.500
88.0 (60 mph)	1.450	18.207	0.526	8.217

the mean values of C_1 at each average speed and the similarity of the corresponding standard deviations. The same effect can also be noted for C_2 at each average speed.

Given such similarities, it was decided to combine all of the data (i.e., eliminate average speed as a parameter). The resulting overall mean values of C_1 and C_2 are 1.72 and 20.47 sec with the corresponding values of the standard deviation being 0.639 and 9.62 sec, respectively. One measure of the effect of parameter variation is provided by the coefficient of variation which is the ratio of the standard deviation to the mean value of the parameter of interest. Here, the coefficients of variation for the vehicle gain and time constant are 37.3% and 47.4%, respectively.

There is no doubt that some of the variation noted here can be attributed to environmental effects; however, in view of the restricted experimental environment, it appears that most of the variation resulted from the nonlinear properties of the vehicle-roadway combination. Thus, even under the greatly restricted conditions considered here, Eq. (8) is not a valid model for vehicle longitudinal dynamics, and in all likelihood, neither is Eq. (5).

The implications of this finding for individual vehicle control are obvious, and it is highly desirable to reduce such variations so that one is dealing with a plant which can be characterized by a constant-coefficient differential equation. This goal can be accomplished by the use of control compensation as is demonstrated in the following section.

Experimental Results–Compensated Dynamics

Step-response data for the compensated vehicle dynamics (Fig. 4) were obtained to determine the range of variation for the parameters K_c and τ. The

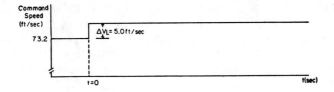

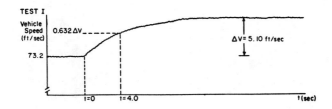

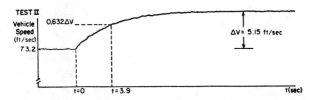

Fig. 8. Step response–compensated vehicle dynamics (50 mph).

compensator design was based on mean values of $C_1 = 1.72$ and $C_2 = 20.47$ sec, δ was chosen as 10 per a previous study,[8] and the design specifications for K_c and τ were 1.0 and 4.0 sec, respectively.

Two typical response curves are shown in Fig. 8 where the average vehicle speed was 73.2 ft/sec and the environmental conditions for each test run were essentially the same. Note from the figure that, in contrast with the previous data, these vehicle response traces are quite similar. Such similarity was present

TABLE II

Statistics of Compensated Vehicle Parameters

Vehicle speed	Mean values		Standard deviations	
ft/sec	K_c	τ (sec)	K_c	τ (sec)
58.6 (40 mph)	1.005	3.947	0.005	0.148
73.2 (50 mph)	1.012	3.957	0.031	0.189
88.0 (60 mph)	1.003	4.107	0.010	0.234

in all of the collected data as can be seen from Fig. 9 where the array of experimentally obtained values of (K_c, τ) for each average test speed is presented. Note from this figure that it was not necessary to obtain a large number of data points to accurately determine the range of variation of these parameters.

The collected data were essentially independent of average speed (see Table II),

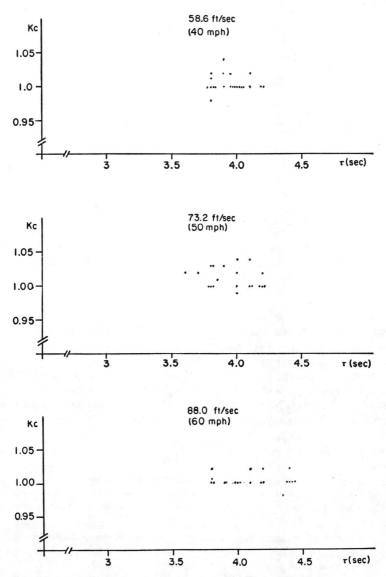

Fig. 9. Experimental data points—compensated vehicle dynamics (K_c vs τ).

and thus it was combined. The resulting overall mean values $\bar{K}_c$ and $\bar{\tau}$ are 1.006 and 4.002 sec, and the corresponding standard deviations are 0.031 and 0.200 sec, respectively. It is clear that a substantial reduction in the variability of the vehicle parameters was obtained via control compensation. [It is interesting to note that similar conclusions were reached in a study of vehicle modeling via a time-series analysis approach. Over the low-frequency range of interest (0–2 rad/sec), the models of the uncompensated vehicle dynamics were characterized by a linear correlation coefficient (ρ^2) which was less than 0.5, while models of the compensated dynamics were characterized by $\rho^2 > 0.95$ over this range. The latter models are contained in Ref. 9.]

To determine the range of validity of this model, additional tests were conducted in which command changes in velocity of ±10 ft/sec and ±15 ft/sec were employed. The system behavior was essentially as predicted by Eq. (10) per the results presented in the Appendix; however, it is anticipated that for larger changes of command velocity the linear capabilities of the propulsion system would be exceeded and Eq. (10) would no longer be valid.

Conclusions

In light of the experimental results, it is concluded that the uncompensated dynamics of a highway vehicle should not be represented by a first-order differential equation with two fixed parameters. This representation is not a particularly good one even under the nearly ideal-road, small-signal conditions considered here, and one would strongly expect that it would not be valid under large-signal conditions.

An examination of the data in the previous section clearly shows that the parameter variations and the effects of disturbance forces can be significantly reduced through the use of control compensation and feedback. Further, essentially the same data were obtained at each test speed, thus indicating that the small-signal model (10) is valid at other average speeds within the range from 58.6–88.0 ft/sec. Hence, it appears that it can be used with considerable confidence for predicting system performance. Given such a model, one can examine various control schemes and obtain realistic estimates of system performance, which would closely approximate those expected under full-scale operating conditions.

Any dynamic improvements achieved by control compensation must ultimately be effected by the propulsion system output, in this case the engine force. One should therefore immediately suspect that large-valued, rapid changes of the input variable (v_i) would place impractical transient requirements on the manipulated variable of engine force as well as on the relevant gears, tires, etc., and a first-order model would no longer be valid. Thus, for practical design considerations, one must insure that the force requirements are within the linear capabilities of the propulsion system if that system is to be represented by a linear relationship.

The results discussed here were obtained from experiments on a single vehicle under nearly steady-state conditions; however, the same general conclusions could be drawn for any automobile although, of course, different statistics for C_1 and C_2 would probably be obtained in any particular case. Further, similar conclusions would probably also be valid for vehicles with unconventional power plants.

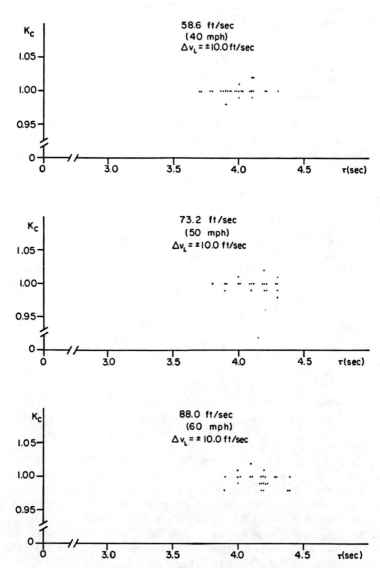

Fig. 10. Experimental data points—compensated vehicle dynamics (K_c vs τ).

Appendix: Additional Experimental Work on Compensated Vehicle Model

The validity of the compensated system was further investigated under quasi-steady-state conditions for two additional cases. Here, in contrast with the previous study, these experiments were performed on a 10-mile section of typical interstate highway wherein the roadway geometry was comprised of varying roadgrades and curves as well as straight and level sections. Testing was conducted on days when the highway surface was dry and the wind speed was nominal, i.e., the velocity varied from 0–25 mph with gusts up to 35 mph. Under these conditions, step-response data were collected as previously described at average speeds of 58.6, 73.2, and 88.0 ft/sec (40, 50, and 60 mph) with the corresponding magnitudes of the command step change (v_L) being ±10.0 and ±15.0 ft/sec at each test speed. The compensator design was based on mean values of $C_1 = 1.72$ and $C_2 = 20.47$ sec, δ was chosen as 10, and the design specifications for K_c and τ were 1.0 and 4.0 sec, respectively.

The range of variation of the parameters K_c and τ for the first of these studies ($v_L = \pm10.0$ ft/sec) is shown in Fig. 10 where an array of experimentally obtained values for each test speed is presented. Note from this figure that essentially the same results were obtained at each speed; thus, it was not necessary to obtain a large number of data points to accurately determine the range of variation for these parameters. The resulting overall mean values of K_c and τ are 0.997 and 4.093 sec with the corresponding values of the standard deviation being 0.015 and 0.180 sec, respectively.

Similar data were obtained for the case where v_L was set equal to ±15.0 ft/sec for each test speed. Here, the overall mean values of K_c and τ are 0.992 and 4.10 sec with the corresponding values of the standard deviation being 0.033 and 0.251 sec, respectively.

An examination of the data shows that the effects of both the parameter variations and the disturbance forces (varying roadway geometry, wind, etc.) can be significantly reduced through the use of control compensation and feedback. Furthermore, these data were for large values of v_L with the resulting system behavior being essentially the same as that predicted by the compensated vehicle model. Thus, it appears that engine force requirements were still within the linear capabilities of the propulsion system. However, it is anticipated that for larger changes of command velocity the linear capabilities of the propulsion system would be exceeded and the compensated vehicle model (Eq. 10) would no longer be valid.

Acknowledgments

The authors wish to thank Messrs. J. Houston, R. Ventola, C. Ventola, and Miss Diane Siekierski, who gave generously of their time, interest and support to help complete this research.

References

1. W. S. Levine and M. Athans, On the optimal error regulation of a string of moving vehicles, *IEEE Trans. Auto. Cont.* AC-11-3, 355-361 (1966).
2. A. H. Levis and M. Athans, On the optimal sampled-data control of a string of vehicles,
Trans. Sci. **2**, 362-382 (1968).
3. S. M. Melzer and B. C. Kuo, The optimal regulation of a string of moving vehicles through
difference equations, *Proc. J. A. C. C.*, Georgia Institute of Technology, Atlanta, Georgia
(June 22-26, 1970).
4. E. P. Cunningham and E. J. Hinman, An approach to velocity/spacing regulation and the
merging problem in automated transportation, *Proc. Joint Trans. Eng. Conf.*, Chicago,
Illinois (October 11-14, 1970).
5. D. F. Wilkie, A moving cell control scheme for automated transportation systems, *Trans.
Sci.* **4**, 347-364 (1970).
6. R. E. Fenton et al., One approach to highway automation, *Proc. IEEE* 56 (April 1968).
7. J. G. Bender and R. E. Fenton, A study of automatic car following, *IEEE Trans. VTG*,
VT-18-3, 134-140 (1969).
8. J. G. Bender, Experimental studies in vehicle automatic longitudinal control, The Ohio
State University, Columbus, Ohio, Report Engr. Exp. Sta. 276A-5 (August 1968).
9. R. G. Rule, R. E. Fenton, and J. G. Bender, On modeling vehicle dynamics—A time-
series analysis approach, *Proc. 26th Ann. ISA Conf.*, Chicago, Illinois (October 4-7,
1971).

Car Following and Spectral Analysis

John N. Darroch
The Flinders University of South Australia, Bedford Park, South Australia
and
Richard W. Rothery
General Motors Research Laboratories, Warren, Michigan

Abstract

An analysis is presented here of a linear car-following model using spectral analysis techniques. In particular, it is assumed that the dynamical characteristics of this traffic situation can be described by a weighting function, $h(\tau)$, or equivalently by its Fourier transform, the frequency response function, $H(f)$.

A brief description is also given of the role these functions play in the analysis as well as a presentation of their numerical estimates obtained from car-following data.

Introduction

In one of the earliest papers which attempted to describe the manner in which one vehicle followed another on a single lane of roadway, Chandler, Herman and Montroll[1] suggested the following relation:

$$\ddot{x}_{i+1}(t) = \int_0^\infty [\dot{x}_i(t - \tau) - \dot{x}_{i+1}(t - \tau)] \, h(\tau) \, d\tau, \tag{1}$$

where $x_i(t)$ is the position of the ith vehicle at the time t, and $h(\tau)$ is a weighting function which is defined as the response of the driver to a unit stimulus applied at a time τ earlier. The integral in Eq. (1) represents the total response of the driver executed at a time t. This is a weighted average of all earlier values in the relative speed between the two vehicles. The weighting function, $h(\tau)$, reflects a driver's estimation, evaluation and information processing in the past until the time now to which he then responds at some future time, t. This concept is illustrated schematically in Fig. 1. We remark that the model given by Eq. (1) can be made more general by allowing $h(\tau)$ to depend on such variables as intervehicle spacing, absolute speed and the sign of relative speed (see, e.g., Refs. 4 and 5).

While the general features of the weighting functions shown here in Fig. 1 are likely to be reasonable, there is little information from which to construct it in detail. Certainly what has happened a number of seconds in the past is not particularly relevant to the driver now, and furthermore, for a short time in the past

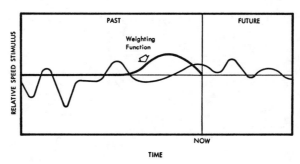

Fig. 1. Schematic diagram of a relative speed stimulus and a weighting function versus time.

a driver cannot evaluate the information available to him. For reasons of simplicity the choice of $h(\tau)$ that has dominated the literature is

$$h(\tau) = \lambda\delta(\tau - T), \tag{2}$$

where $\delta(\tau - T)$ is the Dirac delta function. Equation (1) then becomes

$$\ddot{x}_{i+1}(t) = \lambda\{\dot{x}_i(t - T) - \dot{x}_{i+1}(t - T)\}. \tag{3}$$

We note that in Chandler et al.[1] and Herman et al.[2] the mathematical analysis of the stability of the propagation of disturbances was based on the model given by Eq. (3).

Lee[3] investigated the consequences of using different analytical forms for $h(\tau)$ in the model given by Eq. (1). By contrast, our approach that we shall use here is more of an empirical one. We consider the problem of relating Eq. (1) to an observed data set and, for this, it is necessary to make Eq. (1) more realistic by the inclusion of an error term. Precisely, we assume that $y(t) = \ddot{x}_2(t)$ and $x(t) = \dot{x}_1(t) - \dot{x}_2(t)$ are both *stationary time series* and are linearly related by the equation

$$y(t) = \int_0^\infty h(\tau)x(t - \tau)\,d\tau + z(t), \tag{4}$$

where $z(t)$ is an error term which is assumed to be uncorrelated with $x(t)$. This model is not fully general since it does not allow $h(\tau)$ to depend on the intervehicle spacing, absolute speed or the sign of relative speed as it possibly should (see, e.g., Refs. 4 and 5). However, Eq. (4) is considerably more general than

$$y(t) = \lambda x(t - T) + z(t) \tag{5}$$

which is, apart from the error term, the model that serves as a basis for most of the literature on car-following.

Spectral Analysis Considerations

For the benefit of traffic theorists who are unfamiliar with spectral analysis, we give here a brief introduction to some of the principal concepts. The mathematical aspects are over-simplified and nonrigorous. For a fuller discussion of the same type see Bendat and Piersol.[6] For a rigorous discussion see Jenkins and Watts[7] and the references given in that book.

The initial step towards an analysis of Eq. (4) is to take its Fourier Transform, i.e.,

$$Y(f) = H(f) X(f) + Z(f), \qquad (6)$$

where

$$H(f) = \int_0^\infty h(t) e^{-2\pi i f t} dt$$

and

$$X(f) = \int_{-\infty}^\infty x(t) e^{-2\pi i f t} dt.$$

with similar expressions for $Y(f)$ and $Z(f)$ as functions of frequency, f.

We next rewrite $H(f)$, the "frequency response function" in polar coordinates as

$$H(f) = \lambda(f) e^{-i\phi(f)}, \qquad (7)$$

where $\lambda(f)$ is the "gain factor" and $\phi(f)$ is the associated "phase factor." It is convenient to express $\phi(f)$ in terms of $\tau(f)$, the "time lag" at frequency f. This is defined by

$$\tau(f) = \frac{\phi(f)}{2\pi f} ,$$

and to let $X_{\tau(f)}(f)$ denote the Fourier Transform of $x[t - \tau(f)]$. Then Eq. (6) can be expressed as

$$Y(f) = \lambda(f) X_{\tau(f)}(f) + Z(f). \qquad (8)$$

We note that Eq. (8) above is a direct analog, for frequency, f, of the simple linear model given by Eq. (5). It is also noted that for Eq. (5) to be valid it is necessary that both $\lambda(f)$ and $\tau(f)$ be constants. In the analysis of experimental data we shall be particularly interested in the extent to which both $\lambda(f)$ and $\tau(f)$ change with frequency.

Thus, much of our interest is focused on the polar coordinates of $H(f)$. Now $H(f)$ is related to the spectral density function;

$$S_{xx}(f) = E[X^*(f) X(f)],$$

and the cross spectral density function,

$$S_{xy}(f) = E[X^*(f) Y(f)]$$

by

$$H(f) = \frac{S_{xy}(f)}{S_{xx}(f)} \qquad (9)$$

(see, e.g., Eq. (8.4.5) of Ref. 7). It is noted that $X^*(f)$ is the complex conjugate of $X(f)$.

Equation (9) above is an immediate consequence of the fact that $X(f)$ and $Z(f)$ are uncorrelated since by assumption the processes $x(t)$ and $z(t)$ are uncorrelated. The estimation of $H(f)$ is performed via the estimation of $S_{xy}(f)$ and $S_{xx}(f)$.

In addition to $\lambda(f)$ and $\tau(f)$ there is a third function which we shall use to help describe the behavior of the following driver. This is the coherency function,

$$\gamma^2(f) = \frac{|S_{xy}(f)|^2}{S_{xx}(f) S_{yy}(f)} .$$

This function measures the correlation between the frequency f components of $x(t)$ and $y(t)$ and can take any value between 0 and 1.

One final function which will be estimated is the spectral density function of the error process $z(t)$, namely,

$$S_{zz}(f) = E[Z^*(f) Z(f)]$$
$$= S_{yy}(f) [1 - \gamma^2(f)] . \qquad (10)$$

In the next section, we describe the equations used to calculate estimates of these spectral functions from numerical data obtained in a car-following situation.

Estimates of the Spectral Density Functions

The relative speed and acceleration history measured in the vehicle-following experiment was reduced to a digital format at a sampling rate of 2 per sec. This analog to digital process filtered out all possible frequencies above the so-called Nyquist frequency, f_c, which is given by

$$f_c = \frac{1}{2\Delta} ,$$

where Δ^{-1} is the rate. This is more than an adequate representation of the data in that performance capabilities of present day vehicles prevent all but very small amplitude oscillations in $x(t)$ and $y(t)$ for this high of a frequency, i.e., 1 cycle/sec. Therefore, the physically realizable frequency range considered was from zero to 1 cycle/sec at the discrete frequencies

$$f_k = \frac{k f_c}{m} \qquad k = 0, 1, 2, \ldots m,$$

where m is the maximum number of lag values.

The spectral density function, e.g., in Eq. (9) was estimated by using the following equation for numerical calculations (the notation used here follows Bendat and Piersol[6] in distinguishing an estimate from that which is being estimated by the use of the symbol, ^):

$$\hat{S}_{xx}(f_k) = 2\left\{\hat{R}_{xx}(0) + 2 \sum_{\epsilon=1}^{m-1} D_\epsilon \hat{R}_{xx}(\epsilon) \cos\left(\frac{\pi \epsilon k}{m}\right)\right\} \tag{11}$$

Here $R_{xx}(\epsilon)$ is the auto covariance function which was estimated using

$$\hat{R}_{xx}(\epsilon) = \frac{1}{N-\epsilon} \sum_{n=1}^{N-\epsilon} x_n x_{n+\epsilon} \qquad \epsilon = 0, 1, 2, \ldots m, \tag{12}$$

where x_n's are the digitized values of $x(t)$ at the times $n\Delta, n = 1, 2, \ldots$. Estimates for R_{yy} and R_{xy} are similarly defined. The weighting function, D_ϵ, that was used and which appears in Eq. (11) is given by

$$D_\epsilon = 1 - \frac{\epsilon}{m} \qquad \epsilon = 0, 1, 2, \ldots m,$$

$$= 0 \qquad \epsilon > m. \tag{13}$$

The numerical calculates that were carried out in order to obtain estimates of the coherence function, $\gamma^2(f)$, the gain factor, $\lambda(f)$, and the time lag, $\tau(f)$, used the following expressions:

$$\hat{\gamma}^2(f_k) = \frac{(\hat{C}_k^2 + \hat{Q}_k^2)}{\hat{S}_{xx}(f_k)\,\hat{S}_{yy}(f_k)},$$

$$\hat{\lambda}(f_k) = \frac{|\hat{S}_{xy}(f_k)|}{\hat{S}_{xx}(f_k)} = \frac{\sqrt{\hat{C}_k^2 + \hat{Q}_k^2}}{\hat{S}_{xx}(f_k)}$$

and

$$\hat{\tau}(f_k) = (2\pi f_k)^{-1} \tan^{-1}(\hat{Q}_k/\hat{C}_k),$$

where

$$\hat{C}_k = 2\left[A_0 + 2\sum_{\epsilon=1}^{m-1} A_\epsilon D_\epsilon \cos\left(\frac{\pi \epsilon k}{m}\right) + (-1)k A_m\right],$$

$$\hat{Q}_k = 4\sum_{\epsilon=1}^{m-1} B_\epsilon D_\epsilon \sin\frac{k}{m}$$

and

$$\left\{\begin{matrix} A_\epsilon \\ B_\epsilon \end{matrix}\right\} = \frac{1}{2}\{[\hat{R}_{xy}(\epsilon) \pm \hat{R}_{yx}(\epsilon)]\}.$$

Darroch and Rothery

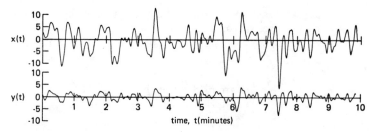

Fig. 2. The relative speed, $x(t)$, and acceleration of a following vehicle, $y(t)$, are shown versus time.

Experimental Techniques

The car-following observations that will be analyzed in this paper were obtained on a freeway facility which is part of the Inter-State highway system of the United States. Two vehicles, a leader and a follower, were used. These vehicles remained in the right lane and traffic was sufficiently light that other motorists did not interfere with their movement.

The driver of the lead vehicle varied his speed "randomly". The only instruction given to the driver of the following vehicle was to "follow" the lead vehicle in a manner which he considered reasonably safe. The duration of the test run was approximately 10 min. During this period the trajectories of the two vehicles were recorded simultaneously on a single magnetic transport by telemetering the position information of the lead vehicle to a receiver mounted in the "follower". The recordings were synchronized by recording a clock channel which was used to gate the information from each vehicle on playback during the data reduction phase. From this information, the relative speed, $x(t)$, and acceleration of the following vehicle, $y(t)$, were calculated. These two variables, for this data set, are shown in Fig. 2.

Analysis

The normalized estimate of the cross correlation function $\hat{\rho}_{xy}(\epsilon)$ between $x(t)$ and $y(t)$ for the data shown in Fig. 2 is plotted against lag time, ϵ, for values up to ± 50 sec in Fig. 3. One notes that this estimate of the cross-correlation function has a maximum of 0.84 at $\epsilon = 0$. In an initial calculation for the cross correlation the maximum value occurred when $\epsilon = 1.5$ sec. The calculation appearing in Fig. 3 has been "aligned." Alignment consists of slipping the time reference of one variable with respect to the other in order that the cross-correlation function peaks at $\epsilon = 0$. The alignment number is equivalent to T. This procedure of aligning the cross-correlation function is carried out in order to improve the accuracy of estimates of the spectral quantities (see, e.g., Jenkins and Watts, Ref. 7).

Using the above values for the lag time and the value of the cross-correlation function for this value of ϵ, the "least-squares" estimates for the parameters λ

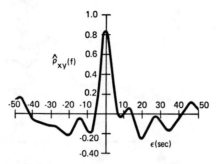

Fig. 3. The estimate for the "aligned" cross-correlation function, $\hat{\rho}_{xy}(\epsilon)$, is shown versus lag time, ϵ, for the relative speed and acceleration data shown in Fig. 2.

and T appearing in the simple linear car-following model are 0.28 sec^{-1} and 1.5 sec, respectively.

The estimates for the spectral density functions for acceleration response, $y(t)$; relative speed, $x(t)$; and the error term, $z(t)$, are shown versus frequency in Fig. 4. As was indicated earlier, the frequency range is low, the main contribu-

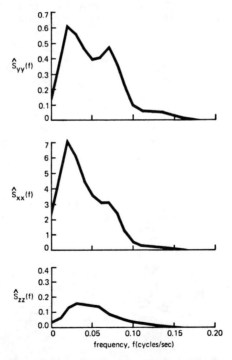

Fig. 4. Estimates for the spectral density function for the acceleration response, $y(t)$; relative speed, $x(t)$; and the error term, $z(t)$, are shown versus frequency (cycles/sec).

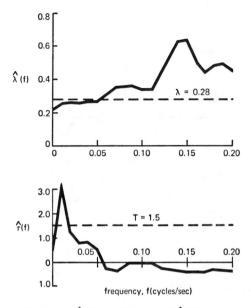

Fig. 5. Estimates for the "gain," $\hat{\lambda}(f)$, and "time lag," $\hat{\tau}(f)$ are shown versus frequency. The dashed line in each graph indicates the "least-squares" estimate of λ and T, assuming $h(\tau)$ is given by Eq. (2).

tions coming from frequencies less than $\frac{1}{10}$ cycle/sec. The error term's contribution while not zero, is small compared to the acceleration response and is not sharply focused in any particular set of frequencies.

Estimates of the "gain" and "time lag" elements of the frequency response function are shown in Fig. 5 where $\hat{\lambda}(f)$ and $\hat{\tau}(f)$ are plotted versus frequency. Superimposed on these graphs are the "least-squares" estimates of λ and T.

Since our weighting function $h(\tau)$ is the Fourier transform of the frequency

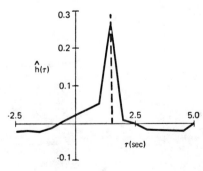

Fig. 6. The estimates of the weighting function, $h(\tau)$, versus τ. The dashed line is the "least-squares" estimate, assuming $h(\tau)$ is given by Eq. (2).

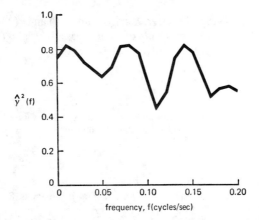

Fig. 7. The estimates for the coherence function, $\gamma^2(f)$ versus frequency (cycles/sec).

response function, i.e.,

$$h(\tau) = \int_0^\infty H(f)e^{-2\pi i\tau f}\, df,$$

it was estimated by using the equation

$$\hat{h}(\tau) = \frac{1}{m}\sum_{k=0}^m \frac{\hat{C}_k}{\hat{S}_{xx}(k)}\cos\frac{\pi\tau k}{m} + \frac{1}{m}\sum_{k=0}^m \frac{\hat{Q}_k}{\hat{S}_{xx}(k)}\sin\frac{\pi\tau k}{m}.$$

The results of estimating $h(\tau)$ are shown in Fig. 6. Superimposed on this figure is the weighting function of Eq. (2) using the "least-squares" estimates for λ and T.

The estimate for the coherence function, $\gamma^2(f)$, is shown in Fig. 7. While this estimate oscillates considerably, it does indicate a relatively high correlation (which tends to decrease with increasing frequency) for each f-frequency component between $x(t)$ and $y(t)$.

Conclusions

Our principal aim in this work has been to point the way towards what appears to be a fruitful method of analyzing car-following models and their associated experiments that heretofore has not been attempted. The data from one test has been presented and analyzed using this technique in order to provide an example. We have not dealt with the question of the accuracy of the estimates and the estimated functions must therefore only be interpreted as suggestive and not necessarily typical. A number of carefully controlled experiments will be required in order to determine whether drivers in this situation can be described by a specific weighting function and an equation of the type given by Eq. (4). The qualitative agreement, however, between λ and T and $\lambda(f)$ and $\tau(f)$ suggest that it might be possible to use these aspects of time series analysis to determine in more detail the characteristics of driver behavior.

Acknowledgments

It is a pleasure to acknowledge with thanks the support of Mrs. Susan Klemmer who programmed the numerical analysis for the digital computer.

References

1. R. E. Chandler, R. Herman, and E. W. Montroll, Traffic dynamics: Studies in car follow-ing, *Opns. Res.* **6,** 165–184 (1958).
2. R. Herman, E. W. Montroll, R. B. Potts, and R. W. Rothery, Traffic dynamics: Analysis of stability in car following, *Opns. Res.* **7,** 86–106 (1959).
3. G. Lee, A generalization of linear-car following theory, *Opns. Res.* **14,** 595 (1966).
4. D. C. Gazis, R. Herman, and R. B. Potts, Car following theory of steady state traffic flow, *Opns. Res.* **7,** 499 (1959).
5. D. C. Gazis, R. Herman, and R. Rothery, Non-linear follow-the-leader models of traffic flow, *Opns. Res.* **9,** 545–567 (1961).
6. J. S. Bendat and A. G. Piersol, *Measurement and Analysis of Random Data,* John Wiley and Sons, New York, 1966.
7. G. M. Jenkins and D. G. Watts, *Spectral Analysis and its Applications,* Holden-Day, San Francisco, 1968.

A Study of Individual Journey Series: An Integrated Interpretation of the Transportation Process

Velibor Vidakovic

Department of Public Works, Urban Research Division, Amsterdam, Netherlands

Abstract

This paper reports the essential findings from an explorative study of the individual series of journeys. The objective of the study was to develop an interpretation of the transportation process that differs from current transportation study techniques by taking into account the linkage of journeys on the individual level. According to the detective character of this study, many of its results are experimental ones, demonstrating the kinds of relationships previously unknown, such as: individual journey frequency as a function of daily linkage of journey ends; relation of the real to the potential travel demand as evaluated from the activities and journey ends linkages; complementary nature of vehicular and walking journeys in relation to the individual journey frequency; journey chains generation and the association of journey purposes; modal distribution in a chain; relation of the travel time to the duration of stay as an aspect of interaction of journey ends within a chain; morphological characteristics of the journey chains.

Evaluating the general meaning of this approach and the specific results obtained, the study concludes with the recommendations of methodological importance, concerned with transportation survey design, integrated interpretation procedure and possible applications to developing an adequate model.

Introduction

The Position of the Study Within the Wider Scope of Urban Research

The transportation study here presented forms a part of an urban research project* which includes: (1) developing the continuous observation of the urban process; (2) analytically expressing the principles of this process; and (3) outlining a systems optimization approach to city planning. Accordingly, the meaning of the study discussed in this paper should be seen within an observation and interpretation program covering the following fields:

(a) socio-functional structure of activity generators (individuals, households, and enterprises);

(b) pattern of space units with regard to building-type and physical state and their geographical distribution;

---

*This project is in its initial phase and it is being supported by the Department of Public Works in Amsterdam. The details on the subject discussed in this paper are reported in the unpublished dissertation of the author "Characteristics of the Urban Transportation Structure," at Technische Hogeschool, Delft.

(c) pattern of channels of communications with regard to potential and actual use and physical state;

(d) time series of activity generation, intensity of space use and communications characteristics;

(e) linkages of activity generators and space units;

(f) activity generators' locational pattern.

As the specific portion of this program, the explorative study of the individual journey series is mainly related to the subjects indicated above as (a), (d), and (e).

The Concept of the Individual Journey Series

Notwithstanding the considerable variety with regard to modeling technique, the current transportation study practice offers a simplified picture of an individual's travel pattern, this latter being assumed to consist of a number of single, independent journeys. Inherent in this approach, each journey is separately described, normally using the data: socio-economic characteristics of travel-maker; location and "land use" of origin and destination; "purpose," mode, distance, duration and time interval of a journey. Although this method provides a correct record of the facts on a given journey, it is here suggested that it is not methodologically satisfying for two reasons: (1) it disregards the succession of journeys as an important aspect of interdependence; (2) it ignores to a great extent the activity pattern as background to the travel pattern.

In order to improve the interpretation of the transportation process, it is necessary to associate the data on both the successive journeys[†] and the sojournments.[‡,§] In that context, the way the individuals produce their series of activities, divide their time into the suitable activity periods, and visit and stay at a number of locations, should be analyzed. Consequently, the journeys should be treated (a) as a part of an individual's time allocation; (b) as an individual journey series, based on either the whole-day period or the period between two sojournments at the same place.

Method, Course and Scope of the Survey

In view of the nature of the study, a method of data collection was required which would provide a detailed record of the individual's use of time and space during the whole observation period. As a technique meeting this requirement,

---

[†]By "journey" is understood the movement of a person from one stationary condition to the following one, whereby those stationary conditions which do not form the destination of the particular movement are not taken into consideration (e.g., transfer points).

[‡]By the expression "sojournment" is understood the stationary condition (i.e., staying at one address) of a person between two successive journeys, no matter how short.

[§]An exception to both these definitions is applied to taking a stroll beginning and ending at the same address. This is seen as a sojournment finding place outside the address, with the journey time between the sojournments taken as zero.

the time-budget survey was applied. In addition, the following modifications have been introduced:

(a) the inquiry for the time-budget notations was designed as a combination of pre-coded and open questions;
(b) during the observation period, the interviewers were in regular contact with the sample persons, checking their records.¶

The survey has been carried out with a 1% sample taken at random from households in Old-South Amsterdam.** All persons aged 15 to 80 years were asked to keep a diary for two consecutive days, noting down:

(1) for every single activity (irrespective of sojournment or activity duration): kind of activity, time interval, kind and full address of the sojournment place;
(2) for every single journey (irrespective of mode and duration): departure and arrival time (based on door-to-door time), and mode(s) of travel (if public transport, also the line numbers).

To insist upon the precision of records, a classification has been introduced, which distinguish between 10 basic travel modes (with the combination thereof), 73 land use categories for the space units (further to identify by address) and 38 kinds of single activities. The returns of the survey embodied the particulars on an average of 23 activity and/or journey intervals per person-day.

The Findings

In accord with the detective character of the study, many of its results are experimental ones, demonstrating the kinds of relationships previously unknown. In the following paragraphs, the findings related to the linkage of journeys, activities and sojournments of the individuals in the sample studied, are discussed.

¶Besides this field-control, a number of other checks of the reliability of responses have been done during coding, such as: (1) checking the reliability of travel duration data, against the limit values of speed per distance and mode, in the case of public transport also taking account of the known frequencies and/or time table; (2) checking the accuracy of the given position of visit places within the sample area, based on the register of shops, services, office buildings, etc.; (3) examining the consistency of the individual time allocation with regard to: (a) known operating periods of various activities such as shops, offices, TV programs, concerts and the like; (b) correspondence of time and other data concerning activities (journeys and sojournments) shared by different members of the same household. Based on all these (and other) checks, about 13% of all time-budget books were found to be unreliable and were eliminated from further analysis.

**For the first stage of the development of the study, this part of Amsterdam has been chosen; it is representative of the pre-war city structure, distinguished by high residential density, the presence of several different socio-economic groups, and a great diversity of economic activities. A point of special interest for planners is that it forms a part of a transition zone designated for general reconstruction. In the further development and extension of the study, which is planned for the period September 1971–May 1972, a 1% sample will be taken from the rest of the urban area of Amsterdam.

Characteristics of the Individual Linkage of Journeys Within a Whole-Day Series

As a general introduction to the survey outcomes, some characteristics of the individual journey series by socio-occupational group are presented (Table I).

TABLE I

Some Characteristics of the Individual Journey Series by Socio-Occupational Group[a]

Socio-occupational group	Mean number per person-day				Mean length of a chain (km)	Mean journey duration (min)	Mean chain size (No. of links)
	Sojournments away from home	Journeys	Home-based chains	Activity transi-tions			
Workers	3.88	5.77	1.89	13.3	6.34	15	3.05
Housewives	3.44	5.22	1.78	16.1	3.46	10	2.93
Students	4.41	6.84	2.43	14.6	4.64	12	2,81
Pensioners	2.07	3.54	1.47	13.8	2.65	13	2,41
Total	3.58	5.44	1.86	14.6	5.52	12	2.92

[a]Mean values Monday through Saturday.

Herein, the mean journey frequency per person-day should be seen as a result of the activity transitions structure and the manner the sojournments away from home are linked. An average of 32% nonhome-based journeys is reported; however, this proportion varied between 17% (pensioners) and 34% (workers). Generally, the higher proportion of nonhome-based journeys is observed at the higher frequency of sojournments away from home.

Individual Journey Frequency as a Function of Daily Linkage of Journey Ends. The significance of the frequency of sojournments away from home for the evaluation of the journey frequency can be observed using the relationship between the number of sojournments per person-day and the mean number of journeys per sojournment (Fig. 1).

As shown, the linking of two sojournments away from home strengthens (i.e., the proportion of the nonhome-based journeys increases) with increasing number of places to be stayed at within a given interval. The same subject can also be described as the ratio of disconnecting the journeys (breaking them into the smaller home-based series) to a given number of sojournments; the experimental results suggest a probabilistic character of this ratio (Table II).

Relation of the Real to the Potential Travel Demand as Evaluated from Activities and Journey Ends Linkages. Another aspect of the individual series is the activity transition matrix which can be seen as the generator of the travel demand. Contextually, the activity matrix has been studied in two ways: as a matrix reduced to "dominant" activities (i.e., the longest ones per sojournment) and as an entire matrix, all activity periods counting equally. Whereas the former approach lends itself to the analysis of the journey purposes within the home-based chain (cf. Journey Chains Generation and the Association of Journey Purposes), the latter one enables the journey generation to be regarded in the

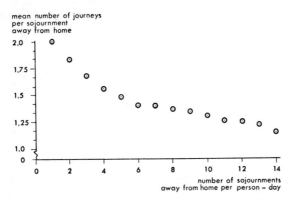

Fig. 1. Relationship between sojournment away from home and journey frequency.

TABLE II

Journey Disconnecting Ratio at Different Frequencies of Sojournment away from Home

Number of sojournments away from home per person-day (n)	2	3	4	5	6	7	8	9	10
Mean value of $\dfrac{J - n - 1}{n - 1}$ (J = number of journeys per person-day)	0.657	0.517	0.446	0.361	0.294	0.326	0.286	0.266	0.204

scope of the whole-day activity series, nondominant and all home-sited activities included.

The result of an analysis of the entire activity transition matrix with regard to the proportion of transitions being attended by a journey, is shown in Fig. 2. Therein, the cumulative percentage frequencies (as ordinates) of activity transitions are plotted against the classes of transitions ranked from the kinds of transitions with lower travel demand to the kinds of transitions with higher travel demand (as abscissas). For example, about 56% activity transitions of various kinds were attended in less than 30% of the cases by a journey, whereas 70% activity transitions were so attended in less than 60% of the cases, and so on.

It can be noted, that a considerable proportion of the total journey generation comes within those kinds of activity transitions which otherwise are distinguished by low and medium level of travel demand (see also Table III). From a great number of kinds of transitions, the following ones can be mentioned, just as an example:

zero level:	reading–watching TV; hobby–eating
low level:	house-keeping–eating; work (study)–taking a rest
medium level:	work (study)–eating; errands–leisure activities
high level:	work–errands; errands–house-keeping
100% level:	shopping–shopping; work–work; shopping–other.

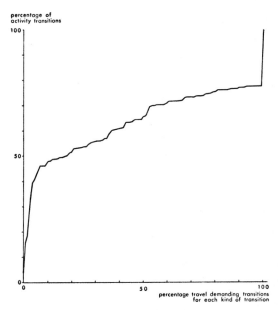

Fig. 2. Cumulative distribution of activity transitions plotted against the percentage of travel demanding transitions.

TABLE III

Distribution of Journey Generation by Level of Travel Demand

Level of travel demand[a]	Percentage activity transitions	Percentage of journey generation	Percentage categories of activity transition[b]
0	4	0	12
0,01–0,39	56	13	17
0,40–0,59	10	13	7
0,60–0,99	8	15	12
1,00	22	59	52
Mean level = 0,37	100	100	100

[a]Ratio between number of activity transitions attended by a journey and total number of activities transitions per category of transition.
[b]Based on 610 categories of activity transition in a 38 X 38 matrix.

Complementary Nature of Vehicular and Walking Journeys Subjected to an Individual Mobility Level. Compared to the conventional home-interview, usually missing a proportion of walking and/or short journeys, the time-budget method of data collection seems to give more insight into the totality of journey

TABLE IV

Distribution of Journey Frequency with Respect to the Basic
Modal Choice (Walk vs Vehicle)[a]

Category of choice per person-day	Percentage person-days	Mean journey frequency per person-day[b]		
		By foot	By vehicle	Total
All journeys by foot	32	4.70		4.70
All journeys by vehicle	20		3.60	3.60
At least one journey by foot resp. by vehicle	44	3.70	3.30	7.00
Made no journey	4			
All categories	100	3.10	2.20	5.30

[a]Mean values Monday through Saturday, all socio-occupational groups.
[b]Journeys between 24.00 and 07.00 excluded.

making (Table IV). Therefore, attention has been paid to the possible substitutive character of vehicular and walking journeys. For each level of mobility (i.e., total number of journeys per person-day) a split between walking and vehicular journeys has been regarded. This goes to show that the higher mobility level is being achieved by means of association of walking and vehicular journeys (see Fig. 3).

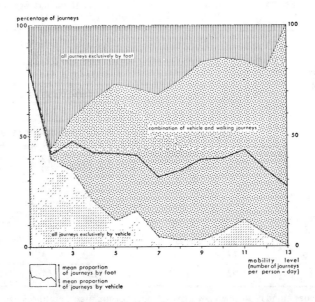

Fig. 3. Complementary nature of vehicular and walking journeys in relation to an individual mobility level.

Characteristics of the Individual Linkage of Journeys Within the Home-Based Series (Chains)

The whole-day journey series of the observed individuals are further studied in relation to the sojournments at home. This latter being taken always as origin and destination of a number of successive journeys, all journeys have been ordered into the home-based series, indicated as "chains". An average chain size of 2.92 links (i.e., journeys) has been found (cf. Table I), while in the frequency distribution by chain size 42.5% journeys formed a part of a two-link chain, followed by 18% and 13% journeys, respectively, within the 3-link and 4-link chains. Furthermore, a considerable proportion of journeys fell in the chains of the size 5 through 10 links.

Journey Chains Generation and the Association of Journey Purposes. The generation of chains by a number of successive sojournments away from home, made for one or more similar or different purposes, proved to be an important aspect of interdependence of journeys. The analysis of the survey data pointed out that almost one half of the sojournments away from home occurred within the two and more purpose chains (Table V). As shown, shopping and walking for pleasure occurred for 60% within a one-purpose chain (e.g., home-shop-shop-home), whereas the sojournments for personal business did so for only 31%. The sojournments away from home for work, education, social visit and other leisure activities came for about 50% within a one-purpose chain.

TABLE V

Distribution of Sojournments within the Chain per Dominant Activity,
by Chain Size and Structure[a]

Dominant activity at the sojournment away from home	Falling in the chain of mean size	Number of different dominant activities in the same chain					total
		1	2	3	4	5	
		The mean chain size					
		2.33	4.01	5.73	7.42	10.33	2.92
		Number of sojournments per thousand					
Work	3.64	117	73	26	12	3	231
Shop	3.32	193	93	28	7	1	322
Education	3.37	20	11	6	1	...[c]	38
Personal business[b]	3.82	37	49	25	5	2	118
Social visit	3.20	67	38	17	5	...[c]	127
Walk for pleasure	3.00	38	13	9	3	...[c]	63
Other leisure activities	3.32	34	31	15	7	1	88
Miscellaneous	4.14	5	5	2	1	...	13
Total	2.92	511	313	128	41	7	1000

[a]Mean values Monday through Saturday, all socio-occupational groups.
[b]Medical and dental service included.
[c]... indicates less than 7%.

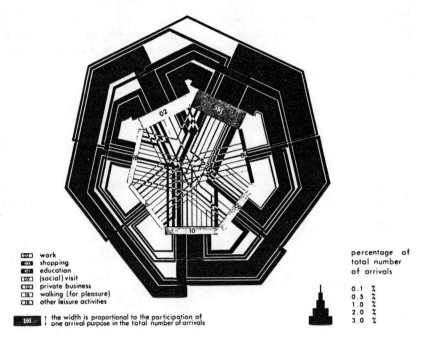

Fig. 4. The frequency wherewith certain associations of two and three arrival purposes occur within the same chain.

Generally, the increasing number of purposes (i.e., different dominant activities) ran concurrently with the increase of the mean chain size. Therein, the combinations of shopping with personal business, social visit or work, the combinations of work with personal business, social visit or other leisure activities and other such combinations totaling 80 in all, made a significant proportion of the journey chain generation (see Fig. 4).

Modal Distribution in a Chain. Analyzing the individual choice of one or more modes of journey-making in the same chain, about 72% of all journeys have been found within a one-mode chain and about 25% within a two-mode chain; the remainder fell in the chains where three and four modes were employed. Contextually, two aspects of modal choice within the chain will be here briefly reported: (a) the simultaneous choice of one mode for a number of journeys; and (b) the use of different modes for journeys within the same chain.

Besides the conventional frequency distribution of journeys per distance by mode, all one-mode chains have been analyzed with regard to the relation between the total chain length and the modes used. The survey data in question suggest a predominance of walk and bicycle chains in the classes of chain length up to 3 km; furthermore, the proportion of the two modes mentioned, declined surprisingly slowly, occurring even in the chain-length-class 5–10 km as a quarter of all chains. Generally, the concurrence of all modes was extended over the long scale of chain distances, from 2 to 15 km.

This indicates that the choice of one mode for a number of successive journeys is made by evaluating the given mode in such a way that its efficiency is related to the journey series rather than to each journey separately. However, with increasing chain size (i.e., number of links), the given mode may ever less satisfactorily correspond with the increasing diversity of journey characteristics, which intensifies a demand for an additional mode. As shown by Table VI, the

TABLE VI

Relation between Chain Size and Number of Journey Modes[a] Employed

Chain size (No. of links)	Percentage of chains made by two or more modes		Mean number of modes used	
	Workers	Housewives	Workers	Housewives
2	7	4	1.07	1.04
3	36	18	1.39	1.20
4	57	29	1.62	1.38
5	75	45	1.89	1.55

[a]Different modes and/or transfer within one journey excluded.

number of modes employed within the same chain increases with increasing chain size. As major combinations of travel modes in the same chain are reported: car-walking; walking-streetcar or bus; (motor)bicycle-walking; car-bicycle-walking and car-streetcar or bus-walking.

The ratio between the journey frequency by a given travel mode and the frequency of all journeys within those chains where at least one journey by that mode occurred, suggests that the streetcar or bus is the most dependently employed mode (ratio 0.47), car and motorcycle being less dependent (0.75), bicycle even less (0.79) and walking the least (0.82).

Relation of the Travel Time to the Duration of Stay as an Aspect of Interaction of Journey Ends Within a Chain. A subject of special interest in this study, the duration of sojournment (i.e., the stay time between two journeys) has been examined in various connections. The "mobilograph" (Fig. 5), based on the total amount of survey data, shows an average time-order of journeys and sojournments in the chains of different size. It suggests a decrease of the mean stay time with increasing chain size, from 148 minutes in a 2-link chain, 82 and 77 min, respectively, in 3- and 4-link chain through to 57 min in a 7-link chain. The mean journey time in chains of different sizes being approximately the same, this implies that the journeys come closer together with increasing chain size; in other words, the journey frequency per time unit spent away from home increases (e.g., 0.69 journeys per hour in 2-link chain against 1.09 journeys per hour in 7-link chain).

The similar information is obtained by comparison of the mean values of the stay time for different purposes in the 2-link chains and the longer ones (Table VII). It should be noted that sojournments which nominally have the same purpose, but which differ in duration, could differ substantially.

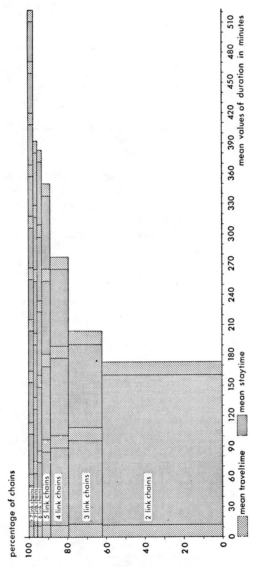

Fig. 5. Mobilograph (mean values Monday through Saturday).

TABLE VII

Mean Duration of Sojournment Away from Home for Selected Activities
and Two Classes of Chain Size

Dominant activity at the sojourn-ment away from home	Mean duration of sojournment (min)	
	In the 2-link chain	In the chain with more than 2 links
Shop	25	13
Social visit	151	40
Work	386	143
Other activities[a]	105	79
All activities	148	67

[a]Walking for pleasure excluded.

Another presentation of the relationship between travel and stay time is given by Figure 6, where, again for the 2-link chains and for the remainder, for each class of total chain duration, the mean ratio travel to total time is given.

Apart from the mean value approaches mentioned, the duration of sojournment has also been studied in its frequency distribution form. Therein, the

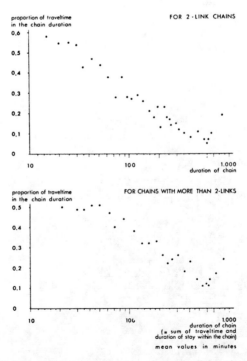

Fig. 6. Relationship between duration of travel and duration of stay.

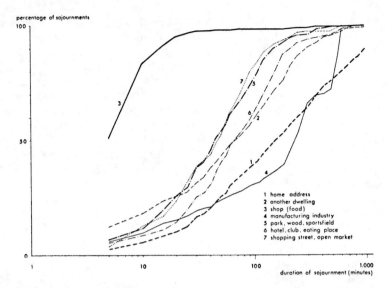

Fig. 7. Cumulative distribution of sojournments with respect to duration of sojournment (per category of place).

characteristic distributions have been found for different kinds of space units (Figure 7). As shown, a considerable proportion of short sojournments is reported, especially in food shops, but also in dwellings and recreation places; surprisingly, even at home still 17% of sojournments lasted no longer than half an hour.

Morphological Characteristics of the Journey Chains. The chains of the sample persons have also been analyzed with regard to their actual shape as represented by straight lines connecting addresses on the map. Thereby, the chains with 4 links and longer are found in two basic shapes: (a) a simple home-based chain, formed as a polygon, in 51% and 49% of cases respectively with and without crossed links; (b) a "bipolar" chain, where an address (usually the own work place, school or a dwelling) twice or more times in the same chain was visited. The latter form has been observed in 18% of all chains counting 4 or more links.

Distinguishing between home-based links and the others, data has been obtained which indicates the increase of spatial interaction of sojournments at increasing chain size. That can be illustrated by the home-based links being somewhat longer and the nonhome-based links being considerably shorter as the chain size increases.

Conclusions

The experimental results of the study indicate the methodological importance of the individual series approach to the transportation process. Based on the modified time-budget technique rather than on the conventional home-interview, the observation of the travel pattern within the succession of individual's journeys and sojournments proves to be useful for the simultaneous explanation of frequency, purpose, mode and spatial distribution of journeys. The first experience in this method suggests a revision of some conventional conceptions. However, to achieve the operational level of the study, further research is needed. Therein, the interest should be sustained for the (dominant) activities transitions matrices, sojournment distribution in space and time, and the journey linkage behavior of different socio-occupational categories. By an integrated interpretation of these elements of the urban process, the interaction of travel demand and space demand could be evaluated in such a way as to enable the development of an urban journey-and-sojournment model.

Modal Choice in Urban Areas

C. G. B. Mitchell
*Transport Research Assessment Group, Road Research Laboratory,
Crowthorne, Berkshire, England*
and
J. M. Clark
*Transport Assessment Group, Cranfield Institute of Technology,
Cranfield, Bedford, England*

Abstract

The common method of analysing modal choice aspects of urban travel survey data is to
assume that the value of time to a traveller is related to his income, and that a function
exists which relates the ratio of the numbers of travellers on two modes to the difference
between the generalized costs (money plus time) for these modes. Values of coefficients in
the function and the relation between the value of time and income can then be found by a
regression analysis. In many practical cases, however, more than two modes are present, a
new mode may be proposed, the data do not include travellers' income, and/or details of
individual journeys are not available. Under such circumstances, the method described
above has proved to be either impossible to apply or to give unsatisfactory results. The
approach described in this paper associates a value of time with each transport mode and
assumes that for journeys on which any two modes carry the same number of travellers the
generalized cost for the two modes is the same. The procedure for finding the values of
time on the modes available is to plot the proportions of travellers on each mode against
journey distance, to assume a value of time for one mode, and at each equal proportion
point on the graph to calculate the value of time on the other mode necessary to give equal
generalized costs. Crossovers between other pairs of modes can be used to check the consistency of the set of values of time. The generalized cost functions obtained by this
method are found to yield good fits to survey data from two dissimilar urban areas (Central
London and Stevenage New Town). Suggestions are made as to how the model can be
applied to situations where a new mode is to be introduced.

Introduction

The large number of transport studies which have been undertaken in urban
areas during recent years has brought sharply into focus the pressing need to
improve urban transport facilities. These improvements frequently take the
form of extending existing facilities, as in the construction of additional urban
motorways, or of developing more efficient operating techniques as is the object of demand responsive bus systems. Less frequently complete new systems
may be introduced, sometimes using existing technologies as with new rapid
transit services, sometimes involving the application of technological innovations
such as pedestrian conveyors. In either case the choice of which mode to improve or introduce is made after the capital investment necessary has been

References p. 85

71

estimated and a prediction made of the benefit to the system operator and to the community as a whole that will result from the investment.

Critical to the estimation of these costs and benefits is the prediction of the ridership of the new and old modes after the changes or innovations have been made. Part of this ridership will be due to trips generated by the changes—the so-called latent demand—but a large proportion will be trips which would have been made anyway using one of the original modes had no improvements been made. The distribution of these latter trips over the available modes is termed the modal split and the estimation of this distribution is the subject of this paper.

The calculation of the proportions of travellers using each mode is commonly made by setting up a model of passenger behavior, in which the choice between modes is assumed to be made on the basis of the relative costs and convenience of travelling by the different available modes. The elements of the cost and convenience of a trip are collected together into a "generalized cost," which in its most general form is the sum of the money costs of the trip, of weighted values of the access (walking and waiting) and riding times for the trip, and of the estimated money values of less easily quantifiable inconvenience aspects of the trip (e.g., difficulty of getting a seat, number of interchanges, unreliability). The form of the model and the weighting factors for time and convenience are determined by fitting the existing travel situation and the model can then be used to predict changes in ridership when the characteristics of the existing modes are changed or a new mode is introduced.

A great many models of modal split have been developed in recent years, and a number of these are mentioned in Refs. 1-10. Perhaps the most simple is the diversion curve, which relates the proportion of car owners travelling by public transport to the ratio of, or difference between, the travel times by private car and public transport. An example of this for London is given by Tresidder.[10] The most commonly used modal split procedure is that of postulating an expression relating the ratio of the numbers of travellers on two modes to the difference or ratio of the generalized costs of using the modes, and then using either linear regression[1] or discriminant analysis[5] to determine the constants in the expression.

Analyses of this type have been successful in determining the weighting factors, or "values of time," that convert to money values the passenger's access time to the mode and ride time on it; in addition they have sometimes been able to derive modal constants that may reflect comfort and convenience factors, but for such an analysis to succeed it is necessary to use travel data selected with great care. For example, it should only include trips in which genuine competition between modes exists, so that car ownership information is necessary; it should also include the traveller's income and the purpose of the trip, as it has been found that these affect the "value of time" in the generalized cost. Only a few cases are known in which the values of inconvenience factors have been determined. It should be noted that the "values of time" determined by these methods are essentially curve fitting parameters in an expression to determine the split of travellers between modes: their significance as indicators of the money value that travellers place on their time is secondary, and the "values of time" necessary to predict modal split correctly may be very different from the money value of saved time used in a cost-benefit analysis.

The summary of existing methods given above is deliberately brief, as these are well known. This paper presents a rather different method of modal split prediction, which is intended for two completely separate problems for which the existing methods may prove to be unsatisfactory. The first, for which it was originally derived by Clark and Seaton,[11] was to handle cases in which the nonquantifiable comfort and convenience aspects of competing modes are as important as their money costs and travel times in determining the modal split between them. The second, for which it has been used by Mitchell, is to make modal split predictions in situations for which the existing travel data are not sufficiently detailed to allow the successful use of the regression analysis described above.

Cases for Which Conventional Methods Fail

The method described in this paper was developed by Clark and Seaton to account for the modal split observed during the journey to work in Stevenage, a town of 58,000 population situated 30 miles north of London. In Stevanage the Development Corporation undertook a comprehensive traffic survey[12] between 1964 and 1966, which showed that the modes used to a significant extent for the journey to work were walk, cycle (including mopeds), bus, car (as driver) and car (as passenger). The ridership split between these modes varied with distance, as is shown in Fig. 1. The generalized cost of making a trip of

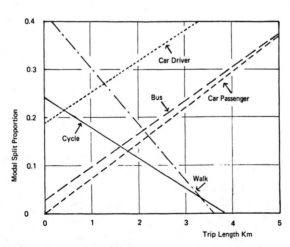

Fig. 1. Observed modal split.

distance x on mode i, $C_i(x)$, was taken to be

$$C_i(x) = P_i x \phi_i + \frac{V x \phi_i}{U_i} + V a_i, \tag{1}$$

where
  P_i = money cost per km of trip on mode i,
  V = "value of time,"
  ϕ_i = ratio of track to airline distance on mode i,
  U_i = block speed for mode i,
  a_i = access time for mode i.

The values of P_i, U_i, ϕ_i, and a_i are given in Table I and were taken from Ref. 13. The "value of time" was set at £0.15 per hour (0.6 d/min*) which closely approximates to the value obtained in orthodox studies and currently used by the Department of the Environment.[14] No values of ϕ_i were available for Stevenage and the values given in Table I were assumed, based on London data.[15]

TABLE I

Values of Constants Used in Stevenage Study

Mode	Money cost P_i (d/km)	Block speed U_i (km/hr)	Access time a_i (min)	Track/airline distance ϕ_i
Walk	0	6.4	0	1.15
Cycle	0	19.3	1	1.15
Car Driver	1.55	38.6	8	1.23
Car Passenger	1.86	38.6	10	1.23
Bus	3.10	22.5	9	1.26

The variation with distance of the calculated generalized cost for the different modes is shown in Fig. 2. It can be seen that the differences in these do not explain the observed modal split, and moreover, no "value of time" could be found which produced generalized costs that did explain the observations. A regression analysis of the data, grouped by trip length but not by origin zone, was made by Mitchell. For this analysis the following expression was used:—

$$n_1/n_2 = \exp - \alpha \{ P_1 x \phi_1 - P_2 x \phi_2 + V_A (a_1 - a_2)$$
$$+ V_R \left(\frac{x \phi_1}{U_1} - \frac{x \phi_2}{U_2} \right) + Q_1 - Q_2 \}, \tag{2}$$

where
  n_1, n_2 = numbers of trips made by modes 1 and 2 respectively,
  α = sensitivity parameter,
  V_A = "value of access time,"
  V_R = "value of ride time",
  Q = convenience factor,

*Throughout the paper, all costs are given in old pence (d), which at the time of writing are exactly equal in value to 1 U.S. cent.

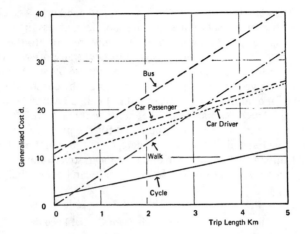

Fig. 2. Estimated generalized costs.

This gave inconsistent and implausible results for different pairs of modes, and a negative overall average value of α.

Another example of poor results from a regression analysis occurred with data for trips within the Central Area of London. These data for the period 8 a.m. to 10 a.m. were derived by Bunce from the London Traffic Survey (LTS) of 1962 and a British Rail Terminal Traffic Survey (BRS).[16] The L.T.S. and B.R.S. provided zone-to-zone matrices of numbers of travellers by the modes, private (cars and taxis), bus, underground rail, and walk, for trips wholly within the central area. These matrices, together with data on money cost, access time and ride time for the various modes (shown in Table II) were processed by a computer program called TRIPMAP to provide a listing of the numbers of trip desire lines starting or finishing in, or passing through, each of 35 regular hexagonal cells covering the central area. These trips were grouped by desire line direction into one of six 60° sectors and by airline trip length into one of

TABLE II

Constants Used in Central London Study

Mode	Money cost $\phi_i P_i$ (d/km)	Block speed $\overline{U}_i$ (km/hr)[a]	Dead time in vehicle b_i(min)	Access time a_i (min)
Walk	0	4.7	0	0
Private	1.5[b]	23.0	3.5	5.0
Bus	2.5	14.5	1.5	13.3
Underground rail	2.5	24.0	2.5	17.1

[a]The overall block speed is $U_i = \phi_i x/(b_i + x/\overline{U}_i)$.
[b]14d parking charge included in areas where this was applicable.

four length brackets (0-2 km, 2-4 km, 4-6 km, and 6-8 km). Associated with each group of trips was a set of average values of cost, access time and ride time. The computer program was intended to give insight into the movement pattern in a town, in order to aid the layout of new transport systems, and the use of its output for modal split studies was attempted on an experimental basis.

A regression analysis was carried out to determine the constants α, V_A, V_R, and Q_i in Eq. (2), for a number of groups of data. In all, 53 estimates of α, V_A and V_R were made. These estimates were not consistent and were often implausible. Slightly more consistent values were obtained for the products αV_A and αV_B than for V_A and V_B, and the attempt to determine the values of Q_i for the four modes was fruitless. The final mean and root mean square (r.m.s.) values of α, αV_A and αV_B are given in Table III, from which it can be seen that the standard

TABLE III

Results of Regression Analysis for Central London

	Mean	Standard Deviation
Sensitivity α	0.05 d^{-1}	0.13 d^{-1}
αV_A	0.085 min^{-1}	0.11 min^{-1}
αV_R	0.05 min^{-1}	0.19 min^{-1}

deviation of the estimates is up to four times their mean, and that the "values of time" are 1.0 d/min. (£0.25 per hour) for ride time and 1.7 d/min. (£0.425 per hour) for access time. The regressions typically accounted for 40 to 60% of the observed numbers of travellers.

In both the cases quoted above it was not, perhaps, surprising that inconsistent results were obtained, as it is known that great care is needed in collecting data for a regression analysis and the data used in these studies did not meet the required standards. In particular, the data used for regression did not include details of income, car ownership or trip purpose. However, a similar study in London by Research Projects Limited[9] in which a sixteen variable discriminant analysis was carried out on data obtained by an intensive socio-economic survey also produced poor results, with only 27% of the population allocated to the correct mode.

New Method

Clark and Seaton proposed a different model[11] to explain the observed modal split in Stevenage. This was based on the concept of finding individual "values of time" for each mode reflecting the different comfort and convenience levels of the mode. Thus, because riding in a car is more comfortable than riding in a bus, time spent in a car should be valued more cheaply than time in a bus.[†]

†The results given later in this paper suggest that comfort is not the primary factor that determines the "value of time" on a particular mode.

The method makes use of the original assumption that a traveller's choice of mode is entirely determined by the generalized costs for the modes. It follows that if at any distance the generalized costs for two modes are the same (in Fig. 1 such a distance is 2.2 km. for car passenger and walk) then equal numbers of travellers will use the two modes for trips of that distance. Thus, at such a distance x the numbers of travellers on the two modes, n_1 and n_2, are equal and

$$P_1\phi_1 x + V_{A1}a_1 + V_{R1}\frac{\phi_1 x}{U_1} = P_2\phi_2 x + V_{A2}a_2 + V_{R2}\frac{\phi_2 x}{U_2}. \qquad (3)$$

Note that no assumption of an expression connecting numbers of travellers with generalized cost, such as equation 2, is required for this method.

A "value of time" for the walk mode was assumed and the "values of time" for the other modes could be determined from equating the generalized costs at the distances for which they had as many passengers as the walk mode. Access time was valued as walking time by Clark and Seaton. The result of this analysis, for a range of values of walking time, is given in Table IV. It can be seen that bus time is valued lower than walk time, while car and cycle time is valued more highly.

TABLE IV

"Values of Time" Derived from Stevenage Data

Mode	Value of time (d/min)			
Walk	0.40	0.60	0.80	1.00
Car Driver	0.54	1.12	1.68	2.40
Car Passenger	0.26	0.68	1.30	2.00
Cycle	1.08	1.56	2.08	2.60
Bus	-0.60	-0.06	0.46	1.00

Mitchell applied this method to the same London data which had proved unsatisfactory for a regression analysis. Of the 35 cells 13 were selected, in which the four modes occurred in at least three of the four distance brackets, and modal proportion figures were plotted for the through trips in each cell. Examples of these plots are given in Fig. 3. Walk time was valued at 1.0 d/min. (£0.25 per hour) and the values of time in the other modes obtained. This was done, firstly valuing access as walk time and secondly giving the same value to access time as to riding time. It was found that the deduced values of times on the different modes were much more consistent between cells in the latter case, as is shown in Table V.[‡]

Further calculations were made starting with values of 0.5 d/min. (£0.125 per hour) and 1.5 d/min. (£0.375 per hour) for walking time, with $V_A = V_R$.

[‡]The values given in Table V are based on data from only 6 cells, while the values in Table VI are based on the full set of 13 cells. The differences between these two sets of estimates are less than the standard deviations for either set.

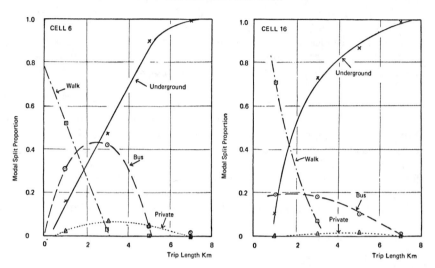

Fig. 3. Observed modal split—morning peak—typical cells.

TABLE V

Estimated "Value of Time" for Six Cells in Central London

Mode	Value of Time (d/min)			
	Total trip time[a]		Access time at 1.0d/min[b]	
	Mean	Standard deviation	Mean	Standard deviation
Walk (defined)	1.00	—	1.00	—
Private	1.72	0.27	2.00	0.44
Bus	0.90	0.16	0.53	0.51
Underground rail	0.72	0.13	−0.03	0.47

[a]With access and ride time given same value.
[b]With access time valued as walking.

The results of all three sets of calculations with $V_A = V_R$ are given in Table VI, and illustrated in Fig. 4. It can be seen that the standard deviations of the estimates are about 0.2 times the mean values and are always less than 0.36 times the mean. It can also be seen that the relative values of walk, car and bus times are similar to those found by Clark and Seaton for Stevenage.

In this method it is, of course, necessary to check that the "values of time" deduced at the proportion curve intersections between the walk mode and the

LONDON CENTRAL AREA

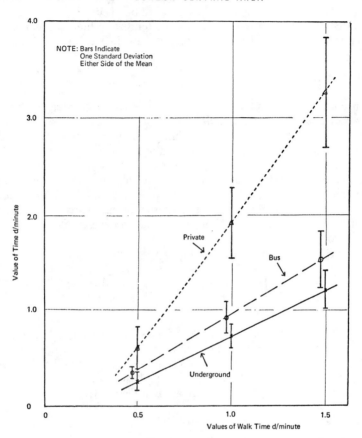

Fig. 4. Derived values of time for the private, bus and underground modes. London morning peak, access and ride times have the same values.

TABLE VI

"Values of Time" Derived from Central London Data for Thirteen Cells

Mode	Value of time (d/min)					
	Mean	Standard deviation	Mean	Standard deviation	Mean	Standard deviation
Walk (defined)	0.50	–	1.00	–	1.50	–
Private	0.60	0.22	1.92	0.36	3.26	0.56
Bus	0.36	0.06	0.94	0.18	1.54	0.28
Underground rail	0.26	0.10	0.76	0.12	1.24	0.24

other modes give equal generalized costs at the intersections of other pairs of modes. If the consistency of estimates based on different mode pairs was less satisfactory than it was in this case an iterative technique could be used to minimize the standard deviation of the estimate differences.

Distribution Functions

Once the "values of time" have been established for the various modes, the distances at which different modes carry equal numbers of passengers as fares or travel times are varied can be calculated without making any assumptions about the expression connecting the ratio of numbers of passengers with the generalized costs of the modes. However, for the assessment of a new mode or of a change to an existing mode it is necessary to determine the passenger loading in greater detail than this. It is, therefore, necessary to assume a modal split expression, and determine the constants for this expression by fitting the original data. The "values of time" in the expression are, of course, those already found from the intersections.

For the London study the following expression was assumed:—

$$n_1/n_2 = \exp - \alpha \left\{ P_1 \phi_1 x - P_2 \phi_2 x + V_1 \left(a_1 + \frac{\phi_1 x}{U_1} \right) - V_2 \left(a_2 + \frac{\phi_2 x}{U_2} \right) \right\}$$

(4)

where V_1, V_2 are the derived "values of time" on modes 1 and 2. α was determined by plotting calculated values of n_1/n_2 for all the pairs of modes against distance for a range of values of α and comparing these with the observed survey values of n_1/n_2. The values of α that best fitted the observed data for the three sets of "values of time" are given in Table VII.

TABLE VII

Values of Sensitivity Constant Obtained
from Central London Data

Values of walk time (d/min)	$\alpha(d^{-1})$
0.5	>0.30
1.0	0.20
1.5	0.15

Where more than two modes are present the proportions of the total travellers on each mode can be determined from the ratios for the pairs of modes that are calculated using Eq. (4). If there are n modes present:—

$$\sum_{i=1,n} \frac{n_1}{n_2} = 1/A \ (1 + 1/A),$$

(5)

where

$$A = \sum_{j=2,n} \frac{n_j}{n_1}.$$

Figure 5 shows the observed proportions of travellers using the four modes, private, bus, underground and walk in the 13 cells used for the London study. Figures 6 and 7 show the predicted proportions using Eqs. (4) and (5) for two different values of V_{walk} (1.0 d/min. and 1.5 d/min.) and the appropriate values of α as given in Table VII and of V_{private}, V_{bus} and $V_{\text{underground}}$ as given in Table VI. It can be seen that for both predictions the general agreement is good, but that in Fig. 6 the proportions of walkers at 3 km and of private travellers at 7 km are over-estimated, in each case at the expense of underground. The proportion of bus passengers is better estimated in Fig. 6 than in Fig. 7 while the reverse is true for walkers. The proportion of underground passengers fluctuates about the observed trend with distance for both predictions. The agreement between the observed and predicted modal split is, as might be expected, less satisfactory for individual cells than for the combination of all thirteen cells.

It is important to notice that the arbitrary choice of a value for V_{walk} has little effect on the overall goodness of fit, provided that the appropriate values for the other modes are used consistently throughout.

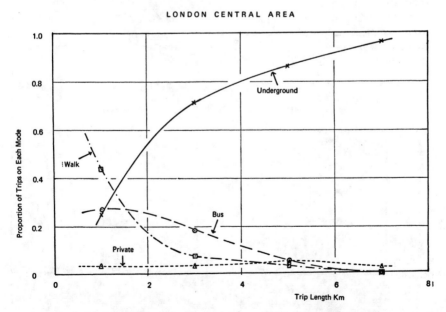

Fig. 5. Modal split for central area trips from L.T.S. & B.R. data 8–10 a.m.

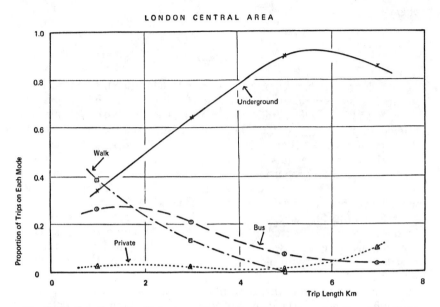

Fig. 6. Predicted modal split for central area trips 8–10 a.m. $\alpha = 0.2$, $V_p = 1.92$ d/min, $V_B = 0.94$ d/min, $V_u = 0.75$ d/min, $V_W = 1.00$ d/min.

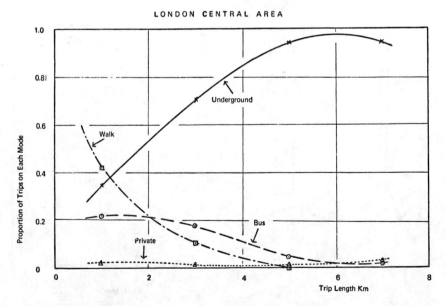

Fig. 7. Predicted modal split for central area trips 8–10 a.m. $\alpha = 0.15$, $V_p = 3.27$ d/min, $V_B = 1.54$ d/min, $V_u = 1.24$ d/min, $V_W = 1.50$ d/min.

Application to a New Mode

Application of the method described above to predict the effect on ridership of fare, speed, or accessibility changes for an existing mode is straight forward, as these involve only changes to P_i, U_i or a_i in Eq. (4). The effect of changes to comfort, such as the introduction of air conditioning in buses, is much harder to predict, as the derived "values of time" do not appear to be related primarily to comfort. If these "values of time" can be interpreted in behavioral terms at all, they would appear to reflect most strongly characteristics such as income and car ownership associated with the traveller, and these dominate the effect of the qualities of the mode. Had it been otherwise, one would have expected for example to find higher values for V_{bus} than for $V_{private}$. However, the derived "values of time" for cyclists in Stevenage are as high as for car drivers, so it appears that for this pair of modes the comfort and income effects counteract each other.

Some work has been done on the application of this method to new modes. The modal splits between each of the existing modes and the new mode are calculated in turn using Eq. (4) with the values of P_i, U_i, ϕ_i, and a_i appropriate to the new mode. It has been assumed that the derived "values of time" for the users of an existing mode are related to aspects of the travellers rather than to aspects of the mode, so that the value for passengers transferring to the new mode is the same as that on the existing mode. Thus

$$n_n/n_e = \exp - \alpha \left\{ P_n\phi_n x - P_e\phi_e x + V_e \left(a_n - a_e + \frac{\phi_n x}{U_n} - \frac{\phi_e x}{U_e} \right) \right\}, \qquad (6)$$

where suffix n refers to the new mode and e to the existing mode. As an example, predictions have been made of the modal split between the existing modes in London and a new mode that has the following characteristics:

Fare	3d to 8d/km airline distance
Access time	5.5 min
Block speed	24 to 36 km/hr
ϕ	1.30

This might, perhaps, be a very fast pedestrian conveyor, although the technical design of the mode is not important for the analysis. Equation (6) was used with $\alpha = 0.2d^{-1}$ and $V_{private} = 1.92d/min$, $V_{bus} = 0.94d/min$, $V_{Ug} = 0.75d/min$, and $V_{walk} = 1.00d/min$. The study was to show the effect of fare level on the diversion of morning peak traffic from existing modes, and the results are shown in Fig. 8. It can be seen that for a 1% change of fare on the new mode the predicted ridership varies by about $\frac{2}{3}$%. The rate of diversion with fare change is somewhat larger than is observed for diversion to car from public transport, but appears plausible for transfer between competing public transport modes. The predicted rate of transfer from private (including taxis) to the new mode is $\frac{1}{2}$% ridership for a 1% fare change.

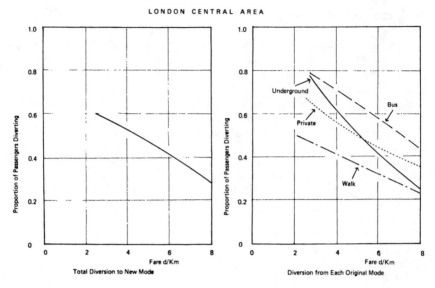

Fig. 8. Predicted diversion of morning peak trips when a novel mode is introduced.

Discussion and Conclusions

The method described in this paper does not appear, from the results so far obtained, to enable the effects of nonquantifiable convenience factors to be included in modal split calculations. However, it does clearly prove able to derive a model from data that do not allow more conventional regression or discriminant analysis. This is not to recommend the collection of data lacking in detail, but to suggest a way of using such data when more rigorous methods fail.

The "values of time" obtained are only curve-fitting parameters in Eqs. (3) and (4), and probably reflect factors such as car ownership, income, and comfort of the mode as well as the value the traveller actually places on his time. They certainly should not be used in the evaluation of the benefits of time saved by a new or modified mode. Whatever interpretation these parameters should have it is significant that similar values are obtained using data from two different surveys in two very dissimilar urban areas.

It might be argued that data can always be better fitted by introducing more parameters (in this case mode-dependent "values of time.") The justification for this procedure lies in the result mentioned above, that these values show consistency in different surveys for different areas. If this result can be demonstrated for several different situations the implication will be that these new values are meaningful and useful.

It is interesting that the analysis suggests that access and ride times have the same values. This is not in agreement with the results of several regression and discriminant analyses; further study is required to resolve this disagreement.

The final agreement between the predicted and observed overall modal split for 13 cells in Central London is certainly good enough to be used for assessment studies. The agreement in individual cells is less good, as might be expected.

Acknowledgement

The authors wish to thank the Director of The Transport Research Assessment Group for permission to present this paper, which is British Crown Copyright and is reproduced with the permission of the Controller, Her Britannic Majesty's Stationery Office. The paper represents the personal views of the authors, which are not necessarily those of the Department of the Environment.

References

1. M. E. Beesley, The value of time spent in travelling: some new evidence, *Economica*, New Series, **32**, 174 (1965).
2. U.S. Dept. of Commerce, Bureau of Public Roads, Modal split-documentation of nine methods for estimating transit usage (1966).
3. R. G. McGillivray, Demand and choice models of modal split, *J. Transport Econ. and Policy* **4**, 192 (1970).
4. R. E. Quandt and W. J. Baumol, The demand for abstract transport modes: Theory and measurement, *J. Reg. Sci.* **6**, 13 (1966).
5. D. A. Quarmby, Choice of travel mode for the journey to work, *J. Transport Econ. and Policy* **1**, 273 (1967).
6. P. R. Stopher, Predicting travel mode choice for the work journey, *Traff. Eng. and Control* **9**, 436 (1968).
7. S. L. Warner, *Stochastic Choice of Mode in Urban Travel: A Study in Binary Choice*, Northwestern University Press, (1962).
8. A. G. Wilson, The use of entropy maximising models, *J. Transport Econ. and Policy* **3**, 108 (1969).
9. Research Projects Ltd, Modal choice—studies of the use and non-use of public transport in the Greater London area (1969).
10. J. O. Tresidder, D. E. Meyers, J. E. Burrell, and T. J. Powell, The London transportation study—methods and techniques, *Proc. Inst. Civ. Eng.* **39**, 433 (1968).
11. J. M. Clark and R. A. F. Seaton, Modal split—A general approach, paper presented to Urban Traffic Model Research Symposium organized by P.T.R.C. Co. Ltd. and held at London School of Economics (1970).
12. Stevenage Deveopment Corporation, Stevenage traffic survey (1964-1966).
13. N. Lichfield and Associates, Stevenage—Public transport cost-benefit analysis, Stevenage Development Corporation (1969).
14. Department of the Environment, The preparation of traffic and transport plans, Tech. Memo. T7/68 (1968).
15. E. M. Holroyd and D. A. Scraggs, Journey times by car and bus in Central London, *Traff. Eng. and Control* **6**, 169 (1964).
16. British Rail Regions, Annual Terminal Censuses (1962).

Minimizing Economic Segregation Through Transit System Changes: A Goal Programming Approach

John W. Dickey
Department of Civil Engineering, Virginia Polytechnic Institute and State University, Blacksburg, Virginia

Abstract

In this study, a goal program was developed for use in determining how transit services (defined in terms of interzonal travel times) should be modified in order to minimize economic segregation (defined by level of family income). The goal programming technique combined the EMPIRIC land use model with mathematical programming techniques in order to determine the municipal service modifications that led to the greatest zonal land use changes in the direction of certain pre-specified goals. The EMPIRIC model, developed for utilization in the Eastern Massachusetts region, is a simultaneous equation regression model which predicts the number of families by income class and the number of employees by job type that will settle in various zones of a region at some future date. These settlement patterns are a function of past land use patterns, future population totals for the region, and various types of municipal services, including interzonal transit services.

In applying the goal programming technique, the desirable proportion of families in each income class (four broad classes were used in this study) was set in each zone at the same proportion of that income class for the region as a whole. In short, a population distribution by family income, based on demographic characteristics for the region as a whole, was established for each zone. Then, after defining slack and surplus variables in terms of the number of families which either were not met or were exceeded by these established proportions, an attempt was made to minimize the total slack and surplus. The constraints in the program are the EMPIRIC model equations, limitations on the range in which transit travel times can be varied, and the equations for the proportions for each family income type in each zone.

Results obtained from a test of the goal programming model using a six-zone subregion of the Eastern Massachusetts region appear to be of consequence. Segregation of families by income (total slack and surplus) was reduced by approximately 40% over the least desirable situation and by 20% over the situation in which no transit travel time changes were made. The changes in transit travel time needed to achieve *maximum desegregation* involve the system into and from Boston—the center city zone—whereas the changes that would bring about *minimum desegregation* involve the system in suburban zones. Further, if no changes are made, some interesting zonal impacts can be noted that do not occur if either "positive" or "negative" ("good" or "bad") changes are made.

Introduction

Throughout cities in the United States and in other countries, one of the major problems is that of segregation, by race, income, religion, or by other characteristics. Of course, not all segregation is undesirable, but in many instances it has

---

References pp. 106-107

meant relatively poorer housing, poorer schools, poorer jobs, and in general a poorer existence for those who happen to be in the minority. And, on a more personal scale, segregation oftentimes leads to a lack of understanding and consideration for those groups that find themselves on the outside.

Naturally the elimination of undesired segregation is a far from simple matter. It probably involves, among other things, changing parts of the structure and operation of government, the housing market, employment practices, and education and training programs. In addition, it also involves changing transportation services, for, as demonstrated by the many angry protests and counterprotests over school busing, transportation (or the lack of it) can be a critical element in maintaining or altering patterns of segregation.

Unfortunately, relatively little is known about the impacts of various transportation related decisions on the extent of segregation. We should have been able to guess long ago, for instance, that changes in the mass transit system which would make it easier for low-income people to reach suitable jobs would lead to lowered job discrimination. But this fact either was not recognized or not felt important until investigations after riots such as those in Watts found that transportation (especially mass transit) to jobs was not available in these areas. (The report *Tomorrow's Transportation*[1] by the Urban Mass Transportation Administration of the U.S. Department of Housing and Urban Development shows several examples of this type.)

This paper focuses on another aspect of the relationship between transportation and segregation—that of the long term impact of mass transit facility changes on locational decisions and thus on the mix of families by income in various zones of a semihypothetical urban area. Use is made of the EMPIRIC land use model[2,3] developed for the Eastern Massachusetts region of the United States. This simultaneous regression equation model predicts the number of families by income class that will live in a given zone and the number of employees by industry type that will work in a given zone over a ten-year period. In the model the predicted number of families or employees is a function of present and expected levels of the other dependent variables and of the highway, mass transit, water, and sewerage systems. It will be our objective to determine how the mass transit system, which has a relatively large effect on the location choices of low-income families, can be changed so as to cause the greatest reduction in family segregation by income.* This investigation will be carried out for a six-zone area abstracted from the Eastern Massachusetts region.† To determine how to minimize segregation, we will accept the EMPIRIC model equations as good indicators of locational choice preferences and will include them within a special mathematical programming technique.

---

*It probably would have been more desirable to look at racial segregation, but families were not classified this way in the EMPIRIC model. Also, to some extent, race and income classifications are synonomous.

†The entire Eastern Massachusetts region could have been used as an example, but this would have required large data processing endeavor beyond the resources of this study.

The EMPIRIC Land Use Model

It will not be our purpose in this section to describe the background, development, and calibration of the EMPIRIC model or of land use prediction models in general. These subjects have been presented elsewhere[4] in great detail. Suffice it to say that a great deal of effort has been exerted in the creation of the model, both for the Eastern Massachusetts region[2,5,6] and presently for the Washington, D.C. metropolitan area.[7] Moreover, at this point in time, the EMPIRIC model represents one of the few workable land use models available. This shortage implies that there is a great deal more to be learned about land use development behavior, but, as noted before, we are not primarily concerned in this paper with the validity and reliability of the EMPIRIC model and so will restrict ourselves here to a brief description of its inputs and outputs.

In any land use model, future zonal development should, in general, be a function of:

(1) Past and present levels of zonal land use,

(2) Past, present, and proposed levels of municipal services such as water, sewerage, and educational facilities,

(3) Past, present, and proposed levels of transportation system variables, and

(4) Projected levels of future *regional* land use development.

TABLE 1

Basic Inputs to the Empiric Model

Present zonal land use variables (zone h), $V_{ih}(t)$

$V_{1h}(t)$ = Number of families with an annual income less than $5000

$V_{2h}(t)$ = Number of families with an annual income between $5000 and $9999

$V_{3h}(t)$ = Number of families with an annual income between $10,000 and $14,999

$V_{4h}(t)$ = Number of families with an annual income equal to or greater than $15,000

$V_{5h}(t)$ = Number of persons in manufacturing and construction employment (S.I.C.[a] Codes 15-39)

$V_{6h}(t)$ = Number of persons in wholesale, transportation, communication, utilities and government employment (S.I.C. Codes 1-14, 40-50, 91-99)

$V_{7h}(t)$ = Number of persons in retail employment (S.I.C. Codes 52-59)

$V_{8h}(t)$ = Number of persons in service employment (S.I.C. Codes 70-89)

$V_{9h}(t)$ = Number of persons in finance, insurance, and real estate employment (S.I.C. Codes 60-67)

Present zonal land area variables (zone h)

$NAP_h(t)$ = Net residential area, in acres

$NAM_h(t)$ = Net manufacturing area, in acres

$NAR_h(t)$ = Net retail area, in acres

$ODA_h(t)$ = Other developed area, in acres

$DA_h(t)$ = Developable area, in acres

$UA_h(t)$ = Total used area = $NAP_h(t) + NAM_h(t) + NAP_h(t) + ODA_h(t)$

$GA_h(t)$ = Gross area = $UA_h(t) + DA_h(t)$

[a]The letters S.I.C. stand for the *S*tandard *I*ndustrial *C*lassification of land use activity taken from: Bureau of the Budget, *Standard Industrial Classification Manual,* U.S. Government Printing Office, Washington, D.C., 1957.

The EMPIRIC model utilizes as its inputs most of the present (time t) and projected or proposed (time t + 1) variables listed above. In Tables I–III it can be seen that land use variables for each zone h, $V_{ih}(t)$, are expressed in terms of: (1) the number of families within different yearly income classes, (2) the number of employees in various types of employment categories, and (3) the area of land devoted to different functions. (These land area variables, as seen in Table I, actually are not treated as land use variables but instead are designated as "present land area variables." This differentiation conflicts with the usual one in that land area commonly is utilized as an indication of intensity of land use whereas population and employment variables are considered as demographic, not land use, characteristics. However, the distinction between the two types of characteristics is somewhat semantic.) The transportation system, both present and future, is represented by interzonal travel times by automobile and by transit, while the water and sanitary services, both present and future, are represented by water and sewage disposal indices for each zone. These indices, presented in Table II, range from 1–7 and 1–5, respectively, with the lower numbers corresponding to the least sophisticated type of system. The final set of inputs,

TABLE II

Basic Inputs to the Empiric Model

Present transportation system variables
   $TA_{gh}(t)$ = travel time by automobile from zone g to h, in minutes
   $TT_{gh}(t)$ = travel time by transit from zone g to h, in minutes
Present municipal service variables (zone h)
   $w_h(t)$ = water supply index, ranging from 1 through 7, indicating type of water supply service
   $s_h(t)$ = sewage disposal index, ranging from 1 through 5, indicating type of sewage disposal service

Code Number	Type of System
	Water supply index
1	Individual wells
2	Combination of individual wells and municipal supply
3	Municipal surface supply
4	Combination of municipal surface and ground supply
5	Municipal ground supply
6	Municipal and Metropolitan District Corporation supply
7	Metropolitan District Corporation supply
	Sewage disposal index
1	Septic tank
2	Combination septic tanks and municipal system
3	Municipal system
4	Combination of septic tanks and/or municipal system and Metropolitan District Corporation system
5	Metropolitan District Corporation System

TABLE III

Basic Inputs to the Empiric Model

Future municipal service variables (zone h)

$w_h(t + 1)$ = Water supply index, ranging from 1 through 7, indicating type of water supply service

$s_h(t + 1)$ = Sewage disposal index, ranging from 1 through 5, indicating type of sewage disposal service

Future transportation system variables

$TA_{gh}(t + 1)$ = Travel time by automobile from zone g to h, in minutes

$TT_{gh}(t + 1)$ = Travel time by transit from zone g to h, in minutes

Future regional land use variables, $T_i(t + 1)$

$T_1(t + 1)$ = Total number of families with an annual income less than \$5000

$T_2(t + 1)$ = Total number of families with an annual income between \$5000 and \$9999

$T_3(t + 1)$ = Total number of families with an annual income between \$10,000 and \$14,999

$T_4(t + 1)$ = Total number of families with an annual income equal to or greater than \$15,000

$T_5(t + 1)$ = Total number of persons employed in manufacturing and construction employment

$T_6(t + 1)$ = Total number of persons in wholesale, transportation, communication, utilities, and government employment

$T_7(t + 1)$ = Total number of persons employed in retail employment

$T_8(t + 1)$ = Total number of persons employed in service employment

$T_9(t + 1)$ = Total number of persons employed in finance, insurance, and real estate employment

---

as displayed in Table III, is the projected future regional levels, $T_i(t + 1)$, for each land use variable in Table I.

Given these inputs, the EMPIRIC model takes the projected future level of each land use in the region and distributes it among the zones in that region. This procedure gives the future number of families in different income categories and number of employees of different types in each zone. These outputs are designated by $V_{ih}(t + 1)$ and are synonomous with the inputs in the top part of Table I.

With the large number of basic inputs to be taken into account by the EMPIRIC model, it can be anticipated that the relationships between them would be fairly complex—and they are. Adding further to this complexity is the rather complicated way in which most of these basic inputs are transformed and combined in order to create more useful and realistic variables for the model. For instance, it was felt by the developers of the EMPIRIC model that the influence of the transportation system (i.e., travel time) on intraregional movement of families and industries decreased exponentially as travel time itself increased (see Ref. 3 for a fuller discussion of this idea). In other words, a zone probably would not develop very rapidly if it were located at distances which made it relatively inaccessible for most family and work trips. This feature subsequently was incorporated through the use of terms such as $\exp(-B \cdot TA_{gh})$ and $\exp(-B \cdot TT_{gh})$ where B was an empirically-derived coefficient (varying for different types of families, industries, and transportation) and TA_{gh} and TT_{gh} were travel times between zones g and h by auto and transit, respectively.

The model developers also speculated that, in conjunction with transportation, two other factors would tend to increase a given zone's growth: (a) the extent to which land within the zone itself is intensively developed, and (b) the extent to which large pools of families and potential employees are in close proximity to the zone. Figure 1 portrays this situation diagrammatically for a given zone h surrounded by the remaining G-1 zones in a designated region. The extent of land development in h is represented by UA_h, the present amount of total used area in the zone. The number of families and employees (of type x) in another zone (g) is represented by U_{xg}. And the connection between the two is the transportation system, represented, as noted before, by either exp $(-B \cdot TA_{gh})$ or exp $(-B \cdot TT_{gh})$. These three factors, when considered together, should be

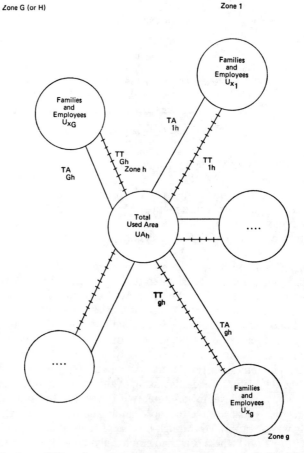

Fig. 1. Accessibility of zone h as related to the other zones in the region.

multiplicative since higher levels of UA_h, U_{xg}, and $\exp(-B \cdot TA_{gh})$ or $\exp(-B \cdot TT_{gh})$ imply the close juxtaposition of families, employees, and developed land which, in turn, should mean greater growth in h. Moreover, a zero level of any of the three factors should mean zero growth in h. This thinking leads to terms of the form $UA_h \cdot U_{xg} \cdot \exp(-B \cdot TA_{gh})$ or $UA_h \cdot U_{xg} \cdot \exp(-B \cdot TT_{gh})$, which then are added for all G zones to indicate the presence, relative to zone h, of families and potential employees in the other zones in the region. The sum which evolves from this series of operations can be thought of as an access measure for zone h.

The equations in the lower part of Table IV are the six different measures for present access actually utilized in the EMPIRIC model. The four equations above them, for $U_{1h}(t)$ to $U_{4h}(t)$, define four different groups of family and employment "activities" pertinent to the access measures $W_{3h}(t)$ through $W_{8h}(t)$. It can also be seen in Table IV that the water and sewage disposal indices have been transformed $[W_{1h}(t)$ and $W_{2h}(t)]$ through multiplication with the total

TABLE IV

Transformed Inputs to the Empiric Model

Present family and employment "activities" (zone h), $U_{xh}(t)$

$U_{1h}(t) = V_{1h}(t) + V_{2h}(t) + V_{3h}(t) + V_{4h}(t) =$ total families

$U_{2h}(t) = V_{1h}(t) + V_{2h}(t) =$ number of families with an annual income less than \$10,000

$U_{3h}(t) = V_{3h}(t) + V_{4h}(t) =$ number of families with an annual income greater than \$10,000

$U_{4h}(t) = V_{5h}(t) + V_{6h}(t) + V_{7h}(t) + V_{8h}(t) + V_{9h}(t) =$ total employment

Present access variables (zone h), $W_{kh}(t)$

$W_{1h}(t) = UA_h(t) \, s_h(t)$

$W_{2h}(t) = UA_h(t) \, w_h(t)$

$$W_{3h}(t) = \sum_{g=1}^{G} UA_h(t) \, U_{1g}(t) \exp[-B \cdot TA_{gh}(t)]$$

$$W_{4h}(t) = \sum_{g=1}^{G} UA_h(t) \, U_{4g}(t) \exp[-B \cdot TA_{gh}(t)]$$

$$W_{5h}(t) = \sum_{g=1}^{G} UA_h(t) \, U_{3g}(t) \exp[-B \cdot TA_{gh}(t)]$$

$$W_{6h}(t) = \sum_{g=1}^{G} UA_h(t) \, U_{1g}(t) \exp[-B \cdot TT_{gh}(t)]$$

$$W_{7h}(t) = \sum_{g=1}^{G} UA_h(t) \, U_{4g}(t) \exp[-B \cdot TT_{gh}(t)]$$

$$W_{8h}(t) = \sum_{g=1}^{G} UA_h(t) \, U_{2g}(t) \exp[-B \cdot TT_{gh}(t)]$$

used area of the zones to which they correspond to form zonal access measures for these municipal services. Future access measures are the same, except that *future* levels of highway, transit, water, and sewerage variables are utilized.

Several other transformations of variables are performed before the actual variables employed in EMPIRIC are established. In fact, EMPIRIC does not predict future levels of land use in each zone directly but instead predicts the change (from present to future) in the zone-to-region share of each variable. In any case, when all of the present or exogenously set or predicted variables (except the future interzonal transit travel times—the decision variables) are inputted, the resulting general form of the EMPIRIC model equations becomes:

$$\sum_{i=1}^{9} a_{ip} V_{ih}(t+1) + \frac{\sum\limits_{g=1}^{H} UA_h(t) \, U_{xg}(t) \, \exp\left[-B \cdot TT_{gh}(t+1)\right]}{\sum\limits_{g=1}^{H} \sum\limits_{h=1}^{H} UA_h(t) \, U_{xg}(t) \, \exp\left[-B \cdot TT_{gh}(t+1)\right]} = C_{ph}$$

$$\text{(all } p, h; \text{ given } x\text{'s)}, \quad (1)$$

where a_{ip} is a coefficient relating land use variables i and p, and C_{ph} is the right hand side constant incorporating the values of all exogeneous factors.

It should be noted that if the future interzonal transit travel times were set at given values, the resultant set of equations would be linear in the $V_{ih}(t+1)$'s, so that these could be determined through the usual simultaneous linear equation solution procedures.

Now, for simplification, let us set:

$$r_{ghx} = UA_h(t) \, U_{xg}(t) \quad \text{(all } g, h; \text{ given } x\text{'s)} \quad (2)$$

and set

$$Y_{gh}(t+1) = \exp\left(-B \cdot TT_{gh}(t+1)\right) \quad \text{(all } g, h). \quad (3)$$

This would leave Eq. (1) in the form:

$$\sum_{i=1}^{9} a_{ip} V_{ih}(t+1) + \frac{\sum\limits_{g=1}^{H} r_{ghx} Y_{gh}(t+1)}{\sum\limits_{g=1}^{H} \sum\limits_{h=1}^{H} r_{ghx} Y_{gh}(t+1)} = C_{ph} \quad \text{(all } p, h; \text{ given } x\text{'s)}. \quad (4)$$

It will be the objective of this study to use Eq. (4) to find the values of the $Y_{gh}(t+1)$'s so as to minimize segregation of families by income, with segregation being defined as a function of the $V_{ih}(t+1)$'s.

Formulating Constraints and an Objective Function

If, as in this investigation, our objective is to minimize segregation by income in zones within a region, we can interpret this to mean that within each zone the distribution of families by income should be as close as possible to that for the region as a whole. Thus, we would want to minimize the difference between the desired distribution for a zone and the regional one. This type of problem comes under the general heading of goal programming, a procedure introduced by Charnes and Cooper.[8] Let us start by letting D_{ih} be the desired level for family type (by income) i in zone h, that is, the level for $V_{ih}(t + 1)$ at which there is no segregation. Thus

$$D_{ih} = V_{ih}(t + 1) + s_{ih} - e_{ih} \quad \text{(all } i, h\text{)}, \tag{5}$$

where s_{ih} is the slack that must be added to $V_{ih}(t + 1)$ if it is less than D_{ih} and e_{ih} is the surplus that must be subtracted if $V_{ih}(t + 1)$ exceeds D_{ih}.

Unfortunately, we would not know the values for the D_{ih}'s beforehand since we do not know how many families in total will end up in zone h. But we do know the percentage that D_{ih} should be of all families in the zone—that being, for minimum segregation, equal to the analogous percentage for the region. Put another way, we should know how many families of type i there should be in a zone in relation to, say, the number of families of type 4 (income $> \$15,000$) in that zone. Thus, if P_i is that ratio, we have

$$D_{ih}/D_{4h} = P_i \quad (i = 1, 2, 3; \text{all } h). \tag{6}$$

Substituting Eq. (5) into (6), we obtain

$$\frac{D_{ih}}{D_{4h}} = \frac{V_{ih}(t + 1) + s_{ih} - e_{ih}}{V_{4h}(t + 1) + s_{4h} - e_{4h}} = P_i \quad (i = 1, 2, 3; \text{all } h) \tag{7}$$

or

$$V_{ih}(t + 1) + s_{ih} - e_{ih} - P_i\left[V_{4h}(t + 1) + s_{4h} - e_{4h}\right] = 0 \quad (i = 1, 2, 3; \text{all } h). \tag{8}$$

These equations will form part of the constraint set for the forthcoming goal program.

The objective function for the goal program is a simple one. Since the D_{ih}'s are the desired (or goal) levels for complete desegregation of families of type i in zone h and the $V_{ih}(t + 1)$'s are the actually obtained levels, we would like to minimize the difference between the two, whether that difference be positive or negative. In other words, we would like to minimize the total of slacks and surpluses for all family types in all zones in the region, or

$$\min \phi = \sum_{i=1}^{4} \sum_{h=1}^{H} (s_{ih} + e_{ih}). \tag{9}$$

If desired, we could also weight each slack or surplus depending on whether that type of segregation of a given family type in a given zone were more or less of concern, although that has not been done in this study.

The Minimum Segregation Goal Program

We now can gather together all of the elements of the goal program to minimize segregation by income. The objective function is as in Eq. (9) and the family zonal segregation constraints as in Eq. (8). Also, we must constrain land use growth to take place as the EMPIRIC model predicts it will, so that relationships of the type in Eq. (4) must be included in the constraint set. Finally, we can add upper and lower bounds on the interzonal transit travel times, indicating that they cannot be less than at present (a service level constraint) and more than that corresponding to a speed of, say, 70 mph‡ (a technology constraint). If the upper level on travel time (and thus the lower level on $Y_{gh}(t+1)$) is symbolized by $LL_{gh}(t+1)$, and the higher level on $Y_{gh}(t+1)$ by $UL_{gh}(t+1)$, then the full goal program can be presented as

$$\min \phi = \sum_{i=1}^{4} \sum_{h=1}^{H} (s_{ih} + e_{ih}) \tag{10}$$

s.t.

$$V_{ih}(t+1) + s_{ih} - e_{ih} - P_i [V_{4h}(t+1) + s_{4h} - e_{4h}] = 0 \qquad (i = 1, 2, 3; \text{ all } h),$$

$$\sum_{i=1}^{4} a_{ip} V_{ih}(t+1) + \frac{\displaystyle\sum_{g=1}^{H} r_{ghx} Y_{gh}(t+1)}{\displaystyle\sum_{g=1}^{H} \sum_{h=1}^{H} r_{ghx} Y_{gh}(t+1)} = C_{ph} \qquad (\text{all } p, h; \text{ given } x\text{'s}),$$

$$LL_{gh}(t+1) \leqslant Y_{gh}(t+1) \leqslant UL_{gh}(t+1) \qquad (\text{all } g, h),$$

$$V_{ih} \geqslant 0, \quad s_{ih} \geqslant 0, \quad e_{ih} \geqslant 0 \qquad (\text{all } i, h).$$

To reiterate, our aim is to find the values of the $Y_{gh}(t+1)$'s, the interzonal transit travel times, so as to minimize ϕ, the level of segregation by income throughout the zones in the region.

Solution of the Goal Program

As can be seen, the goal program in Eqs. (10) is linear with the exception of the second set of constraints. We have chosen to linearize these by setting the denominator equal to a constant, F_x, and then solving the corresponding linear program. This linear program can be stated as

$$\min \theta = \sum_{i=1}^{4} \sum_{h=1}^{H} (s_{ih} + e_{ih}) \tag{11}$$

s.t.

$$V_{ih}(t+1) + s_{ih} - e_{ih} - P_i [V_{4h}(t+1) + s_{4h} - e_{4h}] = 0 \qquad (i = 1, 2, 3; \text{ all } h)$$

---

‡This speed actually is fairly high, although some exceptional cases of transit service may have speeds approaching this magnitude.

$$\sum_{i=1}^{4} a_{ip} V_{ih}(t+1) + \frac{1}{F_x} \sum_{g=1}^{H} r_{ghx} Y_{gh}(t+1) = C_{ph} \qquad \text{(all } p, h; \text{ given } x\text{'s)},$$

$$\sum_{g=1}^{H} \sum_{h=1}^{H} r_{ghx} Y_{gh}(t+1) = F_x \qquad \text{(all } x\text{)},$$

$$LL_{gh}(t+1) \leqslant Y_{gh}(t+1) \leqslant UL_{gh}(t+1) \qquad \text{(all } g, h\text{)},$$

$$V_{ih} \geqslant 0, \quad s_{ih} \geqslant 0, \quad e_{ih} \geqslant 0 \qquad \text{(all } i, h\text{)}.$$

A linear formulation of this type has several advantages. First, we can be sure, as per Charnes and Cooper,[7] that s_{ih} and e_{ih} cannot both be nonzero, i.e., that there is not both a slack and surplus of a given type of family in a given zone. Second, there is a fairly clear interpretation of F_x—it is the total transit access of a given type (x) in the region as a whole—so that as we vary F_x we can see how much increased total transit access affects segregation (θ). At this point we also should mention that there actually are three types of transit access. Referring back to Table IV, we see that the EMPIRIC model was formulated with transit times being associated with total families (U_{1g}), families with an income less than \$10,000 (U_{2g}), and total employment (U_{4g}). It is the regional access totals for three types of "activities" that we must properly estimate if min θ in Eq. (11) is to equal min ϕ in Eq. (10). Moreover, because of the nonlinearity of Eq. (10), we must expect that a global minimum for ϕ may not be found through the process utilizing Eq. (11).

Briefly, the process employed to find the best F_x values was as follows. First, it was noted that, because of the upper and lower bounds on each $Y_{gh}(t+1)$, F_x could vary only in the range between

$$\sum_{g=1}^{H} \sum_{h=1}^{H} r_{ghx} LL_{gh}(t+1) \quad \text{and} \quad \sum_{g=1}^{H} \sum_{h=1}^{H} r_{ghx} UL_{gh}(t+1).$$

A search then was performed within these ranges. At that time it was found that the three F_x values must vary "together," that is, at about the same relative location within their ranges. Otherwise, the problem is infeasible (has incompatible constraints). The resulting optimal level of F_x, F'_x, was found to be:

$$F'_x = 0.375 \left[\sum_{g=1}^{H} \sum_{h=1}^{H} r_{ghx} UL_{gh}(t+1) - \sum_{g=1}^{H} \sum_{h=1}^{H} r_{ghx} LL_{gh}(t+1) \right] \quad \text{(all } x\text{)},$$

$$\tag{12}$$

although the value for min θ did not seem to be overly sensitive to the actual percentage of the range utilized. In any case, the F'_x values found above were the ones used in the semihypothetical example to be presented in the next section.

Data for the Example

Before an example of the minimum segregation goal program can be presented, it is necessary to display some of the data which will form the basis for the example. These data were taken mainly from the 1960 Census of Population.[9] (At the time this research was being undertaken, the Census for 1970 had not been completed. As a result, the reader will continually find references to 1970 as a "future" date.) The test zones selected were located in the metropolitan Boston area, the region for which the EMPIRIC model originally was calibrated. For simplicity, only six zones were selected, corresponding to the minor civil divisions of (1) Boston, (2) Lexington, (3) Natick, (4) Weymouth, (5) Peabody, and (6) Stoneham. Figure 2 depicts the region and corresponding zones.

Family income and employment data were obtained from the Census. The gross area (GA_h) of each zone was measured using a planimeter takeoff of an accurate regional map. Total used area (UA_h), other developed area (ODA_h), net

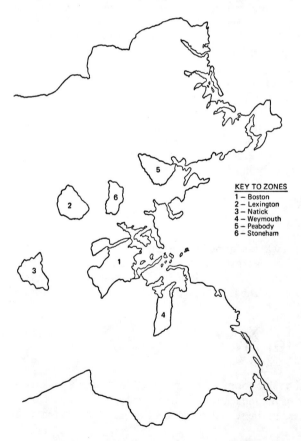

KEY TO ZONES
1 — Boston
2 — Lexington
3 — Natick
4 — Weymouth
5 — Peabody
6 — Stoneham

Fig. 2. The eastern Massachusetts study region.

manufacturing area (NAM_h), and net retail area (NAR_h) were calculated utilizing ratios and percentages developed by Harland Bartholomew in his *Land Use in American Cities*.[10] Water supply and sewage disposal indices were assigned for the 1960 base year by judging the apparent extent of development within each zone as indicated on current local maps and assigning a reasonable index. The indices for both water and sewage were increased by one unit during the forecast period. Interzonal highway travel times were calculated by referring to highway maps of the region from which distances between zones and road conditions could be noted. Transit times in 1960 were arbitrarily equated to highway travel times—admittedly an unrealistic procedure.

Tables V–VII contain the derived data along with hypothetical values for fu-

TABLE V

1960 Levels of Zonal Land Use, 1970 Regional Totals, and 1960 and 1970
Zonal Water Supply and Sewage Disposal Indices[a]

Variable	Zone (h)						Future total (T_i)
	1	2	3	4	5	6	
V_{1h}	63,946	817	1,259	2,505	2,148	956	77,832
V_{2h}	77,955	2,976	3,936	7,106	5,026	2,614	129,721
V_{3h}	16,869	1,662	1,392	1,821	950	715	36,322
V_{4h}	5,445	1,094	531	422	344	197	15,556
V_{5h}	82,716	3,092	3,523	6,405	6,200	2,609	108,262
V_{6h}	77,294	1,736	2,086	3,523	2,016	1,428	90,686
V_{7h}	41,793	1,145	2,087	2,670	2,080	964	47,775
V_{8h}	59,191	2,550	2,595	2,814	1,888	1,389	96,647
V_{9h}	18,224	587	612	1,046	401	443	25,726
$w_h(t)$	7	5	5	4	5	5	
$s_h(t)$	5	3	4	3	4	4	
$w_h(t+1)$	7	6	6	5	6	6	
$s_h(t+1)$	5	4	5	4	5	5	

[a]See Tables I through III for identification and definition of variables.

TABLE VI

1960 Levels of Land Area Development (Acres)

Variable	Name	Zone 1	Zone 2	Zone 3	Zone 4	Zone 5	Zone 6
NAP_h	Net residential area	12,190	1,312	1,226	997	1,648	616
NAM_h	Net manufacturing area	1,893	267	274	200	332	124
NAR_h	Net retail area	1,514	124	116	94	156	58
ODA_h	Other developed area	15,338	5,301	4,958	5,835	6,118	2,285
UA_h	Total used area	30,934	7,004	6,574	7,126	8,254	3,083
DA_h	Developable area	1,628	4,669	5,099	4,751	3,532	1,320

TABLE VII

1960 and 1970 Interzonal Highway Travel Times (min)[a]

Zonal pair	1960 Automobile	1970 Automobile
1-1	15	15
1-2	25	25
1-3	25	25
1-4	20	20
1-5	45	45
1-6	20	20
2-1	25	25
2-2	6	6
2-3	25	25
2-4	60	60
2-5	30	30
2-6	12	12
3-1	25	25
3-2	25	25
3-3	5	5
3-4	40	40
3-5	45	45
3-6	30	30
4-1	20	20
4-2	60	60
4-3	40	40
4-4	6	6
4-5	52	52
4-6	40	40
5-1	45	45
5-2	30	30
5-3	45	45
5-4	52	52
5-5	6	6
5-6	15	15
6-1	20	20
6-2	12	12
6-3	30	30
6-4	40	40
6-5	15	15
6-6	6	6

[a]$B = 0.10$ for all exponential terms.

ture (1970) travel times by auto, municipal service (water and sewer) indices, and levels of regional land use development. There is no doubt that these data are limited because of the restricted region chosen. On the other hand, the individual zonal data are fairly realistic so that the forthcoming example will not be completely hypothetical, and we can place some confidence in the results that are obtained.

An Example Application

Three cases are investigated in this example, the first being that in which an attempt is made to minimize segregation. The second case is the "no change" (to the transit system) case, and the third is the maximum segregation case. These latter two situations provide bases by which to judge the relative impact of transit on segregation. The last case is particularly relevant since it will show the absolute worst condition and thus can be used as a yardstick for measuring the success or failure of the other two situations.

The outcomes of the three investigations are displayed in Tables VIII–X. The top part of the first table shows the land use developments that take place as a result of efforts to minimize segregation. The bottom part shows the desired levels of development, the D_{ih}'s, if the *regional* proportions of families of type i to families of type 4 of 5.00/1, 8.33/1, and 2.33/1 are adhered to and the total number of families is within the range between the EMPIRIC predicted totals and the projected totals. The differences between the two sets of figures are the slacks and surpluses, which are at a minimum. The regional land use totals from the goal-programming approach generally will not be equivalent to either the EMPIRIC predicted or the projected total. This is because the total amounts of slack and surplus do not necessarily balance.

TABLE VIII

Minimum Segregation and Desirable Levels of Land Use Development

Variable	Zone (h)						Total	Pro-jected
	1	2	3	4	5	6		
Minimum segregation levels of land use development								
V_{1h}	71,222	404	927	1,671	2,593	643	77,460	77,832
V_{2h}	95,834	4,061	6,114	8,035	9,443	3,342	126,829	129,721
V_{3h}	22,266	2,609	2,865	2,800	3,093	1,333	34,966	36,322
V_{4h}	9,512	1,990	1,367	993	1,394	537	15,793	15,566
Total	198,834	9,064	11,273	13,499	16,523	5,855	255,048	259,441
Desirable levels of land use development								
V_{1h}	68,902	2,919	4,393	5,777	6,789	2,403	91,185	77,832
V_{2h}	95,834	4,061	6,114	8,035	9,443	3,342	126,829	129,721
V_{3h}	22,639	960	1,444	1,898	2,231	789	29,961	36,322
V_{4h}	7,612	323	485	639	750	265	10,074	15,566
Total	194,987	8,263	12,436	16,349	19,213	6,799	258,049	259,441

The values of the slack or surplus for all three cases and for each family type and zone are presented in Table IX. A plus value indicates a surplus and a minus a slack. Also to be noted is that the "error" between the EMPIRIC predicted and the projected totals for each type of family is relatively small, so that this discrepancy will not be an overly important consideration.

TABLE IX

Slack (−) and Surplus (+) of Families in Each Income Class in Each Zone Resulting
from Minimum Segregation, No Change, and Maximum Segregation Solutions

	Stoneham (6)		Peabody (5)	
	$0–5,000	$10,000–15,000	$0–5000	$10,000–15,000
Minimum	−1,760	+544	−4,196	+862
No Change	−1,575	+681	−4,067	+957
Maximum	−2,502	+486	0	+2,491
	$5,000–10,000	$15,000+	$5,000–10,000	$15,000+
Minimum	0	+272	0	+644
No Change	+675	+311	+459	+670
Maximum	0	+235	+7,219	+1,168

	Lexington (2)		Boston (1)	
	$0–5,000	$10,000–15,000	$0–5,000	$10,000–15,000
Minimum	−2,515	+1,649	+2,320	−373
No Change	−2,236	+1,258	+1,047	−1,306
Maximum	0	+2,282	+8,515	+107
	$5,000–10,000	$15,000+	$5,000–10,000	$15,000+
Minimum	0	+1,667	0	+1,900
No Change	+1,014	+1,725	−4,631	1,636
Maximum	+5,419	+2,030	0	2,206

	Natick (3)		Weymouth (4)	
	$0–5,000	$10,000–15,000	$0–5,000	$10,000–15,000
Minimum	−3,468	+1,421	−4,106	+902
No Change	−3,212	+1,609	−3,683	+1,213
Maximum	−4,661	+1,328	−5,991	+756
	$5,000–10,000	$15,000+	$5,000–10,000	$15,000+
Minimum	0	+882	0	+354
No Change	+933	+936	+1,541	+442
Maximum	0	+823	0	+262

The total amount of slack and surplus for the minimum segregation case is
29,842. This can be compared to 37,317 for the "no change" case and 49,021
for the worst segregation case. The magnitude of these three totals provides
some interesting implications. First, and probably of foremost interest, is the
fact that transit changes can and do have an effect on segregation. The difference between the best and worst cases is 19,189 families, which represents about
7.5% of the families in the region. This difference is not insignificant, especially
since it is almost equivalent to the total amount of what might be called "inherent" segregation (the 29,842 families) that would be left if transit travel
times were improved to their most desirable levels. Rephrased, this means that

TABLE X

1960 Zone-to-Zone Transit Travel Times and 1970 Times Needed to Achieve Minimum or Maximum Segregation[a]

Zonal pairs	1960 Transit travel time	Transit times, min. segregation	Transit times, max. segregation	Min. possible transit time
1-1	15	9		6
1-2	25	17		13
1-3	25	13		13
1-4	20	11	18	9
1-5	45			21
1-6	20			10
2-1	25		13	13
2-2	6		4	4
2-3	25		13	13
2-4	60		35	35
2-5	30		17	17
2-6	12		7	7
3-1	25		13	13
3-2	25		13	13
3-3	5			4
3-4	40		21	21
3-5	45		26	26
3-6	30		17	17
4-1	20		9	9
4-2	60		35	35
4-3	40		21	21
4-4	6		4	4
4-5	52		30	30
4-6	40		21	21
5-1	45		21	21
5-2	30		17	17
5-3	45		26	26
5-4	52		30	30
5-5	6		4	4
5-6	15		9	9
6-1	20		10	10
6-2	12		7	7
6-3	30		17	17
6-4	40		21	21
6-5	15		9	9
6-6	6		4	4

[a]$B = 0.10$ for all exponential terms.

the proper transit improvements could reduce the amount of segregation by income in the region by approximately 40% as compared to what it would be if the transit changes with the worst impacts were implemented.

The *actual* effect of transit probably is not quite as significant, however. Comparison with the "no change" case, for instance, shows that the maintenance of the status quo insofar as transit is concerned would lead to a segregation level roughly in the middle between the best and worst situations (37,817 vs 29,842 and 49,021). The overall effect that transit might have in this situation would thus amount to a level of 7,975 families which would be approximately 20% of the segregation by income that would exist if the "no change" alternative were followed.

As an aside, there would seem to be a great deal of political palatability in the "no change" alternative. This is because the impact on segregation is fairly neutral, not putting great pressure on communities to overcome defacto segregation in the housing market but at the same time not allowing for further segregation. Moreover, since no transit changes need to be made, the public would not get upset by the many alterations that would be required to accomplish the minimum segregation solution. The maintenance of the status quo does not appear to be sufficient, however, if one is seriously interested in reducing segregation.

One other aspect of interest at this junction is that, even if all necessary transit changes were made, there still would be a discrepancy of 29,842 families from the most desirable levels. This amount possibly could be reduced by changes in other municipal services (highways, water systems, and sewer systems) shown to be relevant in the EMPIRIC model. But it is our feeling that such changes probably would not cause a significant reduction. To reduce segregation to its absolute minimum probably would require the action of laws and policies not currently operative in urban areas (and thus generally not incorporated in the EMPIRIC model).

While the total slack and surplus for each case is important, the distribution of these totals also is relevant. It should be noted first that in all three cases the amount of slack exceeded the surplus, with the maximum segregation case having the biggest difference of 23,713 families. From this result comes the not too surprising inference that the process by which segregation is heightened involves the restriction of movement of families into a zone rather than an overexcessive migration of families. However, this conclusion is not upheld in all zones and for all types of families. As should be expected, there are many variables which prevent generalities from being accepted unconditionally. Referring to Table IX, we see that the impacts on various zones can be quite different, with both large and small slacks and surpluses. The effect, though, on a per population basis is somewhat more noticeable. Boston, the zone having the largest number of families by far, is left with about the same amount of slack and surplus to overcome as the suburban zones, which means in essence that, no matter what the transportation changes, the suburban zones will still be faced with a rather difficult task if they desire to overcome segregation.

The results, when viewed in terms of the impacts on each type of family, are a bit more explicit. In particular, families with annual incomes above $15,000 are always in surplus in every zone, and so, with two exceptions, are the families with annual incomes between $10,000 and 14,999. The opposite holds true for the very poor families. The latter problem could be partially overcome by somehow moving the excess of poor families in Boston out into the suburbs, where there is a dearth. But the first two difficulties would not be as easy to overcome, mainly because the only apparent way to alleviate them would be to get rid of many of the wealthier families in the region, and this certainly is not feasible nor desirable from other standpoints.

Another aspect of interest relative to the distribution of slacks and surpluses is that they vary to some extent according to the particular case under study: minimum segregation, "no change," or maximum segregation. Although large differences between the three are not common, one example does bring out an extraordinary result: if the "no change" strategy is chosen, there will be a large slack in the number of families with incomes between $5000 and $9999 in Boston, whereas if either of the other two extreme strategies is chosen, there will be no slack or surplus. Similar results hold for zones three, four, and six, except that there will be large surpluses instead of slacks. What this means is that, in certain situations, conservative policies, such as the "no change" one, may produce much more exaggerated conditions than if a policy involving some changes were followed.

The patterns of transit travel time reductions needed to bring about the impacts discussed above are very distinctive, as can be seen in Table X. There are only four reductions needed to obtain minimum segregation, and these all relate to Boston. In contrast to this, maximum segregation is obtained by reducing most of the travel times within and between the suburban zones, and by leaving Boston alone. These results are reasonable. To get minimum segregation, one must get the low-income families out of Boston, and this is done by providing better transportation in those directions. On the other hand, to increase segregation, one simply isolates Boston travel time-wise and spends available funds on the intersuburban transit system. One result is perplexing, however: decreasing transit travel time from zone one to four helps both to decrease and increase segregation!

As a final remark, it should be mentioned that the improvements needed to achieve minimum segregation really are not as extensive as were at first imagined. Only four changes are needed, and only one of these is to the lower travel time limit, although all four changes would be in Boston where improvement costs would be highest. What might be a significant finding at this point, though, is that attempts by low-income groups to hinder or even stop construction and operation of transit facilities in the inner cities would only do harm to the cause of integration, because the needed travel time improvements in and from the city then would not be realized. We would also expect the same results insofar as urban highways are concerned, although at this point both these conclusions rest on rather scant evidence.

Conclusions

The intent of this paper has been (1) to demonstrate the use of a goal-programming approach in connection with the EMPIRIC model and (2) to investigate the effect of the transit system on segregation by income in a region. Since the example is an overly simplified one, no firm conclusions can be drawn, but several findings with interesting implications have been identified.

(1) Transit does have an effect on segregation by income. Differences in overall effect when transit is used first to minimize segregation and then to maximize it, are great. In the example case, segregation, as measured by the number of slack and surplus families, could be reduced by 40% in going from the worst to the best set of transit changes.

(2) The case in which no changes are made in transit travel times brings about a segregation level midway between the best and worst cases.

(3) There is some "inherent" segregation remaining even after the optimum transit changes have been made. This probably could be reduced only by creation of laws and policies not currently operative in metropolitan areas (and thus incorporated in the EMPIRIC model).

(4) The amount of slack exceeds the surplus in all three cases under study. This indicates that there is a dearth of families, particularly low-income ones, in the region. Greater integration probably could be obtained by decreasing the total number of wealthy families.

(5) The patterns of impact, both by zone and by type of family, evolving from the goal programming application do not appear to be overly distinctive. However, in certain situations it has been found that a "no change" policy may create more exaggerated local impacts than would occur if certain transit modifications were made.

(6) Minimum segregation is obtained by changing transit travel times in and from Boston. Maximum segregation occurs when the intersuburban travel times are reduced and Boston is isolated travel time-wise. These latter changes apparently prevent low-income families from spreading into the suburbs where they are needed in order to reduce segregation.

Further research is needed on an entire region with carefully collected data before any of these conclusions can be substantiated. Moreover, it is important to do additional investigations on the solution of the original nonlinear goal program, especially if large scale programs are to be solved for global optima.

Acknowledgments

The author would like to express his thanks to the National Science Foundation for supporting much of this research endeavor under Grant GK-4747. Thanks also go to Martin Azola, Jim Johns, and Dave White, who helped prepare and run the computer programs involved.

References

1. U.S. Department of Housing and Urban Development, Urban Mass Transportation Administration, *Tomorrow's Transportation,* U.S. Government Printing Office, Washington, D.C., 1968.
2. Traffic Research Corporation, *EMPIRIC Land Use Forecasting Model, Final Report: Development and Calibration of the Model for 626 Traffic Zones,* Boston, February 1967.

3. D. M. Hill, D. Brand, and W. B. Hansen, Prototype development of a statistical land use prediction model for the greater Boston region, *Highway Res. Board Record* 114 (1965).
4. Highway Research Board, National Academy of Sciences, *Urban Development Models,* Special Report 97, Washington, D.C., 1968.
5. Traffic Research Corporation, *EMPIRIC Land Use Forecasting Model, Impact Analysis: The Model as a Planning Tool,* Boston, February 1967.
6. Traffic Research Corporation, *EMPIRIC Land Use Forecasting Model, Reliability Report: 626 Traffic Zones,* Boston, January 1967.
7. Metropolitan Washington Council of Governments, *"EMPIRIC" Activity Allocation Model Study Design,* Technical Memos 1 to 7, Washington, D.C., 1969 (mimeo).
8. A. Charnes and W. W. Cooper, *Management Models and Industrial Applications of Linear Programming,* Vol. 1, John Wiley and Sons, Inc., New York, 1961.
9. U.S. Dept. of Commerce, Bureau of the Census, *1960 Census of Population, Characteristics of the Population,* Vol. II, Part I, U.S. Summary, 1963.
10. H. Bartholomew, *Land Use in American Cities,* Harvard University Press, Cambridge, 1955.
11. CONSAD Research Corp., *Impact Studies: Northeast Corridor Transportation Project,* Vol. II, Federal Clearinghouse, PB-177611, Springfield, Virginia, January 1968.

A Critique of Entropy and Gravity in Travel Forecasting

Martin J. Beckmann
Technische Hochschule, Munich, Germany
and
Thomas F. Golob
Transportation Research Department, General Motors Laboratories, Warren, Michigan

Abstract

Entropy maximization has been explored in recent literature as a theoretical approach to the principle of gravity in traffic forecasting. In this paper it is argued that identical results can be obtained without such metaphysical methods. The alternative proposed is utility maximization by travelers, which ties in neatly with consumption theory in economics. The utility approach, moreover, provides a more general framework for prediction and evaluation and allows a unified approach at all levels of transportation systems analysis.

The gravity formula is derived from a utility maximization model, with distance entering either as a exponential or power function. The derivation illustrates the theoretical limitations of this approach. Also considered is the problem of aggregation over households. The modifications to the functional form that are required to meet more stringent tests of economic theory are identified.

General Model

Consider a household h residing at location i. In making decision on travel to various destinations k, both benefits and costs of trips are relevant. In line with general economic theory, let benefits be represented by a utility function. While costs might be included in this utility, we prefer to separate and subtract them. One may either measure utility in money terms or one may multiply money costs by an appropriate factor to convert dollars to "utiles." For simplicity, the former convention will be observed. Net benefits to a household derived from travel are then given by

$$U(X_1, \ldots, X_k, \ldots, X_n) - \sum_k C_{i,k} X_k, \qquad (1)$$

where U is utility, X_k is the number of trips by household h to destination k during some time period and $C_{i,k}$ is the generalized cost of a round trip from i to k.

Households are assumed to behave rationally (i.e., to maximize net utility). In view of the nonnegativity of X_k, a necessary condition for a maximum is that

$$\frac{\partial U}{\partial X_k} - C_{i,k} \begin{cases} \leq 0 \\ = 0 \end{cases} \quad \text{if } X_k > 0. \qquad (2)$$

If the utility function U is concave, as is usually assumed in economic analysis (this is the operational meaning of diminishing marginal utility) then these conditions are also sufficient, and the set of maximizing X_k is convex. If the U function is strictly concave then the maximizing X_k are unique.

Specific Utility Functions

In all applications it is necessary to specify a utility function or class of utility functions. One specification explored here because of its close relationship to entropy is the integrated logarithm. Concretely, we let

$$U = U_0 + \sum_k [\beta_k X_k - \beta_0 \int_1^{X_k} \log \zeta \, d\zeta]$$

or

$$U = U_0 + \sum_k [(\beta_k + \beta_0)X_k - \beta_0 (X_k \log X_k + 1)] \cdot \tag{3}$$

It is easily verified that this function is increasing for $0 < X_k \lesseqgtr \exp(\beta_k/\beta_0)$ and strictly concave in all X_k. As X_k approaches 0, the marginal utility $\partial U/\partial X_k$, approaches infinity. For large values of X_k the marginal utilities approach minus infinity. Substituting in Eq. (2), we have

$$\beta_k - \beta_0 \log X_k = C_{i,k} \qquad \text{if } X_k > 0. \tag{4}$$

Thus

$$X_k = \exp(\beta_k/\beta_0) \cdot \exp(-C_{i,k}/\beta_0)$$

or

$$X_k = B_k \exp(-C_{i,k}/\beta_0). \tag{5}$$

Aggregating over all households residing at location i, we obtain total travel from (residential) origin i to destination k:

$$X_{i,k} = P_i B_k \exp(-C_{i,k}/\beta_0), \tag{6}$$

where P_i is the number of households at location i. Notice the important assumption that all households at i have identical logarithmic utility functions and, if the formula is to be applied to all locations, utility functions of households everywhere are identical. (For a relaxation of this assumption see section on Aggregation.)

Formula (6) is that variant of the gravity model which can be derived from entropy maximization.[1,2] The entropy approach involves not only a budget constraint on all travel but assumes that originations and terminations of traffic flows are given. Presumably entropy is a measure of diversity, but its operational meaning in connection with travel decisions is obscure. The derivation usually given is a noncritical adaptation of an argument from Kinetic Gas Theory to an economic situation. If entropy is to play the role of a utility function it should be labeled as such, and the decision process and the aggregations involved should be brought into the open.

The flexibility of adaptability of the utility approach becomes apparent in connection with other variants of the gravity formula that cannot be derived from entropy. A function that has been much in use in the economic study of production (and also utility) is the constant elasticity of substitution (C.E.S.) function. Except for a monotone transformation which is irrelevant for purposes of utility maximization, this function is given by

$$U = \sum_k \alpha_k X_k^\rho \qquad 0 < \rho \leq 1$$

or

$$U = \alpha_0 - \sum_k \alpha_k X_k^\rho \qquad \rho < 0. \tag{7}$$

The necessary and sufficient conditions for a maximum are then

$$\alpha_k |\rho| X_k^{\rho-1} = C_{i,k}. \tag{8}$$

Thus

$$X_k = (\alpha_k |\rho|/C_{i,k})^{1/1-\rho}$$

or

$$X_k = A_k / C_{i,k}^{1/1-\rho}, \tag{9}$$

where

$$A_k = [\alpha_k |\rho|]^{1/1-\rho}.$$

Aggregating over all households at location i,

$$X_{i,k} = P_i A_k / C_{i,k}^\gamma, \tag{10}$$

where

$$\gamma = \frac{1}{1-\rho} \geq 0.$$

γ may assume any positive value except $\gamma = 1$. This exception is encountered however when the utility function is assumed logarithmic (and corresponds to $\rho = 0$):

$$U = \sum_k \alpha_k \log X_k, \tag{11}$$

an important type of function known as the addi-log utility function.

The case $\rho = 1$ is also inadmissable but can occur in the case of budget constraints (see next section). Notice that in Eq. (10) the demand for travel is factored into one term depending on the origin, one term depending upon the destination, and one term depending upon the cost of travel or economic distance between origin and destination. Moreover, the origin term is proportional to the number of households (i.e., population at the origin). Note, moreover, that the particular gravity formula

$$X_{i,k} = \frac{A_i C_{i,k}^{-\gamma}}{\sum_j A_j C_{i,k}^{-\gamma}} \tag{12}$$

is a special case of Eq. (10) with

$$P_i^{-1} = \sum_j A_j C_{i,k}^{-\gamma}. \tag{13}$$

The present approach is easily extended to the general case of separable, additive utility functions. Let

$$U = \sum_k \alpha_k \, \phi(\beta_k X_k)$$

with increasing concave functions ϕ. The necessary and sufficient conditions for a maximum are

$$\alpha_k \beta_k \, \phi'(\beta_k X_k) \lesseqgtr C_{i,k},$$
$$\alpha_k \beta_k \, \phi'(\beta_k X_k) = C_{i,k} \quad \text{if } X_k > 0. \tag{15}$$

When

$$\lim_{x \to 0} \phi'(X_k) = \infty,$$

then the second alternative applies for all k, and one has

$$X_k = \frac{1}{\beta_k} \, \psi\left(\frac{\alpha_k \beta_k}{C_{i,k}}\right), \tag{16}$$

where $X = \psi(1/y)$ is the inverse function of $y = \phi'(X)$. (The function ϕ' is monotonically decreasing since ϕ is concave, and therefore X is monotonically decreasing.) However, the resultant demand function (16) can be factored into a term depending on the destination k and a term depending on distance $C_{i,k}$ if, and only if, ψ is a power function. That is, if and only if, ϕ is a power function or logarithm. In any case, the variables associated with the origin are contained in a separate factor obtained through aggregation:

$$X_{i,k} = \frac{P_i}{\beta_k} \, \psi\left(\frac{\alpha_k \beta_k}{C_{i,k}}\right), \tag{17}$$

where P_i is the population at origin i.

Expenditure Constraints

So far we have disregarded any constraints on travel expenditure. This is appropriate as long as travel constitutes a small portion of total household expenditure. However, in a more general approach in line with consumer demand theory limitations on expenditure imposed by the consumers' income must be recognized explicitly. Alternatively, one might subject travel expense to a specific budget limit (e.g., a fixed proportion of the household's income). Another case is that when the limitation of travel is in terms of total time spent rather than money costs. Of course, both elements (and additional limitations as well) may be present. In this section we wish to show how the utility approaches are modified when one or several budget constraints are considered. Mathematically, it makes no difference whether the limitation is on all consumer expenditure or on travel expenditure only, provided we assume as before that the utility function is separable in the various commodities.

Let M be the budget limit. The consumer's object is now

$$\max U(X_1, \ldots, X_k, \ldots, X_n), \tag{18}$$

subject to

$$\sum_k C_{i,k} X_k \leq M. \tag{19}$$

Assume that the constraint is binding. Applying the Lagrangean multiplier λ, we have as conditions for a constrained utility maximum

$$u_k = \lambda C_{i,k}, \quad k = 1, \ldots, n \tag{20}$$

and

$$\sum_k C_{i,k} X_k = M. \tag{21}$$

In the case of a strictly concave utility function, the $n + 1$ conditions (20) and (21) determine X_k and λ uniquely.

For specific results let U be an additive power function as given in Eq. (7). The condition for a maximum then assumes the form

$$|\rho| \alpha_k X_k^{\rho-1} = \lambda C_{i,k}. \tag{22}$$

From which we have

$$X_k = \left(\frac{|\rho|}{\lambda} \alpha_k\right)^{1/1-\rho} C_{i,k}^{-1/1-\rho}. \tag{23}$$

The multiplier λ, whose meaning is that of a marginal utility of money, depends on the budget limit M, the family income level. Equation (22) is identical to Eq. (8) when λ is set equal to unity. By convention we have measured utility in money terms, so that the marginal utility of money is always 1.

The dependence of λ on income M is clearly seen when X_k is substituted in Eq. (21) to obtain an equation for determing λ:

$$\sum_k \left(|\rho| \alpha_k \cdot \frac{1}{C_{i,k}^\rho}\right)^{1/1-\rho} = M \lambda^{1/1-\rho} \tag{24}$$

The left side of (24) is a constant $= 1/q$, say. Thus

$$\lambda = (qM)^{\rho - 1},\tag{25}$$

so that λ is a power function of M with negative exponent $\rho - 1$. Upon substitution of this expression for λ in the demand function (23), it is easily verified that travel demand X_k to every destination k is strictly proportional to M:

$$X_k = M \frac{\alpha_k^{1/1-\rho} \, C_{i,k}^{-1/1-\rho}}{\sum_j \alpha_j^{1/1-\rho} \, C_{i,j}^{-\rho/1-\rho}}.\tag{26}$$

It follows that travel from origin i to any destination k is proportional to aggregate income M_i at origin i. This is a consequence of homogeneity in the utility function.[3] This is no longer true for nonhomogeneous utility functions [i.e., any $U(x)$ where U is not a power function].

Consider now that travel is restricted both by a money and a time budget limit. A household then seeks to

$$\max U(x_1, \ldots, x_n),\tag{27}$$

subject to the constraints

$$\sum_k C_{i,k} \, X_k \leqq M\tag{28}$$

and

$$\sum_k t_{i,k} \, X_k \leqq T,\tag{29}$$

where $t_{i,k}$ is travel time from i to k and T is total time allocated to travel. Introducing Lagrangean multipliers λ and μ, we obtain as conditions for a maximum

$$U_k = \lambda C_{i,k} + \mu t_{i,k}, \qquad k = 1, \ldots, n,\tag{30}$$

where $\lambda, \mu \geqq 0$.

$$\lambda C_{i,k} + \mu t_{i,k} = \eta_{i,k}\tag{31}$$

can be considered a generalized travel cost. Limitations on visits to particular destinations or to sets of destinations may also be treated in the same way, the respective multipliers and cost terms being positive only for the destinations so rationed and zero for all others. Thus, a restriction on certain trips may be translated into a toll charge or supplementary travel cost.

Aggregation

In demonstrating the flexibility of the utility approach we have limited ourselves to those situations in which a formula of the gravity type emerges as a predictor of travel demands. It is equally important, however, to show the limitations of this gravity formula and the restrictions which its use imply on the assumed economic structure underlying this demand. In this section we wish to apply the utility approach to the more general problem situations in which households are no longer subjected to the uniformity assumptions made above.

In the most general case (which, however, is hardly operational) each household has a specific utility function U^h and income level M^h. The result of constrained utility maximization is then given by

$$U_k^h(X_k) = \lambda^h C_{i,k}, \quad k = 1, \dots, n. \tag{32}$$

In the case of separable additive utility functions, weights α and β may also vary from household to household:

$$U^h = \sum_k \alpha_k^h \phi^{(h)}(\beta_k^h X_k). \tag{33}$$

The condition for a utility maximum is then

$$\beta_k^h \alpha_k^h \phi^{(h)\prime}(\beta_k^h X_k) = \lambda^h C_{i,k}$$

or

$$X_k = \frac{1}{\beta_k^h} \psi^{(h)}\left(\frac{\alpha_k^h \beta_k^h}{\lambda^h C_{i,k}}\right), \tag{34}$$

where $X = \psi^{(h)}(1/y)$ is again the inverse function of $y = \phi^{(h)\prime}(X)$.

Total traffic from i to k is hence given by

$$X_{i,k} = \sum_{h \in H_i} \frac{1}{\beta_k^h} \psi^{(h)}\left(\frac{\alpha_k^h \beta_k^h}{\lambda^h C_{i,k}}\right), \tag{35}$$

where H_i denotes the set of households at location i. Letting ϕ be a power function:

$$\phi^{(h)}(X_k) = X_k^{\rho^h} \tag{36}$$

then (dropping β_k)

$$X_k = \left[\frac{\lambda^h C_{i,k}}{\rho^h \alpha_k^h}\right]^{1/\rho^h - 1} \tag{37}$$

and

$$X_{i,k} = \sum_{h \in H_i} \left[\frac{\rho^h \alpha_k^h}{\lambda^h C_{i,k}}\right]^{1/1-\rho^h}. \tag{38}$$

For practical purposes it is too cumbersome to estimate the household parameters α_k^h and ρ^h. The parameter λ^h could, of course, be replaced by the income level M using Eq. (13).

As a first step towards greater operational usefulness, assume that the exponent ρ is the same for all households. Then

$$X_{i,k} = \left[\frac{\rho}{C_{i,k}}\right]^{\gamma} \sum_{h \in H_i} \left[\frac{\alpha_k^h}{\lambda^h}\right]^{\gamma}, \tag{39}$$

where $\gamma = 1/1 - \rho$ as before. This demand formula can be factored into one term depending on household h at origin i traveling to destination k and another term involving only distance $C_{i,k}$ (i.e., a negative power function of distance). Although each household's travel demand is proportional to income, but with a different proportionality factor depending on the α_k^h, aggregate demand is not proportional to aggregate income at location i.

As a further step, assume that all households at a location i possess the same utility function given by

$$U = \sum_k \alpha_k X_k^{\rho}. \tag{40}$$

Assume, however, that the households have different incomes M^h. The trip distribution is given by

$$X_k = \left(\frac{\rho}{\lambda} \alpha_k\right)^{\gamma} C_{i,k}^{-\gamma}, \tag{41}$$

where again $\gamma = 1/1 - \rho$.

Aggregate travel from i to k is consequently proportional to aggregate income at location i:

$$X_{i,k} = \frac{M_i A_k}{C_{i,k}^{\gamma}}. \tag{42}$$

If utility functions are allowed to differ by location then the destination term A_k must be replaced by a term $A_{i,k}$ depending on both origin i and destination k, since A_k is a function of the utility parameters α_k.

Conclusions

We have developed an alternative theoretical explanation of the gravity formula. At the same time we have pointed out the economic restrictions which are implicit in its use: the identity of utility functions for all households at all locations and their specification as logarithmic or power functions. An important consequence of the utility approach is the ability to model trip demand, modal split and destination split within the same analytical framework.[4] The importance to traffic flow theory of such an analysis of demand functions is to show how demand responds to flow conditions, here denoted by generalized travel costs $C_{i,k}$. Also, the consequences of poor flow conditions may be translated into lost utility. It is in this latter respect that the utility approach is clearly superior to an entropy model.

References

1. A. G. Wilson, A statistical theory of spatial distribution models, *Transportation Res.* **1**, 3 (1967); reprinted in R. E. Quandt, Ed., *The Demand for Travel: Theory and Measurement*, pp. 55–82. D. C. Heath, Lexington, Mass., 1970.
2. A. G. Wilson, The use of entropy maximizing models, *J. Transport Econ. Policy* **3**, 1 (1969).
3. J. H. Niedercorn and B. V. Bechdolt, An economic derivation of the "Gravity Law" of spatial interaction, *J. Regional Sci.* **9**, 2 (1969).
4. T. F. Golob and M. J. Beckmann, A Utility Model for Travel Forecasting, Presented at the Second World Congress of the Econometric Society, Cambridge, England, 1970.

Estimation of Person Trip Patterns Through Markov Chains

Tsuna Sasaki
Kyoto University, Kyoto, Japan

Abstract

In a previous paper, the author described the movement of cars as a Markov chain and applied it to the estimation of future trip distributions. In this paper, the person trip pattern will also be described as a Markov chain. Here, a base is defined as the starting place of the first person trip in a day, principally, home, another's home (for a guest), hotel, office (for people who stayed at work overnight) or terminal (for people who slept in a train overnight, or who took the first trip from a terminal in the survey area, though they had started at their own homes), or possibly a hospital (for patients and their attendants).

After people start from their bases, they will complete a chain of trips to achieve their objectives. If the end point of the final trip coincides with the starting place of the first trip, a chain of trips forms a cycle. The distribution of trip-ends which is determined from a land use plan is dependent upon trip purpose.

If the distribution of various bases according to the land use plan, the number and purposes of the first trips, and the transition matrix for trip purposes are known, the person trip pattern will be determined.

In the choice of mode of travel, it is assumed that the rate autos are used depends on the kind of base, purpose of the first trip, change of purposes, and characteristics of destinations.

Introduction

In a study of the person trips in the metropolitan area, we propose to consider the trip pattern as a stochastic process.

In a previous paper, the author described the movement of cars as a Markov chain and applied this to the estimation of trip distributions by means of entropy maximization. Here, the person trip pattern will also be described as a Markov chain. A chain of trips starting from a base is treated as an absorbing Markov chain, which includes various trip purposes and in most cases forms a loop.

A base is defined as the starting place of the first trip in a day, principally, home, another's home, hotel, office or terminal, or hospital, because of the convenience of application of absorbing Markov chains.

After people start from their bases, they will complete a chain of trips and return to their bases according to transition probabilities of trip purposes. There are only two modes of travel in our model, one of them is auto (drivers and passengers) and the other is mass transit. We assume that the rate of using autos decreases with each successive trip in a chain of trips.

In 1970 a person trip survey was performed in the Keihanshin area (Osaka, Kobe, Kyoto and some adjacent prefectures). The analysis of the survey and the

References p. 130

estimation of the future demand for various kinds of transportation facilities is now under way.

The model described below will be taken as one of the simple models to establish a systematic estimation method, although it contains various questionable points that may be clarified in a few years.

Concept of Bases and Cycles

The author is not sure when the phrase "bases" originated. It has been in use since the transportation surveys performed in American cities during 1945–1948. The expression of "from home" or "to home" appeared in Bulletin 203 (Highway Research Board) in 1958.[1]

In the U.S., when either of the trip ends is home, the trip is called a "home based" trip. In this paper, for the convenience of describing the movement of persons through a Markov chain, the base is defined as the starting place of the first person trip in a day. Then we can classify bases into several groups according to the sleeping places of persons, e.g. (1) home, (2) another's home (as a guest), (3) hotel (as a guest), (4) office (as a night worker or a night watch), (5) a train or airplane, etc. (as a passenger), (6) hospital (as a patient or his attendant), (7) a prison (as a prisoner), and so forth.

The places described above should be origins of the first trips in the day. However, the number of trips originating from hospitals and prisons will be negligible among all person trips.

Therefore, the important bases in the urban transportation planning are as follows: (1) home, (2) hotel, (3) office (including factories), (4) terminal (stations, airports, etc.).

An example of a person trip pattern is shown in Fig. 1. Mr. A started at his home, went to his working place, then to another office for a meeting, saw his guest off at a station after the meeting and returned to his office. After work, he went to a hotel to see his friend and then returned to his home. His traffic pattern has two loops with six trips in the day. To treat his pattern of movement as a Markov chain from a base of his home, we should consider his trip of

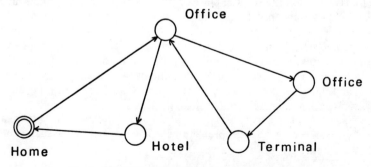

Fig. 1. Illustration of complete trip with two loops.

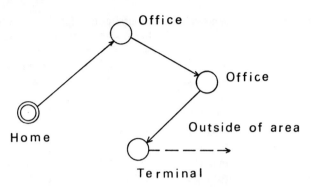

Fig. 2. Illustration of an incomplete trip.

returning to his office as a business trip regardless of its direction. Then his traffic pattern in the day will be treated as a Markov chain with one cycle from his home base. If it has two trips from the base, it should be treated as a Markov chain with two cycles, namely, two Markov chains around the base.

On the other hand, Mr. B went outside of the study area from a terminal after the meeting and could not return to his home that day. His traffic pattern is shown in Fig. 2. Mr. B's traffic pattern will be called an incomplete trip pattern. The pattern of Mr. A shown in Fig. 1 will be called a complete trip pattern.

Mr. A's friend at the hotel might have come into Mr. A's city in the morning from his home outside of the study area. For the transportation planning in Mr. A's city, the trip pattern of his friend in the day should be treated as a terminal based trip if he came by train.

In most cases, a Markov chain from the terminal base will not be more than two cycles, because most people will not enter Mr. A's city more than twice in the same day. Part of the visitors to the city may be assigned to the terminal bases, i.e., railway stations, bus stops, harbors and airports, according to their travel modes. The rest of the visitors will be auto drivers and passengers, and their trips in the city will be considered to belong to highway bases. The terminal from which they leave the study area may not be the same terminal from which they started. However, we assume that both the terminals are the same one and the terminal based trips are complete trips.

In short, for the transportation planning of an area, a base is defined as the starting place of the first trip in the study area and is fixed only for that day for

TABLE I

Home-Based Traffic Patterns in Kobe (on March, 1967)

No. of cycles	0.5	1					2			3	Miscellaneous	Total
No. of trips	1	2	3	4	5	6	4	5	6	6		
No. of samples	54	3699	626	112	34	15	453	48	17	71	76	5179

each person. The number of cycles is the number of trips returning to the same base. An incomplete trip is omitted in this model but the number of incomplete trips is so small as to be negligible, as shown in Table I.

The detailed information of terminal based trips will be obtained by interview surveys at the terminals and the cordon station, as well as home interview surveys.

Description of Person Trip Pattern Through Markov Chains

We assume all person trip patterns form complete trip patterns about every base. After people start from their bases, they will complete a chain of trips and return to their bases according to the transition probabilities of trip purposes. If the returning place of the final trip coincides with the starting place of the first trip, a chain of trips is completed through a Markov chain.

We shall consider several kinds of trip purposes $1, 2, \ldots, q, \ldots, Q$. The number of first trips of each trip purpose in the whole area is described by a vector

$$L^{(1)} = (l_1^{(1)}, l_2^{(1)}, \ldots, l_Q^{(1)}), \tag{1}$$

where the superscript (1) denotes the first trip for every purpose. We also form the following matrix,

$$\mathbf{L}^{(1)} = \begin{pmatrix} l_1^{(1)} & & & 0 \\ & l_2^{(1)} & & \\ & & \ddots & \\ 0 & & & l_Q^{(1)} \end{pmatrix} \tag{2}$$

We distinguish trip production from trip generation. Trip production concerns a whole area and does not include information of subareas or zones. On the other hand, the concept of trip generation expresses information for every zone in the whole area.

The proportions of first trips generated with purpose q for each zone, $1, 2, n, \ldots, N$ are expressed by a vector

$$\mathbf{u}^q = (u_1^q, u_2^q, \ldots, u_n^q, \ldots, u_N^q), \tag{3}$$

where

$$\sum_{i=1}^{N} u_i^q = 1.$$

If we consider the following matrix,

$$U = \begin{pmatrix} u^1 & & & 0 \\ & u^2 & & \\ & & \ddots & \\ 0 & & & u^Q \end{pmatrix}$$

$$= \begin{pmatrix} u_1^1, u_2^1, \ldots, u_N^1 & & & 0 \\ & u_1^2, u_2^2, \ldots, u_N^2 & & \\ & & \cdots & \\ 0 & & & u_1^Q, u_2^Q, \ldots, u_N^Q \end{pmatrix}, \qquad (4)$$

the number of first trips generated for each trip purpose is expressed by the matrix

$$L^{(1)}U = \begin{pmatrix} l_1^{(1)} u_1^1, l_1^{(1)} u_2^1, \ldots, l_1^{(1)} u_N^1 & & 0 \\ & \cdots & \\ 0 & & l_Q^{(1)} u_1^Q, l_Q^{(1)} u_2^Q, \ldots l_Q^{(1)} u_N^Q \end{pmatrix}. \qquad (5)$$

If people make trips according to the transition probabilities between zones, the number of first trips attracted to zones is shown by

$$L^{(1)}UP = \begin{pmatrix} l_1^{(1)} u^1 & & & 0 \\ & l_2^{(1)} u^2 & & \\ & & \ddots & \\ 0 & & & l_Q^{(1)} u^Q \end{pmatrix} \begin{pmatrix} P_1 & & & 0 \\ & P_2 & & \\ & & \ddots & \\ 0 & & & P_Q \end{pmatrix}$$

$$= \begin{pmatrix} l_1^{(1)} u^1 P_1 & & & 0 \\ & l_2^{(1)} u^2 P_2 & & \\ & & \ddots & \\ 0 & & & l_Q^{(1)} u^Q P_Q \end{pmatrix}, \qquad (6)$$

where P_q is the transition matrix between zones for trip purpose q, and N is the number of zones,

$$P_q = \begin{pmatrix} p_{11}^q, p_{12}^q, \ldots, p_{1N}^q \\ \cdots\cdots\cdots\cdots\cdots \\ p_{N1}^q, p_{N2}^q, \ldots, p_{NN}^q \end{pmatrix}, \quad \sum_{j=1}^{N} p_{ij}^q = 1. \qquad (7)$$

The generation of second trips of various trip purposes is assumed to be made through a transition matrix between trip purposes. If the probability that people will take the next trip with purpose j from the previous trip with purpose i is denoted by y_{ij}, then the transition matrix between purposes is expressed by the canonical form

$$Q = \begin{pmatrix} I & 0 \\ R & Y \end{pmatrix} = \begin{pmatrix} 1 & \vdots & 0 & \ldots, 0 \\ \cdots & \vdots & \cdots\cdots\cdots \\ y_{10} & \vdots & y_{11}, \ldots, y_{1Q} \\ \cdot & \vdots & \\ \cdot & \vdots & \\ \cdot & \vdots & \\ y_{Q0} & \vdots & y_{Q1}, \ldots, y_{QQ} \end{pmatrix}, \quad \sum_{j=0}^{Q} y_{ij} = 1, \qquad (8)$$

where the trip purpose 0 designates the return trip to the base, Y is the matrix of moving about, and R is the returning vector.

$$Y = \begin{pmatrix} y_{11}, y_{12}, \ldots, y_{1Q} \\ y_{21}, y_{22}, \ldots, y_{2Q} \\ \cdots\cdots\cdots\cdots \\ y_{Q1}, y_{Q2}, \ldots, y_{QQ} \end{pmatrix}, \quad R = \begin{pmatrix} y_{10} \\ y_{20} \\ \cdot \\ \cdot \\ \cdot \\ y_{Q0} \end{pmatrix}. \qquad (9)$$

The submatrix Y in the canonical form of Q of absorbing Markov chain satisfies the relation

$$I + Y + Y^2 + \cdots + Y^k + \cdots = (I - Y)^{-1}.$$

The inverse matrix $(I - Y)^{-1}$ is called the fundamental matrix for the absorbing Markov chains. The ij-entry of the fundamental matrix shows the expected total number of trips with purpose j after starting with trip purpose i. Table II and III show the values of Y and $(I - Y)^{-1}$, respectively, for the citizens of Kyoto in 1969.

Now we consider an expanded matrix form of the matrix of Y,

$$\overline{Y} = \begin{pmatrix} Y_{11}, Y_{12}, \ldots, Y_{1Q} \\ Y_{21}, Y_{22}, \ldots, Y_{2Q} \\ \cdots\cdots\cdots\cdots \\ Y_{Q1}, Y_{Q2}, \ldots, Y_{QQ} \end{pmatrix}, \qquad (10)$$

TABLE II

Transition Matrix of Trip Purposes (Kyoto, 1969)

	Return	Work	School	Shopping	Recreational
Return	1.0000	0.0000	0.0000	0.0000	0.0000
Work	0.9157	0.0137	0.0009	0.0485	0.0212
School	0.9619	0.0005	0.0067	0.0233	0.0076
Shopping	0.6795	0.0062	0.0022	0.3047	0.0074
Recreational	0.7782	0.0060	0.0024	0.0851	0.1283

TABLE III

The Values of $(I - Y)^{-1}$ (Kyoto, 1969)

	Work	School	Shopping	Recreational
Work	1.0205	0.0008	0.0671	0.0245
School	0.0009	1.0209	0.0387	0.0109
Shopping	0.0098	0.0035	1.4705	0.0140
Recreational	0.0058	0.0058	0.1049	1.4524

where

$$Y_{ij} = \begin{pmatrix} y_{ij} & & & 0 \\ & y_{ij} & & \\ & & \cdot & \\ & & & \cdot \\ 0 & & & y_{ij} \end{pmatrix} \qquad (11)$$

is an $N \times N$ matrix.

The number of second trips generated in each zone after the first trips is expressed by

$$L^{(1)}UP\overline{Y}.$$

Then the number of second trips attracted to each zone is given by

$$L^{(1)}UP\overline{Y}P.$$

In the same way, the number of third trips generated in each zone is described by

$$L^{(1)}UP\overline{Y}P\overline{Y} = L^{(1)}U(P\overline{Y})^2.$$

The number of kth trips generated in each zone is also given by

$$L^{(1)}U(P\overline{Y})^{k-1}.$$

Therefore, the total number of generated trips from the first trips to the kth trips is expressed by

$$\mathbf{L}^{(1)}\mathbf{U}\sum_{l=1}^{k}(\mathbf{P}\overline{Y})^{l-1}.$$

As the value of k becomes infinite, we can evaluate the total number of generated trips before the absorption by the equation

$$\mathbf{A} = \mathbf{L}^{(1)}\mathbf{U}(I - \mathbf{P}\overline{Y})^{-1} \tag{12}$$

$$= \begin{pmatrix} 1\text{-}1, & 1\text{-}2, & \dots, & 1\text{-}Q \\ 2\text{-}1, & 2\text{-}2, & \dots, & 2\text{-}Q \\ \multicolumn{4}{c}{\dotfill} \\ Q\text{-}1, & Q\text{-}2, & \dots, & Q\text{-}Q \end{pmatrix}, \tag{13}$$

where i-j in Eq. (13) shows a $1 \times N$ vector whose entries are the number of trips generated in each zone from chains of trips of which the purpose of the first trips is i and of the final trips is j.

Next we shall consider the OD pattern of these trips. Now arranging $\mathbf{L}^{(1)}\mathbf{U}$ of Eq. (5) in the form

$$\overline{\mathbf{L}^{(1)}\mathbf{U}} = \begin{pmatrix} l_1^{(1)}u_1^1 & & & & & & 0 \\ & \ddots & & & & & \\ & & l_1^{(1)}u_N^1 & & & & \\ & & & \ddots & & & \\ & & & & l_Q^{(1)}u_1^Q & & \\ & & & & & \ddots & \\ 0 & & & & & & l_Q^{(1)}u_N^Q \end{pmatrix}, \tag{14}$$

the total pattern of trips from the first to the kth is described by

$$\overline{\mathbf{L}^{(1)}\mathbf{U}}\mathbf{P} + \overline{\mathbf{L}^{(1)}\mathbf{U}}\mathbf{P}\overline{Y}\mathbf{P} + \overline{\mathbf{L}^{(1)}\mathbf{U}}\mathbf{P}\overline{Y}\mathbf{P} + \cdots + \overline{\mathbf{L}^{(1)}\mathbf{U}}\mathbf{P}(\overline{Y}\mathbf{P})^{k-1}.$$

Therefore, we have the OD pattern of persons before the absorption in the form

$$\overline{\mathbf{L}^{(1)}\mathbf{U}}\mathbf{P}(I - \overline{Y}\mathbf{P})^{-1} = \begin{pmatrix} M_{11}, & M_{12}, & \dots, & M_{1Q} \\ M_{21}, & M_{22}, & \dots, & M_{2Q} \\ \multicolumn{4}{c}{\dotfill} \\ M_{Q1}, & M_{Q2}, & \dots, & M_{QQ} \end{pmatrix}, \tag{15}$$

where M_{ij} gives the OD pattern of persons whose trip purpose of the final trip just before absorption is j, after starting initially with trip purpose i.

In addition we should apply this method to the trip pattern from each kind of base.

In our model, there are some questionable assumptions. In the example of trips to work, the length of the first trips might be longer than the kth trips, because the first trips will occur earlier in the day than the kth trips. However, we assume the transition matrix between zones is independent of k.

The transition matrix between trip purposes is also assumed to be independent of k.

These assumptions will be examined for a few years during the Keihanshin (Osaka, Kobe, and Kyoto) metropolitan transportation survey.

Estimation of OD Pattern

In order to determine the person trip pattern, we must investigate the values of $L^{(1)}$, U, P, and Q.

In the evaluation of $L^{(1)}$, the number of households and the type of occupation will be important. This is the problem of trip production. The values of U are given by the land use plan. Then one of the important factors is the distribution of employees for each zone. In order to determine the value of P, the number of first trips of each trip purpose will be investigated for each zone.

The present value of Q will be given by the trip survey. Although future values of Q may be different, they will be assumed to be the same, as the present values.

Choice of Mode of Trip

There are many factors which influence the modal split. The effect of differences of trip purposes and characteristics of zones will be important to the choice of mode.

We now consider in the modal split model the effect of variation of trip purposes and zones.

To simplify the consideration, we have two kinds of modes, auto and mass transit.

If the fractions of trips by auto are given by

$$\{\mu_i^j\} = \begin{pmatrix} \mu_1^1, \mu_2^1, \ldots, \mu_N^1 \\ \cdots\cdots\cdots\cdots \\ \mu_1^Q, \mu_2^Q, \ldots, \mu_N^Q \end{pmatrix}, \quad \begin{pmatrix} i = 1, 2, \ldots, n, \ldots, N \\ j = 1, 2, \ldots, q, \ldots, Q \end{pmatrix}, \quad (16)$$

then, through consideration of the characteristics of car trip generation by purposes j of the first trips in each zone i, Eq. (5) is replaced by

$$\begin{pmatrix} l_1^{(1)}\mu_1^1 u_1^1, \ldots, l_1^{(1)}\mu_N^1 u_N^1 & & 0 \\ & \cdot & \\ & \cdot & \\ & \cdot & \\ 0 & & l_Q^{(1)}\mu_1^Q u_1^Q, \ldots, l_Q^{(1)}\mu_N^Q u_N^Q \end{pmatrix}. \qquad (17)$$

In the transition from the previous trip purpose i to next trip purpose j, the fractions of trips using autos will be reduced by a factor ϕ_{ij}, because of the inconvenience of parking or the avoidance of drunk driving, etc.

Therefore, the matrix of $\overline{Y}$ of Eq. (10) is generalized to

$$\overline{Y} = \begin{pmatrix} \phi_{11} Y_{11}, & \phi_{12} Y_{12}, & \ldots, & \phi_{1Q} Y_{1Q} \\ \cdots\cdots\cdots\cdots\cdots\cdots\cdots\cdots\cdots\cdots \\ \phi_{Q1} Y_{Q1}, & \phi_{Q2} Y_{Q2}, & \ldots, & \phi_{QQ} Y_{QQ} \end{pmatrix}, \qquad (18)$$

if we consider the transition from auto to mass transit as well as the transitions between trip purposes.

Hence, the OD pattern of autos is expressed by Eq. (15) using Eq. (17) and Eq. (18).

Finally, we consider the returning pattern. The return pattern from the first trips is denoted, by the use of Eq. (14), by

$$[\overline{\mathbf{L}^{(1)}\,\mathbf{U}\mathbf{P}\overline{R}}]^t, \qquad (19)$$

where the superscript t means of the transpose matrix and

$$\overline{R} = \begin{pmatrix} R_{10} \\ R_{20} \\ \cdot \\ \cdot \\ \cdot \\ R_{Q0} \end{pmatrix}, \qquad R_{q0} = \begin{pmatrix} y_{q0} & & & 0 \\ & y_{q0} & & \\ & & \cdot & \\ & & & \cdot \\ 0 & & & y_{q0} \end{pmatrix}. \qquad (20)$$

The returning pattern from the second trips is also written as

$$[\overline{\mathbf{L}^{(1)}\,\mathbf{U}\mathbf{P}\,\overline{Y}\mathbf{P}\overline{R}}]^t. \qquad (21)$$

Similarly, the returning patterns from the third and the kth trips are denoted by

$$[\overline{\mathbf{L}^{(1)}\,\mathbf{U}\mathbf{P}(\overline{Y}\mathbf{P})^2\overline{R}}]^t, \quad [\overline{\mathbf{L}^{(1)}\,\mathbf{U}\mathbf{P}(\overline{Y}\mathbf{P})^{k-1}\overline{R}}]^t, \qquad (22)$$

respectively. The sum of all returning patterns gives the OD pattern

$$[\overline{\mathbf{L}^{(1)}\,\mathbf{U}\mathbf{P}(I - \overline{Y}\mathbf{P})^{-1}\overline{R}}]^t = \begin{pmatrix} S_{10} \\ S_{20} \\ \cdot \\ \cdot \\ \cdot \\ S_{Q0} \end{pmatrix}, \qquad (23)$$

where S is an $N \times N$ matrix of the returning pattern and q is the purpose just before absorption

The fractions of trips by auto are introduced into $\overline{R}$ as

$$\overline{R} = \begin{pmatrix} \phi_{10}\ R_{10} \\ \phi_{20}\ R_{20} \\ \cdots\cdots \\ \phi_{Q0}R_{Q0} \end{pmatrix}. \tag{24}$$

In such a way, we can describe the OD pattern of autos (auto drivers and auto passengers), though the OD pattern of car trips is not described.

An illustration of modal split of the citizens of Kyoto is shown in Table IV.

TABLE IV

Modal Split of Persons by Trip Purposes
(Kyoto, 1969)

	Mass transit	Auto	Taxi
Work	0.724	0.256	0.020
School	0.960	0.037	0.003
Shopping	0.588	0.301	0.111
Recreational	0.467	0.322	0.191
Sightseeing	0.760	0.144	0.096
Business	0.120	0.836	0.044
Total	0.583	0.369	0.048

Introduction of ϕ_{ij} in Eq. (18) is supported by the assumption that the rate of using autos will be influenced by the change of purpose, because the duration of parking is affected by the parking purposes and has a great influence on parking costs.

However, if the parking capacities in each destination zone have a great effect on choice of autos, we should have variable coefficients on zones instead of Y_{ij} in Eq. (11), i.e.,

$$Y_{ij} = \begin{pmatrix} \psi_1 y_{ij} & & & 0 \\ & \psi_2 y_{ij} & & \\ & & \cdot & \\ & & & \cdot \\ 0 & & & \psi_N y_{ij} \end{pmatrix}. \tag{25}$$

Choice of mode of trips is strongly affected by the trip purposes and the characteristics of zones. Therefore, the values of μ, ϕ, and ψ should be carefully determined through factor analysis, etc.

Conclusion

In the transportation planning in an urban area, the distribution of bases and the number of the first trips of each trip purpose are important. The OD pattern of person trips described as a Markov chain will be useful in estimation of the future trip pattern from the land use plan. Though we assume that the future values of the transition matrix between purposes are the same as present values, this needs to be verified.

To observe the trip pattern from all kinds of bases, surveys of guests at hotels, lodgers at home, and passengers at terminals should be performed at the same time as dwellers at home.

Acknowledgment

The author wishes to express his sincere appreciation to Prof. Eiji Kometani for his valuable suggestion, and to research assistant Toshiaki Okamoto, a humble gratitude, for his cooperation in calculation.

References

1. Gordon B. Sharpe, Walter G. Hansen, and Lamelle B. Hamner, Factors affecting trip generation of residential land-use, *Highway Res. Board Bull.* 203, 20–36 (1958).
2. Tsuna Sasaki, Category of person trips pattern, *Traffic Engineering* (in Japanese), Vol. **4**, No. 1, 3–9 (1969).

A New Look at the Traffic Assignment Problem*

I. Jeevanantham

University of California, Berkeley, California

Abstract

Traffic assignment models in use presently assume that the cost of using a link in a network is the same to all the users of the link. There is no doubt that this assumption does not correctly describe the actual situation. For example, the value of time and the operating cost of the vehicle, two of the components of the cost of travel, are not the same to all the users. The existence of traffic assignments when the above assumption is not made is examined for some very simple networks. We show that stable traffic assignments exist and that these assignments are quite different from the assignments obtained with the present models.

Introduction

The models developed for assigning traffic to a network are based on the assumption that all the users evaluate the network in exactly the same manner. There is always a "cost" (time, money, etc.) associated with the use of the network; in the models it is assumed that the users who are physically alike (example, all those in buses, or all those in automobiles) agree on the cost of using the network, at least on those elements of the network which they use.

If we consider the elements that constitute the cost, we realize that the above assumption is an oversimplification. For example, suppose we consider a traffic of automobiles only. There are cars of different sizes, age, models on our highways; the operating costs of these are not obviously the same. Secondly, there is the question of the value of the time spent on a trip. There is no doubt that each individual has his own value of time; some people drive slowly and spend a longer time on a trip than others. At issue here is not what should be the cost of a trip; the point we wish to make is that given a certain number of users (flow) on a route, it is neither obvious nor necessary that the cost to every one of them is the same.

It is perhaps more realistic to assume that each individual has his own estimate of the cost of a trip (perhaps based on his experience) for different traffic conditions. We will assume that this is a continuous function of the flow on the route and that every individual bases his decision on his cost function. We will further assume that the traffic using a network are physically alike (i.e., all automobiles

*This research was supported in part by the National Science Foundation under Grant GP-9323.

References p. 153

153

131

or buses, etc.); thus the cost is a function of the single variable f which is the flow on a route. This assumption is made basically to simplify the mathematics; in fact, it enables us to obtain very simple graphical solutions.

Given that the users do not agree on the costs, how does this influence the traffic assignment? We investigate here the existence of solutions for such a problem for some very simple networks.

The Model

The networks we examine have a single origin O and a single destination D. No two routes between O and D have a common link.

There is a cost for using any route, this cost being a function of the number of users (flow) on the route. With a route flow f_j on the jth route ($j = 1, 2, 3, \ldots$) between O and D, the cost to the users varies; we group the users by cost into types A, B, C, etc. Thus the cost of a trip on the jth route to a person of type A is $C_{Aj}(f_j)$. There are similar cost functions for every type of user on all the routes. We will assume that the cost functions are all monotone increasing.

Given a total flow F between O and D, our problem is to investigate the existence of a stable assignment of the flow to the routes between O and D. This assignment naturally depends on the objective of the users. We will consider the following objectives:

(i) The users try to minimize their individual costs by using the route which is cheapest to them.[1,2]

(ii) The users try to minimize their total cost.[1,2]

(iii) A combination of (i) and (ii); for example, A type may attempt to minimize their total cost while the others desire the minimization of their individual cost.

Part I

Minimize Cost to Every User

To achieve the above objective, the assignment of the users should be such that everyone occupies the route that is cheapest to him. For the simple networks examined a graphical solution that provides all the necessary information is given. This solution is a generalization of the graphical solution when the users are all alike (i.e., $A \equiv B \equiv C$, etc.). We will therefore take this case as our first example and use the solution in the other examples.

Example 1—All Users Alike

We have a flow F between O and D, the flow being all of one type, say A. The solution is given for two routes between O and D; it is easy to see that it could be extended to any number of routes between O and D.

For a given F, if an assignment with flows on both routes is possible, it will be stable only if the costs on the two routes are equal. If this were not so, some of the users on the costlier route will transfer to the cheaper route tending to equalize the costs. Thus the route flows f_j must satisfy

$$C_{A1}(f_1) = C_{A2}(f_2)$$

where

$$f_j \geqslant 0 \quad \text{and} \quad \sum_j f_j = F.$$

The graphical solution of the above set of equations is shown in Fig. 1. We first construct the composite curve $C_A(F)$ as the horizontal sum of the $C_{Aj}(f_i)$; that is, $TW = TU + TV$ and is the total flow between O and D when the cost to the users is OT. Since the $C_{Aj}(f_i)$ are monotone increasing, TU and TV are both single valued and increase as OT increases. Thus TW is also single valued and increasing, i.e., $C_A(F)$ is monotone increasing. For T such that $OY \leqslant OT \leqslant OX$, $C_A(F) \equiv C_{A2}(f_2)$ as $C_{A1}(f_1) = 0$.

Having constructed $C_A(F)$, to obtain the cost and route flows for some F, we first locate W for the given F and then U and V; f_1 and f_2 are then the route flows and OT the cost of a trip between O and D.

The existence and uniqueness of the solution follows from the properties of the cost functions. Since $C_A(F)$ is monotone increasing, W, the intersection of $C_A(F)$ and the vertical through F, always exists and is unique. Similarly, given W, U and V, which are the intersections of the $C_{Aj}(f_i)$ and the horizontal through W, exist and are unique.

For $F \leqslant F_1$, the construction gives all the flow on route 2. This is consistent with the objective; for example, if $F = F_2$, the cost of introducing some

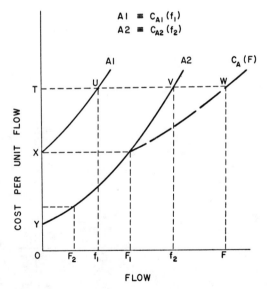

Fig. 1. Construction to determine route flows. Two routes, all users alike.

flow on route 1 is more than the cost on route 2 with $f_2 = F_2$. Therefore, nobody will choose to use route 1.

In practice, there is an upper bound to the capacity of a route and the cost functions are all asymptotic. If F_j is the capacity of the jth route, then $C_A(F)$ is asymptotic to the vertical through $F = \Sigma_j F_j$. If $F > \Sigma_j F_j$, then $W, U,$ and V do not exist and there is no optimal assignment.

Example 2—Two Routes with Two Types of Users

We are given the cost functions $C_{Aj}(f_j), C_{Bj}(f_j)$. Using the construction in Example 1 we obtain f_j^A as the unique route flows if the total flow F between O and D had been all of A (Fig. 2). Similarly f_j^B would be the unique route flows if F was all of B.

To determine the assignment when $F = F_A + F_B$ where F_A, F_B are the number of A and B, respectively, we first load the network with a flow F all of A and gradually substitute B for A until F has the required composition. We will continue the substitution till all the A have been replaced, thereby obtaining solutions for all possible combinations of F_A and F_B.

Referring to Fig. 2, with route flows f_j^A (F all of A).

$$C_{B1}(f_1^A) > C_{B2}(f_2^A). \tag{1}$$

Route 2 is thus the cheaper route to B and if we replace ϵ of A by B, the latter would choose route 2. If $\epsilon < f_2^A$, this substitution does not alter the route flows; the cost to A remains unchanged and equal on the two routes. B of course is using the cheaper route; therefore, this is the optimal assignment.

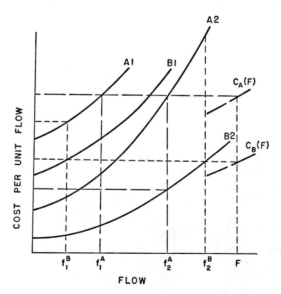

Fig. 2. Construction to determine route flows. Two routes, two types of users.

As we continue to replace A, B continues to use route 2 with A on both routes till $F_B = f_2^A$. Up to this stage the route costs have not changed; route 2 is yet cheaper to B. If we further increase F_B by replacing A on route 1, B will prefer route 2 and transfer to it, thereby reducing route 1 flow and the cost to A on route 1. Since route 2 flow increases $(> f_2^A)$, $C_{A2}(f_2) > C_{A2}(f_2^A)$. We now have A and B occupying routes with the cheaper cost to them; again this is the optimal assignment.

The above assignment with A on route 1 and B on route 2 and the route flows changing continuously as F_B is increased prevails till (i) the route flows become f_j^B, or (ii) route 1 flow is zero which implies that all the A have been replaced. We note that (i) and (ii) are mutually exclusive.

When the route flows become f_j^B, route cost to B on the two routes are equal. [From Eq. (1) we see that the route flows at which cost to B will be equal on the two routes, i.e., f_j^B, must be such that $f_1^B \leqslant f_1^A$ and $f_2^B > f_2^A$. Thus the results are consistent.] Now as we continue to replace A (on route 1), B will choose route 1; once again the route flows and the route costs remain constant. This assignment with A on route 1 and B on both routes is maintained till all the A have been replaced by B.

The occurrence of the second possibility is shown in Fig. 3. For a total flow F_3, $f_1^B = 0$ and $f_2^B = F_3$, and when the A are completely replaced, B uses only route 2 which is cheaper (to B) than route 1. This is true for all F such that $F_2 < F \leqslant F_4$.

A further exception, perhaps a trivial one, to the assignments discussed above occurs for small values of F when only one route is occupied. In Fig. 3, for a flow of F_1, only route 2 is occupied—whatever the composition of the flow. This is true for all $F \leqslant F_2$.

The above discussion establishes the existence of a solution. The uniqueness of the solution (at every stage) follows directly from the uniqueness of f_j^A (and f_j^B).

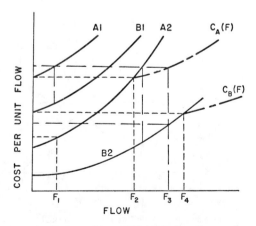

Fig. 3. Repetition of Fig. 2 for low flows.

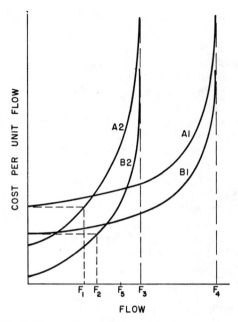

Fig. 4. User cost curves. Two routes, two user types.

   A typical set of cost curves is shown in Fig. 4. By drawing the cost curves asymptotic to the same vertical line for a particular route, we are assuming that all the users agree on the capacity of the route. This is a reasonable assumption as the capacity is a function of the degree of congestion on the route; the latter in turn is a function of the number of vehicles but not the cost of the trip. In Fig. 4, route 1 may represent the cost curve for a typical freeway while route 2 that of an arterial.

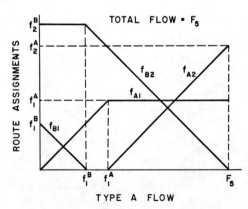

Fig. 5. Variation of route assignments with composition of flow. Users choose cheapest route.

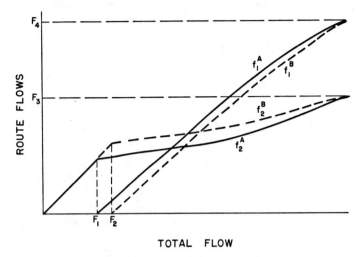

Fig. 6. Variation of critical route flows with total flow. Two routes, two user types.

Figure 5 shows the variation in route assignments as the composition of F varies for $F = F_5$ (Fig. 4). This is for the typical case where both routes are occupied all the time. Points of interest here are the two critical values of F_A (and $\therefore F_B$) which determines the assignment of the users. For $F_A > f_1^A$, A "controls" the flows and route costs by occupying both routes while B is on the cheaper of the two routes. For $F_A < f_1^B$, B controls.

Next we consider the variation in route flows as F varies. This is shown in Fig. 6 for the cost curves of Fig. 4. In view of the variation in route assignments with the composition of F, Fig. 6 really shows how the critical values of F_A (and F_B) vary with F.

In Fig. 6 we note that when F = capacity of the network, $f_j^A = f_j^B$; A and B estimate the costs on the two routes to be equal and may trade places. While this is true in all cases (if the users agree on the capacity of the routes), there is the possibility that $f_j^A = f_j^B$ even at some intermediate flow. We illustrate this in Fig. 7. Figure 7(a) gives the cost curves (a part only); we assume that $F = F_A + F_B$ is such that A occupies both routes. Figure 7(b) gives the variation in $C_{Bj}(f_j)$ for F such that $C_{Aj}(f_j)$ varies from C_1 to C_2. At $F = F^*$, $C_{Bj}(f_j)$ on the two routes are equal [but not necessarily equal to $C_{Aj}(f_j)$] and once again A and B may trade places.

The above phenomenon by itself is not very significant. What is more important is the behavior as F increases through F^*. For $F < F^*$, from Fig. 7(b), route 1 is cheaper while, for $F > F^*$, route 2 is cheaper; thus all the F_B on route 1 will transfer to route 2. (We can also show that a similar situation exists even when B is on both routes with A on the cheaper route.)

The transfer of B creates a turbulent situation as an (almost) equal number of A will transfer from route 2 to route 1 to maintain equilibrium.

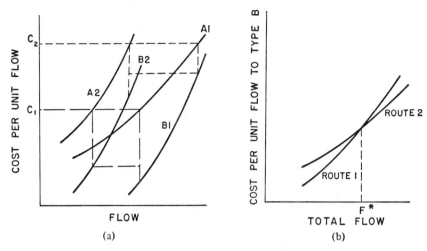

Fig. 7. Nonunique equilibrium. Two routes, two user types.

Flows on transportation networks are not constant; they are subject to fluctu-ations. If the total flow is in the neighbourhood of F^*, one could expect un-settled conditions on the network as the users jockey for the route of their choice.

Example 3—Three Routes with Two Types of Users

We adopt the same procedure as before; first, loading the network with flow F all of A and then gradually replacing A by B.

In Fig. 8, $C_A(F)$ and $C_B(F)$ are the composite cost curves for A and B, re-spectively, being the horizontal sums of the three cost curves for each type. $C^*(F)$ is another composite curve; we will consider the construction and use of it later.

The following outlines briefly the changes in route flows and assignments as the composition of F varies; the changes can be followed in Fig. 8.

(i) With $F_A = F, F_B = 0$, the route flows are f_j^A.

(ii) When route flows are f_j^A, route 3 is the cheapest to B (W below U, V). As B are introduced, they replace A on route 3 with the route flows remaining un-changed till all A on route 3 are replaced by B.

(iii) Now we must replace the A on routes 1 and 2. B yet considers route 3 to be cheaper. Thus route 3 flow increases ($f_3 > f_3^A$) as more B use it than in (ii) while A adjust their flows on the other two routes to equalize their costs. We have $f_1 < f_1^A$ and $f_2 < f_2^A$. Given an F_A ($< f_1^A + f_2^A$), we obtain f_1 and f_2 from the composite curve for A on routes 1 and 2 only with total flow F_A (this is not shown in Fig. 8). Note that the route flows are continuously changing.

(iv) When the route flows in (iii) become f_j^* we see that the cost to B on route 2 is equal to the cost on route 3 which B occupies. Hence, as the B in-crease they replace A on route 2. Once again route flows remain unchanged at

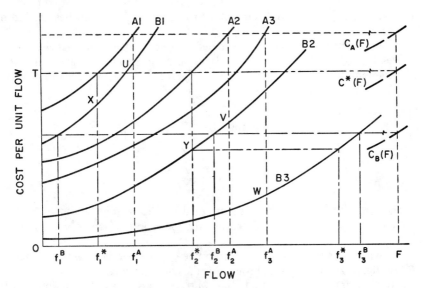

Fig. 8. Construction to determine route flows. Three routes, two user types.

f_j^*. In Fig. 8 we show $f_3^* < f_3^B$ and $f_2^* < f_2^B$ where f_j^B are the route flows when all the flow is B type. It is easy to show that this is a necessary condition.

(v) Once all the A on route 2 have been replaced by B, the route flows once again begin to change, for as we commence replacing A on route 1, B will transfer to the other two routes which are yet cheaper to B. B adjusts the flow on routes 2 and 3 to equalize their cost. Route 1 flow is now less than f_1^* while the flow on routes 2 and 3 increase beyond f_2^* and f_3^*, respectively. Given F_B we obtain flows on routes 2 and 3 from the composite curve of these two routes for B.

(vi) Route flows continue to change till the cost to B on all three routes becomes equal which is at the unique route flows f_j^B (uniqueness follows from Example 1). From here on, the route flows remain constant as B now replaces A on route 1 till all the A have been replaced.

The uniqueness of the solution at every stage follows directly from the uniqueness of f_j^A, f_j^B (these two follow from Example 1) and f_j^* (which we show later).

There are of course some variations to the above behavior. For instance, all three routes may not be occupied always; or in (ii) with route flows f_j^A, two routes may be equally costly but cheaper than the third route to B. It is fairly easy to show that each of these leads to stable unique solutions. Finally, as in Example 2, if $f_j^A = f_j^B$ for all j, flows remain constant whatever the composition of F and the users may choose any route they like.

Construction and Use of $C^*(F)$

In step (iv) of the assignment in the last example we have A and B each occupying two routes but with only one route being common to both. We use the composite curve $C^*(F)$ to determine the route flows f_j^* that satisfies the above condition.

Referring to the example we considered, in step (iii) we know that A occupies routes 1 and 2 and B occupies route 3; it is also clear from our algorithm that the common route is either route 1 or 2, Therefore, the f_j^* are such that

$$C_{A1}(f_1^*) = C_{A2}(f_2^*)$$

$$C_{B3}(f_3^*) = \min\ [C_{B1}(f_1^*),\ C_{B2}(f_2^*)]$$

for

$$\sum_j f_j^* = F,\ f_j^* > 0\ \cdot$$

The composite curve $C^*(F)$ is constructed as follows (refer to Fig. 8):

(i) For some cost OT to A, determine the route flows f_1^* and f_2^*. These are unique.

(ii) With route flows f_1^* and f_2^*, determine the cheaper of these two routes to B. In Fig. 8, Y is below X and route 2 is the cheaper route.

(iii) Now determine f_3^* so that $C_{B3}(f_3^*) =$ cost to B in (ii). f_3^* is unique as (ii) gives a unique cost to B.

(iv) Plot Z such that $TZ = \Sigma_j f_j^*$ determined in (i), (ii), and (iii). TZ is the total flow in the network. Locus of Z is $C^*(F)$.

The uniqueness of the f_j^* follows from the monotonicity of the cost functions. Now if we took some other cost to A greater than OT, we can easily show that all the f_j^* increase and therefore F increases. Thus $C^*(F)$ is also monotone increasing.

Knowing $C^*(F)$ we simply reverse the procedure to obtain the route flows and assignment for some F. Given an F, Z is a unique point as $C^*(F)$ is monotone increasing; uniqueness of Z means there is a unique cost to A when they occupy routes 1 and 2. [Note we may interpret $C^*(F)$ as the cost to A, on routes 1 and 2, when network flow $= F$.] Therefore f_j^* are unique.

The construction of $C^*(F)$ as explained here is only true for those F where A occupies routes 1 and 2 and B occupies route 3 and, either route 1 or 2, i.e., for those F where route 3 is cheapest to B with route flows f_j^A.

Example 4—Two Routes with Three Types of Users

We first obtain f_j^S, $(S = A, B, C)$ the route flows when all the flow on the network is of the Sth type. These route flows are obtained from the composite curve for each type (these are not shown in Fig. 9).

(i) With route flows F_j^A we see from Fig. 9 that route 2 is cheaper to both B and C. Thus, if we start with a flow F all of A (giving route flows f_j^A) and gradually replace A by B and/or C, B, and C would choose route 2. For $F_A \geqslant f_1^A$ $(\because F_B + F_C \leqslant f_2^A)$ the route flows remain constant at f_j^A with A on both routes and B and C on route 2. When $F_A = f_1^A$, A uses only route 1 while B and C use route 2.

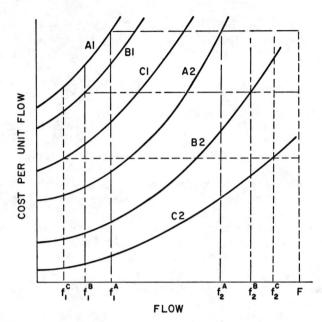

Fig. 9. Construction to determine route flows. Two routes, three user types.

(ii) As we continue to replace A, B and C continue to use route 2 as it is yet cheaper. Therefore route 1 flow decreases and route 2 flow increases continuously, till $F_A = f_1^B$.

(iii) Now B estimates costs on the two routes to be equal, and as more A are replaced, B uses both routes. A continues to use route 1 and C uses route 2. The route flows remain constant at f_j^B. B is able to control the flows only so long as $F_C \leqslant f_2^B$. When $F_C = f_2^B$ (and $F_A + F_B = f_1^B$) A and B use route 1 while C uses route 2.

(iv) If F_C increases beyond f_2^B, C continues to use route 2; A and B remain on route 1 and the route flows change continuously with route 2 flow increasing and route 1 flow decreasing.

(v) When the route flows in (iv) become f_j^C, C will begin to use both routes. Therefore for $F_C > f_2^C$, C uses both routes while A and B use route 1; once again the route flows remain constant.

Again, the uniqueness of the solution at every stage follows from the uniqueness of the f_j^S.

As in the earlier examples, the behaviour may be somewhat different at other values of F. For example, unlike in (i) where with route flows f_j^A route 2 was cheaper to both B and C we could have route 1 cheaper to B and route 2 cheaper to C. This situation is similar to (iii) with renaming of the users, and has a stable solution. If $f_j^A = f_j^B = f_j^C$ for all j the route flows remain constant while the users may choose any route they like. Finally, if $f_j^A = f_j^B \neq f_j^C$, A and B may occupy either or both routes while C occupies the cheaper of the two routes–

say route 2. Then the route flows remain at f_j^A so long as $F_C \leqslant f_2^A$; at $F_C = f_2^A$, A and B use only route 1 with C alone on route 2. The assignment now follows (iv) and (v).

Implications of the Model

It is quite clear from the examples analyzed so far that the traffic assignments under the proposed model are different from those under previous models. From the manner in which the solutions develop, one also feels that there is a stable solution for the general case of n types of users and j routes between the origin and destination.

In formulating the previous models, some regular flow pattern was observed on the routes between two points, and if more than one route was occupied, the conclusion was that the cost on the occupied routes must be equal. The cost to the various users was not measured or estimated; it was inferred from observation. However, from the results under the proposed model, we see that this conclusion is not necessarily true; that people could disagree on the relative costs of the various routes and yet distribute themselves in a stable manner that meets their objectives. Thus we are tempted to question the validity of the previous models.

In transportation planning, it is customary to estimate the distribution of trips between zones, by trip purpose; this is found to give a better fit to the model. There is some indication that a further classification by socio-economic status gives an even better fit. This implies that people are not evaluating similar trips in the same manner. Therefore the assumption that all the people do not agree on the cost of trips influences not only the traffic assignment models but also the distribution models. We could also infer that the modal split, too, is affected.

Part II

Minimize Total Cost

One has to consider two different possibilities in the minimization of the total cost to the users.

(i) We have classified the users into different groups. Each group may attempt to minimize the total cost to the members of the group, regardless of the consequences to the other groups.

(ii) We minimize the total cost to all the users of the network, evaluating the total cost as a function of each person's estimate of his cost.

Suppose we had a network where the traffic was all trucks belonging to two different companies; their cost of operation are not necessarily the same. A reasonable objective of each company would be minimize their total cost; thus this is an example of (i).

Now suppose that these two companies are really subsidiaries or different sections in a larger company. The latter would primarily desire to minimize the overall cost of operation, i.e., objective (ii).

The next two examples analyze a network with two routes and two classes of users for each of the two different objectives. The general principles therein are applicable for networks with more than two routes and more than two classes of users.

Mathematically we pose the problem in the following manner. Let f_{Aj}, f_{Bj} be the flow of A and B, respectively, on the jth route ($j = 1, 2$). Then by definition

$$\text{Total cost to } A = C_A = \sum_{j=1,2} f_{Aj} C_{Aj}(f_j) \qquad (2a)$$

$$\text{Total cost to } B = C_B = \sum_{j=1,2} f_{Bj} C_{Bj}(f_j) \qquad (2b)$$

where

$$f_{Aj} \geqslant 0, \quad f_{Bj} \geqslant 0$$
$$f_j = f_{Aj} + f_{Bj}, \quad \sum_j f_{Aj} = F_A, \quad \sum_j f_{Bj} = F_B \quad \text{and} \quad \sum_j f_j = F.$$

In (i) we want to minimize both C_A and C_B while in (ii) we want to minimize $C_A + C_B$, subject to the constraints in (2).

Example 5—Minimize Total Cost to Each Class of User

Suppose we have some distribution of flow on the network. Then

$$\text{Total cost to } A \text{ on route } 1 = C_{A1} = f_{A1} C_{A1}(f_1),$$

$$\text{Total cost to } A \text{ on route } 2 = C_{A2} = f_{A2} C_{A2}(f_2),$$

and

$$\text{Total cost to } A \text{ on the network} = C_A = C_{A1} + C_{A2}.$$

If we were to add a flow ϵ of A on to route 1, the additional cost to A on route 1 is given by

$$\epsilon \frac{\partial C_{A1}}{\partial f_{A1}} = \epsilon \left[C_{A1}(f_1) + f_{A1} C'_{A1}(f_1) \frac{df_1}{df_{A1}} \right] \quad \text{where} \quad C'_{A1}(f_1) = \frac{dC_{A1}(f_1)}{df_1}$$

$$= \epsilon [C_{A1}(f_1) + f_{A1} C'_{A1}(f_1)] \quad \text{as} \quad \frac{df_1}{df_{A1}} = 1$$

$$\equiv \epsilon C^*_{A1}(f_1).$$

Thus $C^*_{A1}(f_1)$ is the total additional cost A must bear in order to introduce an additional unit of flow on route 1; it is the marginal cost to A on route 1. There is, of course, an increase in cost to the B already on route 1 but A does not consider this. We assumed that A's actions are completely selfish.

$C_{A1}(f_1)$ is always positive; it is also always increasing making $C'_{A1}(f_1)$ positive. Therefore, as $f_{A1} \geqslant 0, C^*_{A1}(f_1)$ is always positive.

If the ϵ of A added to route 1 were from route 2, the change in cost to all the A on the network is given by

$$\Delta C_A = \epsilon \, [C_{A1}^*(f_1) - C_{A2}^*(f_2)]$$

from which it is clear that C_A is a minimum only when $C_{A1}^*(f_1) = C_{A2}^*(f_2)$. If A is to use only one of the two routes, then the marginal cost on that route must be less than or equal to the marginal cost on the other route. Therefore in order to minimize C_A, A should use the route or routes that minimizes their marginal cost.

So far we have only considered the behavior of A. B, too, is faced with the same problem. Therefore, we may restate our assignment problem as the determination of route flows to minimize the marginal cost to all the users; it is very similar to the problem in part one. The only difficulty now is that unlike the cost functions, the marginal cost functions depend on the composition of the route flows; they are therefore functions of the optimal assignment.

To investigate the nature of the solution, we adopt the same technique we have used in the earlier examples. Initially we load the network with a flow F, all of A, and then gradually substitute B in place of A. The solution is as follows:

(i) With $F = F_A$, A uses both routes adjusting route flows to equalize marginal costs on the two routes. We assume that F is large enough to require the use of both routes.

(ii) As B is introduced, flows readjust as A continues to use both routes while B uses the cheaper (in marginal cost) of the two routes.

(iii) As F_B increases flows continually readjust until the marginal cost to B too is equal on the two routes even though they yet occupy only one route.

(iv) If F_B increases further, B uses both routes and we have A and B on both routes.

Further increase in F_B would result in A and B interchanging roles as we retrace (iii), (ii), and (i).

The solution to (i) is quite straightforward. We know that only A is present and the route flows are given by

$$C_{A1}^{**}(f_1) = C_{A2}^{**}(f_2) \qquad f_1, \; f_2 > 0$$

where

$$f_1 + f_2 = F_A = F$$

and

$$C_{A1}^{**}(f_1) = C_{A1}(f_1) + f_1 C_{A1}'(f_1)$$
$$C_{A2}^{**}(f_2) = C_{A2}(f_2) + f_2 C_{A2}'(f_2)$$

as we have

$$f_1 = f_{A1} \quad \text{and} \quad f_2 = f_{A2}.$$

We therefore construct $C_{A1}^{**}(f_1)$ and $C_{A2}^{**}(f_2)$ and use the graphical method in Fig. 1 to solve for the f_j ˙̇ ̇he uniqueness of the solution in example 1 was a consequence of the monotonicity of the cost functions. However, monotonicity of the cost functions does not necessarily imply that the marginal cost functions are also monotone. Perhaps it is reasonable to assume that $C_{Aj}^{**}(f_j)$ is also monotone increasing. Examining the expressions for $C_{Aj}^{**}(f_j)$ it is evident that as long as $C_{Aj}'(f_j)$ is nondecreasing with increasing f_j, then the marginal cost functions are monotone increasing. However, this may not be a necessary condition. If the $C_{Aj}^{**}(f_j)$ are monotone increasing, then there is a unique solution.

It is perhaps more convenient to consider (iv) next; in fact (ii) and (iii) are special cases of (iv). As A and B are present on both routes, the marginal cost on the two routes are equal for A as well as B. If f_j^* is the optimal assignment, they are given by

$$C_{A1}(f_1^*) + f_{A1}^* C_{A1}'(f_1^*) = C_{A2}(f_2^*) + f_{A2}^* C_{A2}'(f_2^*), \qquad (3a)$$

$$C_{B1}(f_1^*) + f_{B1}^* C_{B1}'(f_1^*) = C_{B2}(f_2^*) + f_{B2}^* C_{B2}'(f_2^*), \qquad (3b)$$

where

$$f_{Aj}^* + f_{Bj}^* = f_j^*, \quad \sum_j f_{Aj}^* = F_A, \quad \sum_j f_{Bj}^* = F_B \quad \text{and} \quad \sum f_j^* = F_A + F_B = F.$$

For known F_A we first obtain the f_{Aj} that satisfy (3a) for sets of values of f_j that satisfy $\Sigma f_j = F$. This is shown graphically in Fig. 10. For a given set of f_j, H, θ_1 and θ_2 are all unique. As f_1 increases among these values of f_j ($\therefore f_2$ decreases) H increases; if we assume $C_{Aj}'(f_j)$ to be nondecreasing then θ_1 either in-

$$H = C_{A1}(f_1) - C_{A2}(f_2)$$
$$\theta_1 = C_{A1}'(f_1)$$
$$\theta_2 = C_{A2}'(f_2)$$

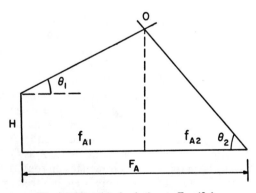

Fig. 10. Graphical solution to Eq. (3a).

creases or remains constant while θ_2 either decreases or remains constant. Therefore M in Fig. 10 moves to the left or moves vertically (when say $f_{A1} = 0$). Therefore (a) the f_{Aj} are unique for each set of f_i and (b) f_{A1} is either decreasing or remains constant (at 0 or F_A) as f_1 increases such that $\Sigma f_i = F$.

The f_{Bj} satisfying (3b) may be estimated in a similar manner and f_{B1} will also be unique and either decreasing or be constant. Therefore, the graph of $f_{A1} + f_{B1}$ against f_1 will be single valued and be decreasing or remain constant. From this graph we read off f_{A1}^* and f_{B1}^* at route flow f_1^* such that $f_{A1}^* + f_{B1}^* = f_1^*$. This is the required solution. The f_{Aj}^* is the intersection of the graph of $f_{A1} + f_{B1}$ and a straight line through the origin of slope 1. As the former is monotone decreasing (or constant) and the latter a monotone increasing function there is a unique intersection; therefore, the solution is unique. Strictly we have only shown that if a solution exists, it is unique [if $C_{Sj}'(f_j)$ is nondecreasing]. A solution may not exist for all F.

In (ii) and (iii) we have A on both routes while B uses only one route. If we assume that B uses route 2, the optimal assignment is given by

$$C_{A1}(f_1^*) + f_1^* C_{A1}'(f_1^*) = C_{A2}(f_2^*) + (f_2^* - F_B)C_{A2}'(f_2^*), \qquad (4a)$$

$$C_{B1}(f_1^*) \geqq C_{B2}(f_2^*) + F_B C_{B2}'(f_2^*) \qquad (4b)$$

The inequality applies for (ii) while the equality holds for (iii). As B occupies route 2 only, $f_{B1}^* = 0$ and $f_{A1}^* = f_1^*$ while since $f_{B2}^* = F_B$, $f_{A2}^* = (f_2^* = F_B)$. Therefore (4a) and (4b) with the equality, is a special case of (iv) and may be solved in exactly the same manner; hence there is a unique solution when the $C_{Sj}'(f_j)$ are nondecreasing. Similarly (ii) is also a special case of (iv) but now because the inequality prevails we need consider only the solution of (3a). [And check inequalities in 4(b). But we show later that for values of F_B less than F_B satisfying (iii), the inequality holds]. It is easily verified that once again there is a unique solution if $C_{Sj}'(f_j)$ are nondecreasing.

A direct method of solving (4a) is shown in Fig. 11. Rewriting (4a) as

$$C_{A1}(f_1^*) + f_1^* C_{A1}'(f_1^*) = C_{A2}^*(f_2^*) + f_2^* C_{A2}'(f_2^*) - F_B C_{A2}'(f_2^*),$$

we recognize that we can construct the marginal cost curves for the two routes (F_B is known). We then determine the optimal route flows using the composite curve. For the f_j^* to be a feasible solution they must also satisfy (4b). It is also clear from Fig. 11 that as F_B increases (for a given F), f_2^* increases and ultimately the inequality in (4b) becomes an equality.

It is apparent that there are some exceptions to the four step assignment we have considered. We have already mentioned that route assignments with A and B using both routes may not always exist. For example, as F_B increases, if before (iii) is true either $f_1 = 0$ or $f_2 = F_B$ is realized, neither (iii) nor (iv) is a possibility. In the former case both A and B use route 2; as F_B increases there is a possibility of B using both routes while A remains on route 1. If $f_2 = F_B$, we have A on route 1 and B on route 2. As more A are replaced by B, once again there is the possibility of B using both routes while A now remains on route 1.

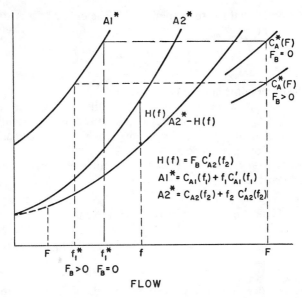

Fig. 11. Graphical solution to Eq. (4a).

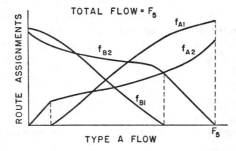

Fig. 12. Variation of route assignments with composition of flow as the total cost to each user type is minimized.

The results of a numerical example is shown in Fig. 12. It gives the route assignments at $F = F_5$ for the cost curves in Fig. 4 as the composition of F is varied. (Note that Fig. 12 is not to the same scale as Fig. 4.) The value of F chosen is sufficiently large to follow the typical behavior discussed here.

Example 6—Minimize Total Cost to all the Users

We have already discussed the difference between this objective function and the objective function in the last example. Perhaps another significant difference is that unlike all the earlier examples, we have one objective function (encompassing all the users) that has to be minimized. The optimal flows are given by the minimization of

$$(C_A + C_B) = \sum_{\substack{j=1,2 \\ S=A,B}} f_{Sj} C_{Sj}(f_j) \tag{5}$$

subject to the linear constraints in (2).

All the terms in (5) are nonnegative; $(C_A + C_B)$ therefore has a lower bound, a minimum of $C_A + C_B$ exists and there is a solution to this assignment problem.

Perhaps one may treat (5) as a programming problem. We could also continue to use the concept of marginal cost.

With a flow f_{A1}, f_{B1} on route 1, if we were to add ϵ of A on to route 1, the change in total cost to all the users is given by

$$\Delta(C_{A1} + C_{B1}) = \epsilon \frac{\partial}{\partial f_{A1}} [f_{A1} C_{A1}(f_1) + f_{B1} C_{B1}(f_1)]$$

$$= \epsilon [C_{A1}(f_1) + f_{A1} C'_{A1}(f_1) + f_{B1} C'_{B1}(f_1)]$$

$$= \epsilon_A C_1^*(f_1).$$

$_A C_1^*(f_1)$ is the price that society must pay to add unit flow of A to route 1. We note that the main difference between marginal cost here and marginal cost in the previous example is that at every stage the increase in cost to all the users must be considered. Strictly $_A C_1^*(f_1)$ is the marginal cost to society due to increase in A; we will refer to it as 'marginal cost due to A'. We can derive similar marginal cost expressions for B, for both routes 1 and 2.

We can easily verify that $_A C_1^*(f_1)$ is positive. We can further argue that, as in Example 5, the total cost is minimized when A chooses the route or routes of least marginal cost due to A and B chooses the route or routes of least marginal cost due to B.

If we neglect the trivial case when all the flow is on one route, there are only three possibilities we need consider.

(i) Route assignments such that marginal costs are equal on the two routes, due to A as well as due to B.

(ii) Marginal cost due to one of the two, say A, are equal on the two routes but B occupies the route with the lower marginal cost.

(iii) Each type occupies a separate route.

The conditions for (i) are

$$C_{A1}(f_1) + f_{A1} C'_{A1}(f_1) + f_{B1} C'_{B1}(f_1) = C_{A2}(f_2) + f_{A2} C'_{A2}(f_2) + f_{B2} C'_{B2}(f_2), \tag{6a}$$

$$C_{B1}(f_1) + f_{A1} C'_{A1}(f_1) + f_{B1} C'_{B1}(f_1) = C_{B2}(f_2) + f_{A2} C'_{A2}(f_2) + f_{B2} C'_{B2}(f_2), \tag{6b}$$

where the $f_{Sj} \geqslant 0 \quad S = A, B; j = 1, 2$

$$\sum_S f_{Sj} = f_j, \tag{6c}$$

$$\sum_j f_{Sj} = F_S. \tag{6d}$$

From (6a) and (6b) we obtain the condition

$$C_{A1}(f_1) - C_{B1}(f_1) = C_{A2}(f_2) - C_{B2}(f_2) \tag{6e}$$

subject to

$$\sum_j f_j = \sum_S F_S = F.$$

The solution to (6e) (if it exists) is a feasible solution only if the f_{Sj} are feasible. For every set of f_j satisfying (6e) there is a unique set of f_{Sj} satisfying (6c) and (6d). There will be a unique solution to (i) only if there is a unique solution to (6e). The cost functions being monotone increasing does not necessarily give a unique solution to (6e).

If the solution is to be A on both routes with B on the route with the cheaper marginal cost, say route 2, the optimal flows are such that

$$C_{A1}(f_1) + f_1 C'_{A1}(f_1) = C_{A2}(f_2) + f_{A2} C'_{A2}(f_2) + F_B C'_{B2}(f_2), \tag{7a}$$

$$C_{B1}(f_1) + f_1 C'_{A1}(f_1) > C_{B2}(f_2) + f_{A2} C'_{A2}(f_2) + F_B C'_{B2}(f_2), \tag{7b}$$

giving

$$C_{A1}(f_1) - C_{B1}(f_1) < C_{A2}(f_2) - C_{B2}(f_2). \tag{7c}$$

Equation (7a) may be rewritten as

$$C_{A1}(f_1) + f_1 C'_{A1}(f_1) = C_{A2}(f_2) + f_2 C'_{A2}(f_2) - F_B [C'_{A2}(f_2) - C'_{B2}(f_2)] \tag{8}$$

which may be solved by the graphical method in Fig. 11 with

$$H(f) = F_B [C'_{A2}(f_2) - C'_{B2}(f_2)].$$

Once again if the $C'_{Aj}(f_j)$ are nondecreasing, there is a unique solution to (8). This solution is a feasible solution to (ii) only when (7c) is also satisfied.

Finally if the solution is such that A and B each occupy a separate route—say A on route 1 and B on route 2, then F_A, F_B satisfy the following conditions

$$C_{A1}(F_A) + F_A C'_{A1}(F_A) \leqslant C_{A2}(F_B) + F_B C'_{B2}(F_B), \tag{9a}$$

$$C_{B1}(F_A) + F_A C'_{A1}(F_A) \geqslant C_{B2}(F_B) + F_B C'_{B2}(F_B). \tag{9b}$$

If the equality holds in both cases, (9) is a special case of (6).

In the above analysis we have not shown that the assignment problem (i.e., given F_A, F_B what is the route assignment) has a unique solution. We have only investigated the existence of local minima. If (5) is a convex function, any local minimum is also a global minimum. However, even if the cost functions are convex, (5) is not necessarily convex. If (5) is not a convex function and there is more than one local minimum, we can always decide on the optimal solution in particular cases.

Figure 13 shows the route assignments for the numerical example of Fig. 12 with the new objective function. We see that in this particular case an assignment with A and B occupying both routes at the same time does not exist. For the cost curves of Fig. 4, it was found that this assignment never exists.

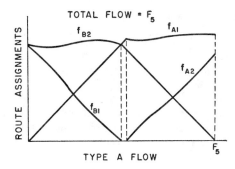

Fig. 13. Variation of route assignments with composition of flow as total cost to all the users is minimized.

However, this does not imply that such a solution does not exist in general. For example, suppose the two routes are identical; then by symmetry both routes must be occupied by A and B always.

Minimize Total Cost to A but Minimize Individual Cost to B

Here we consider a combination of the objectives of Part I and Part II. Suppose A represented a trucking company or a bus company while B represents the collection of private automobiles. It is reasonable to assume that A is primarily interested in minimizing their total cost while the individuals in B prefer to choose the cheapest route. This is a very real problem; perhaps even more realistic than our objective in Part I.

The argument of Example 5 applies to A; therefore their total cost is minimized by the minimization of the marginal cost to A. Hence the optimal route flows are those flows that minimize the marginal cost to A and minimize the (individual) cost to B.

If we commence with a flow F all of A, the route flows are given by

$$C_{A1}^{**}(f_1) = C_{A2}^{**}(f_2),$$

$$f_1, \ f_2 \geqslant 0, \ f_1 + f_2 = F,$$

where we assume that F is such that both routes are occupied. The marginal cost $C_{Aj}^{**}(f_j) = C_{Aj}(f_j) + f_j C_{Aj}'(f_j)$. We have already discussed this solution in Example 5; the f_j are unique if we may assume that the $C_{Aj}^{**}(f_j)$ are monotone increasing.

With the above route flows let route 2 be cheaper to B. If we replace ϵ of A by B, B would choose to use route 2; however, the entry of B into route 2 changes the marginal cost to A who readjust their flows to equalize their marginal cost. The optimal route flows are given by

$$C_{A1}(f_1^*) + f_1^* C_{A1}'(f_1^*) = C_{A2}(f_2^*) + (f_2^* - F_B)C_{A2}'(f_2^*), \qquad (10a)$$

$$C_{B1}(f_1^*) > C_{B2}(f_2^*). \qquad (10b)$$

Equation (10a) is the same as (4a); from Fig. 11 it follows that as F_B increases f_1 decreases and f_2 increases, the solutions being feasible only so long as (10b) is also satisfied. The optimal flows are given by (10a) and (10b) till route flows are such that

$$C_{B1}(f_1^*) = C_{B2}(f_2^*) \tag{10c}$$

with B yet on route 2. From example 2, we know that for a given F there is a unique solution to (10c)—namely, the route flows are constant at f_j^B whatever the composition of the flow. Therefore, it is clear that for a given F, both (10a) and (10c) can be satisfied at most for a single value of F_B. Suppose such a solution exists when $F_B = F_B^*$. From the manner in which the route flows changed as F_B was increased from zero, we see that F_B^* exists only if neither $f_1 = 0$ nor $f_2 = F_B$ at some value of $F_B < F_B^*$. With $F_B = F_B^*$ let the route flows be f_j^{B*}.

As F_B increases beyond F_B^* there are two possibilities we need to consider: (i) B uses both routes, and (ii) B uses only one route.

If B uses both routes, the route flows are constant at f_j^B. If A, too, occupies both routes, the assignment of A must satisfy

$$C_{A1}(f_1^B) + f_{A1}C_{A1}'(f_1^B) = C_{A2}(f_2^B) + f_{A2}C_{A2}'(f_2^B), \tag{11}$$

where

$$f_{A1} + f_{A2} = F_A.$$

Equation (11) is the same as (3a) and may be solved using Fig. 10—now, however, H, θ_1, and θ_2 are constant while F_A varies. There is a unique solution for each F_A and as F_A decreases we can easily see that f_{A1} decreases; further it is easy to show that f_{A2}, too, decreases. Therefore, as F_A decreases, both f_{A1} and f_{A2} decrease until one of them becomes zero. (If H is positive f_{A1} is zero first and if H is negative f_{A2} is zero first.) Suppose $f_{A1} = 0$ first; then all the A are on route 2 and Eq. (11) becomes

$$C_{A1}(f_1^B) = C_{A2}(f_2^B) + F_A C_{A2}'(f_2^B). \tag{11a}$$

As F_A decreases further, (11a) becomes an inequality and A continues to use only route 2 which is the route with the lower marginal cost.

We will next show that (ii) is not an optimal solution. Since both routes were occupied at $F_B = F_B^*$, clearly both routes must be occupied for $F_B > F_B^*$; the solution we have to consider then is where A occupies both routes with B on one route. From (10a) if B is to be on route 2, $f_2^* > f_2^{B*}$ (i.e., route 2 flow with $F_B = F_B^*$); but this means that B is occupying the more expensive route and is not a stable solution—some B will transfer to route 1. Suppose B were to occupy route 1. Then the route flows are given by

$$C_{A1}(f_1^*) + (f_1^* - F_B)C_{A1}'(f_1^*) = C_{A2}(f_2^*) + f_2^* C_{A2}'(f_2^*) \tag{12a}$$

and we have assumed

$$C_{B1}(f_1^*) < C_{B2}(f_2^*). \tag{12b}$$

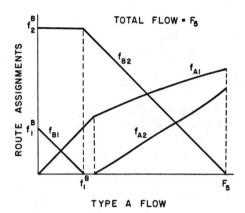

TYPE A FLOW

Fig. 14. Variation of route assignments with composition of flow. Type A user chooses cheapest route while type B users minimize their total cost.

Equation (12b) implies $f_1^* < f_1^B$ and therefore $f_2^* > f_2^B$. Comparing (12a) and (11), marginal cost to A under the present assignment is clearly higher than under (i). Therefore as F_B increases beyond F_B^*, the route flows remain constant at f_j^B.

Now if $f_1 = 0$ at $F_B < F_B^*$, all the flow is on route 2 and we have

$$C_{A1}(0) = C_{A2}(F) + F_A C_{A2}'(F)$$

and

$$C_{B1}(0) > C_{B2}(f).$$

As F_B increases, all the flow continues to use only route 2.
If $f_2 = F_B$ at some $F_B < F_B^*$, we have A on route 1 and B on route 2 such that

$$C_{A1}(F_A) + F_A C_{A1}'(F_A) = C_{A2}(F_B)$$

and

$$C_{B1}(F_A) > C_{B2}(F_B).$$

As F_B increases, f_1 decreases and f_2 increases; A uses only route 1 and B route 2. If the route flows become f_j^B (if F is large enough; otherwise $f_1 = 0$ before route flows are f_j^B) they remain constant and B begins to use both routes.

Figure 14 shows the variation in route assignment under this new objective for the earlier numerical example. F_5 is such that $f_2 = F_B$ before F_B^* is realized. For some higher values of F, F_B^* does exist.

Conclusion

The model that is relevant to the urban transportation scene is that discussed in Part I. We have already discussed the implications of this model.

Of the two models that minimized total cost, we cannot find a meaningful application of the minimization of total cost to each class of user in the context of urban transportation. However, the second model—the minimization of the overall cost to all the users is a typical objective of most planning authorities. There is no doubt to the difference in the assignments by this model and the model of Part I (compare Figs. 5 and 13). Hence, it is doubtful that one may achieve the objective of the planning authority if people are given a free choice of routes.

The model of Example 6 describes best the individual objectives of the different type of users on the urban transportation network. For we do have trucking companies, bus companies, etc., using the network. However, most often they constitute only a small percentage of the total traffic and it is sufficiently accurate to use the model of Part I.

References

1. N. O. Jorgensen, Some Aspects of the Urban Traffic Assignment Problem, I.T.T.E. Graduate Report, Berkeley, Calif., 1963.
2. J. G. Wardrop, Some Theoretical Aspects of Road Traffic Research, Inst. Civil Eng. Road Paper No. 36, London, 1952.

Equilibrium and Marginal Cost Pricing on a Road Network with Several Traffic Flow Types

Maurice Netter

Institut de Recherche des Transports, Arcueil, France

Abstract

In this paper, we prove that classical results concerning equilibrium traffic assignments and road network efficiency tolls are based on an implicit assumption: the homogeneity of flows (or the uniqueness of traffic flow types). The main results are not valid when one considers several traffic flow types.

In the first part of this paper, we recall, under a relatively generalized and mathematical form, what these results are: existence and uniqueness of the equilibrium, efficiency of social marginal cost pricing.

In the second part, we explain why (in the general case of several traffic flow types) the equilibrium is no longer uniquely defined and why social marginal cost pricing is not necessarily efficient (even from a theoretical point of view).

We discuss briefly some relations between pricing policies and the setting up of rules relative to the use of road networks.

Generalized Mathematical Statement of the "Classical" Social Marginal Cost Pricing Theory

The Generalized "Classical" Network Equilibrium Concept

Let us consider a directed graph (Ω, A), where A is the set of links, and $\Omega = \{w_1, w_2, \ldots, w_N\}$ is the set of nodes. This graph represents a road network.

Let

$x_r^{nn'}$ be the flow on the link $(w_n, w_{n'}) \in A$ with destination w_r;

$x^{nn'}$ be the total flow on the link $(w_n, w_{n'}) \in A$;

x_n^r be the traffic with origin w_n and destination w_r;

$g^{nn'} (x^{nn'}, x^{n'n})$ be the private user cost on the link $(w_n, w_{n'})$ if the flow from w_n towards $w_{n'}$ is $x^{nn'}$, and the flow from $w_{n'}$ towards w_n is $x^{n'n}$.

ASSUMPTION (A1). *The functions $g^{nn'}$ with $(w_n, w_{n'}) \in A$ are continuous, positive, strictly increasing and convex.*

This assumption is quite natural, from physical and economic points of view.

By definition, the private (user) cost on a path is the sum of the private costs on the links of this path. We assume that *Wardrop's first principle*[1] is satisfied, i.e., for every origin-destination (w_n, w_r):

1) the private costs on all paths that are actually used have the same value p_n^r,
2) the private costs on the paths that are not used, are not smaller than p_n^r,
3) p_n^r will be the "*price*" on the origin-destination (w_n, w_r).

We assume moreover that the traffic between w_n and w_r is uniquely defined by the price p_n^r. So, for every origin-destination (w_n, w_r):

$$x_n^r = f_n^r (p_n^r).$$

ASSUMPTION (A2). f_n^r *is a strictly decreasing, continuous function, and for* $p_n^r \longrightarrow \infty, f_n^r (p_n^r) \to 0$.

Notations. Let

$x^{::}$ be the table of $x_r^{nn'}$ numbers;
$x:$ be the table of x_n^r numbers;
$x^{\cdot\cdot}$ be the table of $x^{nn'}$ numbers;
$p:$ be the table of p_n^r numbers.

DEFINITION. $(x^{::}, x:, x^{\cdot\cdot}; p:)$ is an *equilibrium* on the graph (Ω, A), for the cost functions $g^{nn'}$ and the demand functions f_n^r if, and only if, simultaneously:

$$x_r^{nn'} \geqslant 0 \quad \text{for } (w_n, w_{n'}) \in A, w_r \in \Omega,$$

$$- x^{nn'} + \sum_r x_r^{nn'} = 0, \quad \text{for } (w_n, w_{n'}) \in A,$$

$$- x_n^r + \sum_{n'} (x_r^{nn'} - x_r^{n'n}) = 0.$$

Wardrop's first principle is satisfied, p_n^r being the price of the origin-destination (w_n, w_r)

$$x_n^r = f_n^r (p_n^r).$$

Generalized "Classical" Results on Network Equilibria: Existence and Uniqueness

ASSUMPTION (A3). *For every set* $\{n, n'\}$ *such that* $(w_n, w_{n'})$ *or* $(w_{n'}, w_n)$ *is a link, there is a strictly convex and differentiable function* $\Gamma^{\{n,n'\}} (x^{nn'}, x^{n'n})$ *(n being assumed smaller than n') such that*

$$\frac{\partial}{\partial x^{nn'}} \Gamma^{\{n,n'\}} (x^{nn'}, x^{n'n}) = g^{nn'} (x^{nn'}, x^{n'n}),$$

$$\frac{\partial}{\partial x^{n'n}} \Gamma^{\{n,n'\}} (x^{nn'}, x^{n'n}) = g^{n'n} (x^{n'n}, x^{nn'}).$$

There are two important particular cases of Assumption (A3). The first is the case where private costs in both directions are equal, and depend on the sum of flows in both directions:

$$g^{nn'}(x^{nn'}, x^{n'n}) = \gamma^{\{n,n'\}}(x^{nn'} + x^{n'n}) = g^{n'n}(x^{n'n}, x^{nn'}).$$

Then

$$\Gamma^{\{n,n'\}}(x^{nn'}, x^{n'n}) = \int_0^{x^{nn'}+x^{n'n}} \gamma^{\{n,n'\}}(t)\,dt.$$

This case has been studied by Beckmann, Winsten, and McGuire[2] in 1956 (uniqueness of the equilibrium and pricing theorem).

In the second important particular case, the directions are independent (separate directions or one-way roads):

$$g^{nn'}(x^{nn'}, x^{n'n}) = \gamma^{nn'}(x^{nn'})$$

Then

$$\Gamma^{\{n,n'\}}(x^{nn'}, x^{n'n}) = \int_0^{x^{nn'}} \gamma^{nn'}(t)\,dt + \int_0^{x^{n'n}} \gamma^{n'n}(t)\,dt.$$

This case has been studied by Bruynooghe and Sakarovitch[3] in 1969 (uniqueness of the equilibrium, and computation algorithm).

In these two cases, the results are the same: the equilibrium problem is equivalent to a convex programming problem. This is the reason why we have generalized them.

Notations. Let h_n^r be the inverse function of f_n^r:

$$x_n^r = f_n^r(p_n^r) \longleftrightarrow p_n^r = h_n^r(x_n^r).$$

Let $G(x:, x^{..})$ be the function defined by:

$$G(x:, x^{..}) = \sum_{(n,n')\in T} \Gamma^{\{n,n'\}}(x^{nn'}, x^{n'n}) - \sum_{n,r} \int_0^{x_n^r} h_n^r(t)\,dt,$$

where T is the set of $\{n, n'\}$ such that $(w_n, w_{n'})$ or $(w_{n'}, w_n)$ is a link of A.

PROPOSITION (P1). *Under Assumptions (A1), (A2), and (A3), G is a convex function*

Remark. The convexity of g functions is not necessary for the validity of (P1).
PROBLEM (II). Minimize $G(x:, x^{..})$ under constraints:

$$(C_r^{nn'}) \quad x_r^{nn'} \geqslant 0, (w_n, w_{n'}) \in A, \quad w_r \in \Omega,$$

$$(C^{nn'}) - x^{nn'} + \sum_r x_r^{nn'} = 0, \quad (w_n, w_{n'}) \in A,$$

$$(C_n^r) \quad -x_n^r + \sum_{n'}(x_r^{nn'} - x_r^{n'n}) = 0.$$

(II) is a convex programming problem.

PROPOSITION (P2). *(x:·, x:, x··, p:) is an equilibrium if, and only if, (x:·, x:, x··) is a solution of Problem (Π) and the p_n^r are the dual variables corresponding to constraints (C_n^r) in this problem.*

This proposition can be easily proved by applying Kuhn-Tucker's duality conditions to Problem (II). Since it is a convex programming problem, these conditions are necessary *and* sufficient. (P2) implies the following theorem.

ASSUMPTION (A4). *There is a positive real number Q such that for every link $(w_n, w_{n'}) \in A$ and every origin-destination (w_v, w_r):*

$$x^{nn'} > Q \Rightarrow f_v^r(g^{nn'}(x^{nn'}, x^{n'n})) < Q.$$

THEOREM (T1). *Under assumptions (A1), (A2), (A3), and (A4) equilibria always exist. The set of these equilibria is such that the $x^{nn'}$, x_n^r, and p_n^r are uniquely defined.*

So the prices (then the traffics by origin-destination) and the total flow on every link in both directions are uniquely defined.

The "Classical" Marginal Cost Pricing Results

The users' surplus, corresponding to a traffic assignment $(x:·, x:, x··)$ is:

$$S(x:, x··) = \sum_{n,r} \int_0^{x_n^r} h_n^r(t)\, dt - \sum_{(w_n, w_{n'}) \in A} x^{nn'} g^{nn'}(x^{nn'}, x^{n'n}).$$

The first term represents the maximal amount of money that the users are ready to pay to use the network. The second term represents the total generalized travel cost. We do not consider other costs (pollutions, accidents, etc.)—these are not important, from the mathematical point of view.

THEOREM (T2). *In Beckmann's and Bruynooghe-Sakarovitch's cases, under Assumptions (A1) and (A2), the maximization of users' surplus S can be achieved through an equilibrium with private cost functions:*

$$c^{nn'}(x^{nn'}, x^{n'n}) = g^{nn'}(x^{nn'}, x^{n'n}) + \bar{x}^{nn'} \frac{\partial}{\partial x^{nn'}} g^{nn'}(\bar{x}^{nn'}, \bar{x}^{n'n})$$

$$+ \bar{x}^{n'n} \frac{\partial}{\partial x^{nn'}} g^{n'n}(\bar{x}^{n'n}, \bar{x}^{nn'}),$$

where $\bar{x}^{nn'}$ is the optimal flow on the link $(w_n, w_{n'})$.

The theorem is proved by applying Kuhn-Tucker's duality conditions. In Beckmann's and Bruynooghe-Sakarovitch's cases, *the surplus function is concave: let us remark that this is the key of the theorem.*

Normative Economic Interpretation of Theorems (T1) and (T2)

In order to achieve the "optimal" traffic assignment $(\bar{x}:·, \bar{x}:, \bar{x}··)$ it is sufficient to perceive at the entrance of each link a toll equal to *the social marginal cost at the "optimum," i.e.,*

$$\bar{x}^{nn'} \frac{\partial}{\partial x'^{nn'}} g^{nn'} (\bar{x}^{nn'}, \bar{x}^{n'n}) + \bar{x}^{n'n} \frac{\partial}{\partial x'^{nn'}} g^{n'n} (\bar{x}^{n'n}, \bar{x}^{nn'}).$$

So, from a rigorous point of view, the process is the following: (1) computation of the "optimal" traffic assignment and the corresponding tolls; (2) perception of tolls; (3) new equilibrium, which is "optimal."

Let us remark (this will be essential in the second part), that the validity of Theorems (T1) (uniqueness of the equilibrium) and (T2) (sufficiency of tolls) are quite necessary for the success of the third step.

The Case of Several Flow Types

Now, let us consider that there are J flow types: cars, trucks, etc. We shall prove that the results relative to the case of one flow type are no longer valid for this more general case.

Mathematical Formulation

Let $x_{rj}^{nn'}, x_j^{nn'}, x_{nj}^r$ be the same concepts as, respectively, $x_r^{nn'}, x^{nn'}, x_n^r$ in the first part of this paper, but relative to the jth flow type.

Let $g_j^{nn'} (x_1^{nn'}, \ldots, x_j^{nn'}, \ldots, x_J^{nn'}, x_1^{n'n}, \ldots, x_J^{n'n})$ be the private (user) travel cost for the jth flow type on the link $(w_n, w_{n'})$, when the flow of jth type in the direction $w_n \longrightarrow w_{n'}$ is $x_j^{nn'}$, and when the flow of the same type in the direction $w_{n'} \longrightarrow w_n$ is $x_j^{n'n}$.

ASSUMPTION (A'1). *The functions* $g_j^{nn'}$, *with* $(w_n, w_{n'}) \in A$ *and* $1 \leqslant j \leqslant J$ *are continuous, strictly increasing and positive.*

Moreover, we assume that Wardrop's first principle is true for every flow type, and every origin–destination. So, for every flow type j, and every origin–destination (w_n, w_r), a *price* p_{nj}^r is defined.

As in the first part of this paper, we assume the validity of the demand function concept:

$$x_{nj}^r = f_{nj}^r (p_{nj}^r).$$

ASSUMPTION (A'2). f_{nj}^r *is a strictly decreasing, continuous function, and*

$$\lim_{p_{nj}^r \to \infty} f_{nj}^r (p_{nj}^r) = 0.$$

We define an equilibrium in the same way as in the first part of the paper, by adding the j subscript to every variable.

Existence and Nonuniqueness of Equilibria

When we consider a road network with only one traffic flow type, the equilibrium problem is equivalent to the maximization of a differentiable concave function constrained by linear inequalities (i.e., to a differentiable concave programming problem). In this section we prove that, in the case of several traffic flow types, no differentiable programming problem is equivalent to the equilibrium problem. As a consequence, the equilibrium is not, in this case, uniquely defined.

PROPOSITION ($P'1$). *In the general case, when there are several flow types, the equilibrium problem is not equivalent to a differentiable programming problem.*

Proof. This equivalence would imply the existence of functions:

$$\Gamma^{\{n,n'\}} (x_1^{nn'}, \ldots, x_j^{nn'}, \ldots, x_J^{nn'}, x_1^{n'n}, \ldots, x_j^{n'n}, \ldots, x_J^{n'n})$$

[for every (n, n'), with $n < n'$ such that $(w_n, w_{n'}) \in A$], such that

$$g_j^{nn'} = \frac{\partial}{\partial x_j^{nn'}} \Gamma^{\{n,n'\}}, g_j^{n'n} = \frac{\partial}{\partial x_j^{n'n}} \Gamma^{\{n,n'\}}$$

So, the minimand of the mathematical programming equivalent problem would be the sum of the G functions given in the previous section on generalized classical results on network equilibria, relative to the different values of j (flows types). By applying the Kuhn–Tucker theorem, the equilibrium conditions would be obtained.

But the existence of these Γ functions would imply, by Schwartz theorem, that

$$\begin{cases} \dfrac{\partial}{\partial x_j^{nn'}} g_{j'}^{nn'} = \dfrac{\partial}{\partial x_{j'}^{nn'}} g_j^{nn'}, \\[2ex] \dfrac{\partial}{\partial x_j^{nn'}} g_{j'}^{n'n} = \dfrac{\partial}{\partial x_{j'}^{n'n}} g_j^{nn'}, \end{cases}$$

for every flow type j and $j' \neq j$, for every (n, n') such that $(w_n, w_{n'})$ is a link.

In the general case, there is no reason why these integrability conditions should be valid; so, there is no equivalence between equilibria and optima of a differentiable programming problem, and Proposition ($P'1$) is proved.

Economic Interpretation of Integrability Conditions. These mathematical conditions mean that users of the jth type are as concerned by a marginal increment of flow of j'th type, as users of the j'th type by a marginal increment of flow of jth type.

Consequences of Proposition $P'1$. It is not possible to prove the existence of equilibria in the same way as in the first part of the paper. But we have proved this existence (Netter and Sender, 1970)[4] by using Brouwer's fixed point theorem. We build a function the fixed points of which are in one to one correspondence with equilibria.

This fixed point theorem is a powerful existence theorem, but not a uniqueness theorem. Actually, we have found numerical examples where there are several equilibria, or even an infinite number of equilibria. Some of these examples are developed in the quoted report.

THEOREM (T'1). *Assumptions (A'1), and (A'2) imply the existence of equilibria, which are not, in the general case, uniquely defined. The equilibria set may be infinite.*

Properties of the Surplus and Social Cost Functions: Local and Global Optima

In Beckmann's and Bruynooghe–Sakarovitch's cases the total travel (social) cost function is convex under Assumption (A1). So, under Assumption (A2), the surplus function is concave. This implies that this surplus function has only one local maximum; so, first-order conditions (conditions on first-order derivatives) are sufficient for global maximality. The Kuhn–Tucker's conditions, which introduce prices, are first-order conditions: they are necessary *and* sufficient. So prices (perception of tolls) are sufficient for (global) optimality.

This very important property is no longer true in the general case when there are several flow types.

PROPOSITION (P'2). *In the general case, when there are several flow types, there is not necessarily one local optimum.*

We give an example in the research report by Netter and Sender (1970)[4] with two local optima (and three local extrema).

The reason of the difference between this case and the Beckmann's and the Bruynooghe–Sakarovitch's cases is the nonconvexity in the general case of the social cost function, i.e., of

$$\sum_{j=1}^{J} \sum_{(w_n, w_{n'}) \in A} x_j^{nn'} g_j^{nn'} (x_1^{nn'}, \ldots, x_j^{nn'}, x_1^{n'n}, \ldots, x_j^{n'n})$$

even if the $g_j^{nn'}$ functions are convex.

Consequences for Pricing Theory

FIRST CONSEQUENCE. *In the general case, the (global) optimal equilibrium will not follow necessarily (even in the theory) social marginal cost pricing.*

Proof. Since there may be several local extrema (Proposition P'2) first-order conditions are not sufficient for optimality. Moreover, Proposition (P'1) implies that several equilibria may correspond to given tolls. So, even if one knows the tolls corresponding to the (global) optimum, the achievement of this optimum will not necessarily follow the enforcement of these tolls.

SECOND CONSEQUENCE. *Social marginal cost pricing does not necessarily improve (in the surplus criterion sense) the use of the network.*

Applying social marginal cost pricing may bring a network use (locally optimal) which may be worse than the initial one.

These two consequences are in contradiction with the results of the "classical" pricing theory.

An Example. In the quoted report, we have studied a numerical example with two roads, and two traffic flow types (for instance, cars and heavy vehicles). The total demand for each type is supposed to be inelastic.

Without tolls, there are three equilibria. In the first equilibrium, all the cars are on road number 1, and all the heavy vehicles on road number 2. In the third equilibrium, all the cars are on road number 2, and all the heavy vehicles on road number 1. In the second equilibrium, the traffic of each type is divided between the two roads.

When one applies social marginal cost pricing, there are still three equilibria. The first and the third one are the same as when no tolls are perceived. In the second equilibrium, the two types of traffic are again divided between the two roads. The corresponding social costs are the following:

$$300.8 \qquad 343.57 \qquad 360.21$$

The first equilibrium is optimal.

Let us note that, in this example:

(1) Marginal cost pricing does not imply that the optimum is reached (two other nonoptimal equilibria are possible).

(2) The optimum can be reached if no tolls are perceived; moreover, social marginal cost pricing does not increase the chances of reaching this optimum.

(3) Still, there are some ways to reach the optimum with certainty. For instance, to forbid heavy vehicles to go on the road they should not take.

Some Relations Between Pricing and Rules Concerning Network Use

When pricing fails to ensure the "optimal" use of a road network, what other means are at disposal? One may think of certain types of rules: for instance, forbidding certain roads to certain flow types in one or both directions.

So, one may hope to restrict the set of admissible traffic assignments in such a way that in this set the surplus function has only one local maximum; so, in connection with these rules, pricing could be efficient.

But, in fact, travel costs depend on these rules: for instance, speeds are higher on a road in one direction if travel in the opposite direction is forbidden.

So, if one wants to take into account possible rules, one must consider that private cost functions depend on them. If there are R possible rules, one may try to maximize the surplus function:

$$S(x:,x^{\cdot\cdot},\rho^{\cdot}) = \sum_{n,r} \int_0^{x_n^r} h_n^r(t)\,dt - \sum_{(w_n,W_{n'})\in A} x^{nn'} g^{nn'}(x_.^{nn'},x_.^{n'n},\rho^{\{n,n'\}})$$

with $\rho^{\{n,n'\}}$: rule applied on the road between w_n and $w_{n'}$,

$$\rho^{\cdot} \text{ set of } \rho^{\{n.n'\}} ((w_n,w_{n'})\in A \text{ or } (w_{n'},w_n)\in A),$$
$$x_.^{nn'} = (x_1^{nn'},\ldots,x_J^{nn'}),$$
$$x_.^{n'n} = (x_1^{n'n},\ldots,x_J^{n'n}).$$

Let $\bar{\rho}^{\cdot}$ be the optimal set of rules: it is not sure that only one equilibrium corresponds to $\bar{\rho}^{\cdot}$, and therefore, it is not sure that social marginal cost pricing is efficient.

Other Results

If we assume that demand functions are constant (inelastic demands), the above propositions and theorems remain valid. The maximization-of-surplus criterion is replaced by the minimization-of-social-cost criterion.

If we consider several periods of time, we may assume that, during each period, Wardrop's principle is fulfilled; so, for every flow type, origin–destination, and period, there is one price. Demands depend on prices of each period. We have proved in our research report the existence of equilibria. Some assumptions are necessary to ensure the existence of the surplus function (integrability) and the efficiency of social marginal cost pricing.

We have built heuristic but efficient procedures to compute equilibria for relatively high-dimensional problems.

Conclusions

The consideration of several flow types on a network forces us to revise traffic assignment theory and network marginal cost pricing theory. Mathematically, it introduces some nonconvexity and uniqueness properties.

From an economic point of view, it forces us to consider jointly pricing policies, and the setting up of rules applied to the use of the road network. But, perhaps, these rules are much more important than pricing (with reference to the surplus criterion). This would be a very strong argument against pricing.

References

1. J. G. Wardrop, Some theoretical aspects of road traffic research, Inst. Civ. Eng. road paper n° 36, London, 1952.
2. M. Beckmann, C. B. McGuire and C. Winsten, Studies in the Economics of Transportation, Cowles Foundation for research in economics at Yale University, 1956.
3. G. Bruynooghe and M. Sakarovitch, Une Méthode d'Affectation du Trafic, Institut de Recherche des Transports, Arcueil, France, 1969.
4. M. Netter and J. G. Sender, Equilibre Offre-Demande et Tarification sur un Réseau de Transport, Institut de Recherche des Transports, Arcueil, France, 1970.
5. M. Netter, Affectations de Trafic et Tarification au Coût Marginal Social: Critique de Quelques Idées Admises (to be published in "Transportation Science," vol. 6, 1972).

A Study of the Travel Patterns in a Corridor with Reference to the Assignment Principles of Wardrop

Sam Yagar

Department of Civil Engineering, University of Waterloo, Waterloo, Ontario, Canada

Abstract

A corridor consisting of a freeway and arterials is examined in an attempt to determine how closely the existing traffic patterns approximate each of the principles of Wardrop,[1] which correspond to individual and total travel cost minimizations, respectively. The existing set of flow values and unit travel times are obtained and from these values is estimated the total vehicle time (in motion and in queue) which is consumed in satisfying the peak period O–D demands on the network.

A representation of the traffic flow pattern which would exist if the peak period users assigned themselves according to Wardrop's first principle is presented. This is obtained through a proposed dynamic assignment technique which is an extension of Homburger's[2] technique for assigning traffic when there is no queuing. The flows, queues, and total time consumed are compared with the existing values.

A set of flows (not necessarily realistic) corresponding to a lower bound on the minimum time required to satisfy the demands is obtained by linear programming. These "flows" and the corresponding queues and total time value are compared to the existing values. The difference between this value of total time and the existing value represents an upper bound on the possible time saving within the corridor in the peak period.

The total travel time in the corridor which was studied is found to be between the total time values corresponding to assignments according to Wardrop's two principles. It is lower than it would be if all the users chose their expected shortest time paths. This is because of an apparent user bias against the freeway with a resultant saving in congestion on the relatively high-capacity freeway. It was not possible to conclude which of Wardrop's principles was emulated more closely in that corridor.

Introduction

The peak-period traffic in a corridor was studied with a view toward determining the type of self-assignment emulated by the users of that corridor. An attempt was made to justify, in turn, each of the principles proposed by Wardrop.[1] The question of the effective self-assignment by network users is an old one which is still not resolved. The answer to this question would aid in the construction of future networks.

In planning new roadway facilities or in the expansion of existing facilities, the designer is faced with the problem of where available finances might best be spent. Given the demand over time for the corridor in terms of origin–destination pairs, he might want to design a system to satisfy these demands with the least amount of total time spent in travel. Various alternatives for facility additions

---

References pp. 180-181

or control schemes might have to be compared with respect to their relative costs of travel. Before testing candidate systems, the planner will likely face the dilemma of choosing a criterion for assigning traffic to the network, for he must know how these systems would be used before he can evaluate and compare them.

Generally, the first principle of Wardrop is used. Under this principle each user minimizes his own individual travel cost (represented here by time), leading to equal times on all routes that are used between any pair of points. This assignment does not necessarily yield the minimum total time for the system, for the user does not bear the entire cost that he brings to the system, but is minimizing only that part which he himself bears. The following is an illustration of such a system where a large part of the marginal cost added by an additional user is borne by other users. Suppose the unit cost is an increasing function of flow as in Fig. 1. This curve is typical of intersection delays. The respective total costs are $n \cdot C_n$ for n users and $(n + 1) \cdot C_{n+1}$ for $n + 1$ users. The cost to the $(n + 1)$th user is C_{n+1}, but the increase in cost to the other n users by his being there is $n \cdot (C_{n+1} - C_n)$. Individuals minimizing only their own costs would tend to introduce large costs into the system as a whole.

There are, however, some such as Taylor[3] and Blunden[4] who feel that Wardrop's second principle describes the actions of network users quite well, if not better, than his first principle. According to the second principle, a network is used in such a manner that the total cost to the system of satisfying the demands of the users is minimized. Although the first principle (where the individual considers only his own cost) seems logical, it can be argued that most people do not know their travel times very accurately, and that many either are not aware of or would rather not compete for the faster routes.

Either of these explanations, though plausible, does not necessarily logically lead to Wardrop's second principle. However, it may cause his second principle to be a better description of traffic than his first principle. In a study of the East Freeway corridor in Columbus, Ohio, Taylor concluded that the traffic

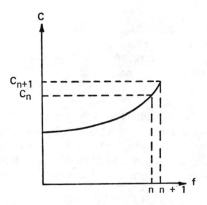

Fig. 1. Cost per user as a function of flow.

there resembles an "optimal" allocation much more closely than it does an "individual minimum travel time" allocation. Blunden has noted that individuals acting independently often assume a near-optimal solution automatically.

Considerable doubt has been expressed by some traffic assignment people concerning the type of self-assignment that drivers tend to emulate. If they are indeed assigning themselves approximately optimally from the point of view of the total network, there is little, if any, assignment control required; and the planner should be using this optimality principle in his modelling assignments. However, if the users are minimizing their own individual travel times, the planner should either (1) design with this minimum individual time assignment in mind, or (2) find the best combination of design and enforceable assignment; then use this design and enforce the desired assignment.

To choose between the above alternatives, given Wardrop's first principle, one should know the gains in system travel time through the enforcement of (2) above, as well as the total cost of enforcing the controls that would be required by (2). The more attractive alternative could then be chosen.

One of the objectives described herein is that of finding a lower bound for the minimum total travel time in the corridor, for comparison with actual total travel time. Subtracting these values yields an upper bound for the difference between the present total time and the theoretical minimum total time. This upper bound in turn gives a limit to spending on traffic control in the corridor that could be justified by savings in time to the users.

Description of the Corridor Which was Studied*

The corridor which was studied is approximately $3\frac{1}{2}$ miles long and is located in Oakland, Emeryville and Berkeley, California. It is shown in Fig. 2 as it was represented by a network of nodes and links for the purposes of traffic assignment. Any flows in the weak (southbound) direction were not included in the model, but their relevant effects were considered in determining the characteristics (such as capacity) of the components of the network.

The peak period demands on the network were obtained by handing out questionnaires to the users of the network in order to determine their origins, destinations and travel paths. From the information obtained a set of origin and destination matrices was developed in terms of the nodes in Fig. 2. Each of these matrices represents the demand for the corridor during one of the consecutive 15-min time slices which make up the afternoon peak period. (Since the demand may vary with time it is broken up into short time slices having different rates of demand. The demand rate is assumed to remain constant over a time slice which is of the order of 15 min. Homburger effectively uses a single time slice covering

*The characteristics of the network components and the demands for the network were derived from data collected by the Bay Area Freeway Operations Study which was conducted for the Division of Highways of the State of California and the United States Bureau of Public Roads by the Institute of Transportation and Traffic Engineering of the University of California.

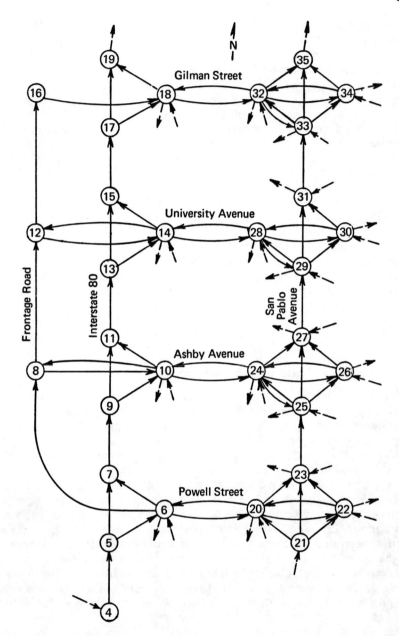

Fig. 2. Network representation of the real corridor studied.

the entire demand period.) The matrices which were obtained are contained in Yagar's dissertation.[5]

Existing Traffic Pattern During the "Typical" Peak Period

It was observed that queuing in the network, due to demands in excess of capacity, started in the time slice between 4:00 and 4:15 p.m. and ended in the slice from 4:45 to 5:00 p.m. Therefore the peak period was defined as extending from 4:00 to 5:00 p.m. This period is not to be confused with a peak hour, as it could have been shorter or longer than an hour, depending on the duration of queuing in the network.

The flows and unit travel times were recorded for each of the time slices in the peak period. These are illustrated for the slice from 4:30 to 4:45 p.m. in Fig. 3. Similar diagrams for the other time slices are presented by Yagar.[5] The 15-min flow on each link is shown followed by the effective unit travel time (in seconds) on that link with the latter in parentheses. If a link is acting as a bottleneck (in this case link (11)——(13)) and causing a queue to form, the effective unit travel time for that link includes the unit delay due to queuing for that link and is indicated for that link after its nonqueuing unit time at capacity flow. The links such as (9)——(11), which store a queue which they do not cause, have associated with them unit travel times corresponding to their flows. The implicit assumption is that a queue stacks up vertically at the entrance to the bottleneck link. For the intersections the flow is given for each type of movement, and to the right of the main diagram are given the effective flows and unit travel times for the approaches to the intersections which were considered. The thick lines represent physical queues on the respective links.

The existing total travel time, including time in motion and in queue, for the peak period was calculated from the flows and unit travel times by summing the products of the flows and corresponding unit travel times on the individual links and adding in the queue delays which were not already represented in the travel time. The existing total time was found to be 1342 vehicle hours. This same accounting procedure was adapted to both of the methodologies which follow so that the results could be compared.

Assignment According to Individual Travel Time Minimization

To observe how the existing travel pattern compares to an assignment based on this principle, it is necessary to know how the flows would look if the users assigned themselves according to minimizing their individual travel times. This in turn requires a technique to emulate such an assignment.

It is felt that any model attempting to represent peak-period travel in a capacitated network should meet the following criteria: (i) flow-dependent costs on each link, (ii) simultaneous assignment of the demands from all origin-destination pairs, (iii) allowance for time-varying demands and resultant queuing

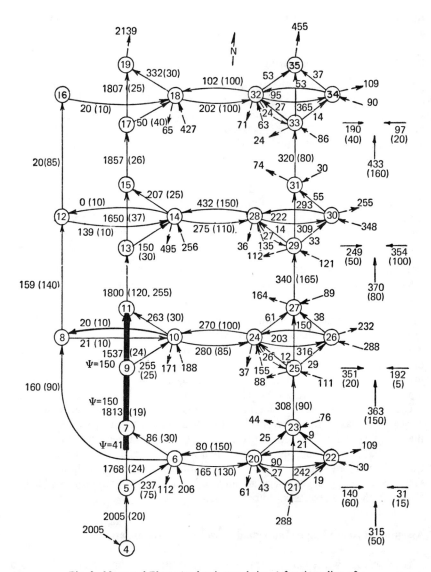

Fig. 3. Measured Flows (and unit travel times) for time slice #3.

considerations when demand exceeds capacity. The assignment methods of Steel[6] and Homburger[2]† satisfy items (i) and (ii) above but consider only time-invariant demands represented by a single matrix. No method could be found which satisfies all the criteria. Both of the above methods assign representative fractions of the whole demand matrix (on successive iterations), incrementing each link's cost as its flow increases. Homburger's method seems to be more efficient than that of Steel, as it assigns, on each iteration, as large a representative fraction of the matrix as it can before having to increase the unit cost on any link. It then increases the costs for all the links whose flows warrant an adjustment and initiates another step to repeat this procedure. The process continues until all the traffic has been assigned.

Extension of Homburger's Method to Allow for Queuing

It is felt that an extension to Homburger's[2] method allowing for queuing could lead to more realistic assignments. A proposed extension is described here. It assigns traffic in much the same way as Homburger's method. The shortest paths are found for all origin–destination pairs which have trip demands. As large a fraction as possible of the whole demand matrix is assigned to the network before changing any link costs. The link costs are then changed, new paths found and more of the demands assigned. The procedure continues until all the demands in a time slice have been assigned. This finishes Homburger's assignment.

Whereas Homburger eliminated a link when its assignment reached capacity, it is proposed here to leave the link in the network but increase the cost of that congested link (one which cannot serve any more vehicles in the present time slice although it will accept further assignment of vehicles which join a queue to be served in the next time slice) so that it reflects the cost of queuing for it. In this way a network user can still choose to use the link after taking into consideration the cost of queuing.

A trip which involves one or more congested links will go only as far as the first point of congestion and then joins a queue at that point, from which it will be assigned to its destination in the next time slice along with the new demand for the next time slice. The above procedure is repeated for each time slice until all the demands have been assigned.

Basic Assumptions and Implications

Queue Storage. A queue is assumed to be stored at a single point immediately prior to the bottleneck which is causing the queue. All users wishing to pass through the bottleneck are thus assumed to effectively join the queue at its rear. Also, any users which do not reach a bottleneck would not feel the effect of the physical queue for that bottleneck, by this procedure.

Queue Dissipation. A queue which dissipates in a certain time slice is assumed to decrease at a constant rate over the entire length of that time slice and thus

†This method was explained to the writer verbally by Wolfgang S. Homburger.

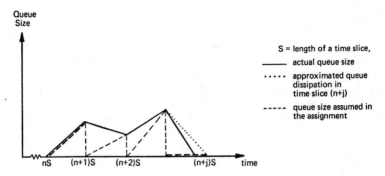

Fig. 4. Evolution of a queue over time.

disappear at the end of the slice. This is illustrated by the dotted line in time slice $(n + j)$ of Fig. 4.

Queue Evolution. A queue which exists at a point in the network at the end of a time slice is assumed to be taken out and fed back into the network at that point at a constant rate over the duration of the following time slice. This causes the queue evolution of Fig. 4 to be approximated by that of the dashed curve. By the assumption that the queue dissipates at the end of time slice $(n + j)$ the total queue time as approximated by the dashed curve is one-half of the actual queue time. To correct for this the total queue time obtained by the outlined procedure is simply doubled.

User's Knowledge of Travel Times. It is assumed that the user who is being assigned knows the unit travel and queue times of all the links for the present time slice but not for the next time slice. This has the effect that the present best path can be chosen for him but if that path leads to a queue he will re-select the remainder of his path based on new information when he is ready to leave the queue.

Approximations

Constant Turning Equivalents. A given type of turning movement at a given intersection is assumed to have a constant through flow equivalent in terms of its effect on the intensity of flow at the intersection, independent of the number of such movements.

Travel Time Functions of Flow. The relations of unit travel time vs. flow are approximated by piecewise-constant components as done by Homburger.[2] This technique effectively replaces a link by a number of "sublinks" in parallel, each having a different constant unit cost as illustrated in Fig. 5. Also shown in Fig. 5 is a linear portion on the unit travel time curve corresponding to queuing $(a > c)$. The added delay due to queuing reflects the number of vehicles already queued up for the link. This added delay is proportional to the size of the queue. When a user considers such a congested link as part of his path he does so by taking the appropriate sublink from those corresponding to $a > c$.

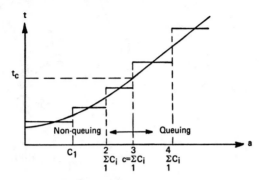

Fig. 5. Unit travel time (including queue time) vs assignment.

Methodology Used in the Dynamic Assignment

The time varying origin–destination demands on the network were assigned using the logic described in Table I. This logic is presented in greater detail by Yagar,[5] as is a computer programme which was written to carry out the logic.

TABLE I

Logic Used in Assigning Traffic to the Network

1. Make the unit travel time on each link correspond to that of its first sublink.
2. Read in the origin–destination demand matrix for the time slice
3. Add any demand queued up from the previous time slice to the demand matrix.
4. *From each node*
   (a) Find the tree of shortest paths to all other nodes which can be reached from that node.
   (b) Mark the first congested link encountered in each path.
5. Assigning each demand to its chosen path, assign as large a fraction of the remaining matrix as possible to the network without over-assigning to any sublink. Any demand reaching a congested link is not assigned further along its path, but is put into a matrix of demand queued up for the next time slice.
6. If there is any unassigned demand left in the matrix, replace any used up sublink by the next component for the link and go to step 4. Otherwise continue to step 7 as the time slice has been completed.
7. If this is not the last time slice go to step 1 for the next time slice. Otherwise the assignment is complete.

Results of the Assignment Using the Outlined Technique

For comparison with the actual observed flows, the results for the third time slice from the assignment by the described method are shown in Fig. 6. The explanations for the items in Fig. 3 apply here.

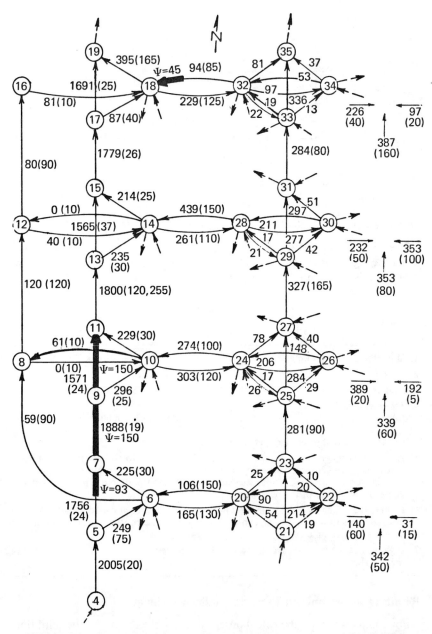

Fig. 6. Flows and unit travel times and queues for the third time slice corresponding to assignment using Wardrop's first principle.

The total travel time of all the network users in the four time slices corresponding to this assignment is 1436 vehicle hours. There is a leftover queue of 185 vehicles for link ⑪——⑬ at the end of the fourth time slice. Correction for this queue brings the total travel time to 1449 vehicle hours. This corrected value can be used for comparison with the total travel time of 1342 vehicle hours calculated for existing traffic.

Discussion of Results

The total travel time corresponding to the assignment representing Wardrop's first principle differs from that of the actual measured traffic by 1449–1342 = 107 vehicle hours, or 8%. A comparison of assigned flows with existing flows shows the main difference between them to be in the greater use of the freeway by the former. This results in more queuing on the freeway in the assignment by Wardrop's first principle. The difference in the respective amounts of queue time spent on the freeway is approximately 100 vehicle hours. This accounts for most of the difference of 107 vehicle hours between them. The other flow differences between these two assignments result in relatively small contributions to difference in total travel time.

The difference between the respective amounts of queue time spent on the freeway results from the difference in the relative amounts of flow assigned to the bottleneck link ⑪——⑬ . A psychological explanation of the model's apparent over-assignment to the freeway is a possible tendency of users to shy away from the freeway in favor of alternate routes which take more time. In so doing these users tend to help the total system in terms of total travel time. This may happen because some users do not know their shortest cost paths and/or do not measure the costs of the alternative paths strictly in units of time. Network users may also tend to avoid[‡] the freeway because of the unreliability of travel time estimates due to the variance in freeway travel times.

A Lower Bound for the Minimum Total Travel Time

As discussed previously, the writer is somewhat skeptical of the oft-accepted theory that the principle of individual travel time minimization describes traffic better than that of total system travel time minimization. The above results tend to reinforce this skepticism.

The objective here was to find a lower bound on the minimum theoretical travel time and thus an upper bound on the difference between the actual total time and the minimum possible total time.

[‡]The writer is presently attempting to quantify the apparent bias by individual users against the use of the freeway and include an appropriate correction into the route choice phenomena.

Method for Finding the Lower Bound[§]

The lower bound was obtained by assigning all the demand matrices using linear programming in order to minimize the total time. In order to keep the linear program to a feasible size, it was necessary to ignore origin–destination pairs and satisfy only the total exogenous flow requirements at the nodes. In linear programming terminology, a multicommodity problem was treated as a single-commodity one. The problem as stated would have yielded a total so low as to be an obvious lower bound, but meaningless. Therefore further sets of constraints had to be added, which would drive the value of the lower bound up and increase the credibility of the "assignment." This is illustrated by the example network shown in Fig. 7.

Suppose that there is a rectangular network with links AB, AC, CA, CD, BD and DB each having a cost of U per unit of flow. There is a demand on the network of Z units of flow from A to D and Z units from C to B. In a single-commodity flow analysis the demands would be treated as Z units of exogenous input flow at each of A and C, and Z units of exogenous output at each of B and D. A linear program would satisfy the single-commodity requirements by assigning flows of Z on links AB and CD and of zero on the other links. The cost would be $2UZ$. However, a realistic solution would have to yield a cost of $4UZ$.

Such a realistic solution can be enforced on the single-commodity analysis in this example by adding flow constraints across certain cuts as follows:

$$F_{CA} + F_{DB} \geqslant Z,$$
$$F_{AC} + F_{BD} \geqslant Z.$$

The addition of flow constraints can be made relatively straightforward by utilizing the special structure of the corridor network. In a corridor with no backward flow demands (such as from right to left in Fig. 7), the addition of these constraints on the flows across cuts can have relatively predictable effects. This property of a corridor is exploited somewhat and single-commodity analysis used with the addition of certain types of flow constraints to force the so-

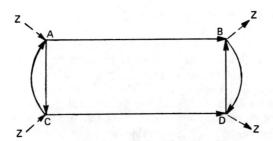

Fig. 7. A single-commodity flow representation.

[§]The assumptions and approximations which were discussed previously are continued in this section with the exception that the assumption regarding "User's Knowledge of Travel Times" does not apply here.

lution value toward the minimum theoretical total time value which can not be obtained in this case by multicommodity analysis due to the prohibitively large linear program required to model the problem exactly which is further complicated by queuing and its associated demand transfers. The single-commodity problem contains no constraints that are not inherent in the multicommodity analysis. Therefore, its total time estimate cannot be greater than that obtainable by the multicommodity analysis, and is a lower bound for it. If sufficient multicommodity constraints can be represented in the problem the lower bound can approach the theoretical minimum. However, due to the micro nature of the description of intersections in our model, it is closer to a general network than a corridor. Therefore it was not possible to impose sufficient constraints to obtain the quality of bound that was hoped for. However, the writer feels that the results are of some benefit.

Results of the Pseudoassignment

The constraints imposed were not intended to achieve a realistic assignment, but merely to force the value of the lower bound to a realistic and useful value. Also some hints were sought from the set of "flows" derived by the optimization technique. These goals were achieved to some extent but the set of "flows" obtained was obviously unrealistic. Time slice 3 of the pseudoassignment is illustrated in Fig. 8. The "flows" for the other slices are presented by Yagar.[5] The reader is cautioned against attaching undue significance to this pseudoassignment.

The lower bound on the minimum total travel time was found to be 993 vehicle hours. Since the existing value is 1342 vehicle hours, an upper bound on the possible saving in total travel time is 1342 – 993 = 349 hours or 26%.

It is also seen in the pseudoassignment that the freeways not only never have queues, but also never have flows which will decrease their velocities much below the free-flow values. Instead, any flows approaching capacity take place on the surface streets.

Discussion of Results

Studying Fig. 8 reveals obvious infeasibilities in the "flows" obtained. Forcing the assignment to a feasible set, if it were possible, would drive the lower bound up by some unknown amount. The minimum total travel time could in fact still be quite close to the measured time. Therefore the possibility of a self-assignment being near-optimal in terms of total travel time has been neither proven nor disproven.

The pseudoassignment gives qualitative hints as to the direction that an optimal solution wants to take, as it seems to want to dump any congestion onto the surface streets and keep the freeway velocities high. This tends to support ramp-metering as practised by freeway-operations people as not only having a high benefit to cost ratio but also finding near-optimal solutions without excessive and complicated efforts. It must be pointed out that care should be exercised in generalizing this idea of clear freeways to areas outside of the

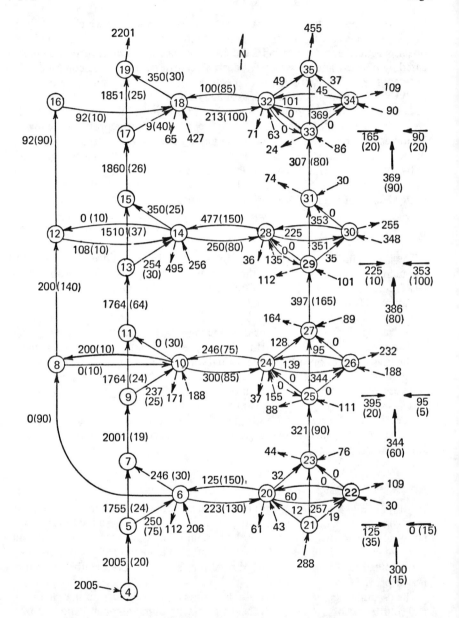

Fig. 8. "Flows" (and unit travel times) for time slice #3, corresponding to lower bound on total travel time.

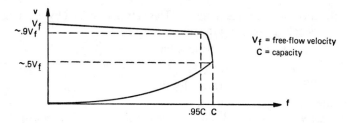

Fig. 9. Velocity as a function of flow on the freeway which was studied.

San Francisco Bay Area. The reason for this is that the velocity-flow curve for the freeway studied was found to be quite insensitive to flows up to about 95% of capacity and to be very sensitive in the range between 95 and 100% of capacity, as illustrated in Fig. 9.** Therefore the linear program found it best to keep the freeway assignment below its capacity as the incremental assignment in this range would appreciably slow down all the traffic already assigned, adding considerably to the total travel time. The corridor studied was also able to accept greater flows on the northbound surface streets whose existing flows were generally considerably below capacity.

Summary of Assignments

In comparing the flows corresponding to individual path minimization and lower bound on total travel time with the existing flows it was observed that the

TABLE II

Breakdown of the Total Travel Times for Each of the Assignments

	Assignment minimizing total travel time	Actual traffic pattern	Minimum individual travel time assignment
Effective travel time on bottleneck link (11)—(13)	126	240	240
Effective time spent in queue on the freeway	0	234	335
All other time	867	868	881
Totals	993	1342	1456

**This curve was obtained from the Bay Area Freeway Operations Study conducted by the Institute of Transportation and Traffic Engineering of the University of California on behalf of the California Division of Highways and the U.S. Bureau of Public Roads.

major differences were on the freeway. This was determined by dividing the travel time for each assignment into three categories:

   (1) the travel time on the bottleneck link ⑪———⑬ ;

   (2) all the time lost due to queuing for this link (this includes all the delay in upstream links due to queuing for link ⑪———⑬);

   (3) all the rest of the travel time.

The summary of these travel times is given in Table II, which shows that the major difference among the three "assignments" is accounted for on the freeway and can be attributed to the bottleneck link.

Conclusions

In the application performed on the existing corridor it was found that in terms of total travel time the self-assignment in effect in the corridor was somewhere between those corresponding to the two principles of Wardrop. Except for a tendency to avoid the freeway, traffic seems to follow quite closely Wardrop's first principle of individual path minimization. This tendency to avoid the freeway may result from the variance in travel times due to unreliability of service resulting from factors such as accidents. Due to this factor the users may be imposing a better assignment on the network as a whole, in terms of total travel time, as suggested previously by Blunden and Taylor. The study did not reveal how closely the self-assignment resembles an "optimal" one, although in the corridor studied the difference in total travel times was somewhere in the range from 0 to 26%.

In corridors where the freeway flows are characterized by curves resembling the one in Fig. 9 it might be advisable to keep the freeway flows less than 95% of capacity.

The corridor which was studied, and its users, may not be typical enough for general acceptance of the above conclusions. It may in fact be felt that further studies along these lines are desirable. If so, it is hoped that the methodologies used would be beneficial to any such studies. In particular, it is felt that the methodology proposed for the assignment of time-varying demands according to Wardrop's first principle could, with some additional work, be made to emulate the self-assignment of peak period users in a corridor. There is much potential for the application of such a model, for example, in estimating the loads on the links of a corridor for projected future demands, or in testing proposed ramp metering or ramp closure schemes to determine their net effects on the network as a whole.

References

1. J. G. Wardrop, Some theoretical aspects of road traffic research, *Inst. of Civil Eng.*, Road Paper #36 (1952).

2. W. S. Homburger, *Traffic Estimation–Computer Programs for Educational Purposes*, 2nd ed., *ITTE Course Notes*, University of California, Berkeley, Calif., 1969.

3. W. C. Taylor, Optimization of traffic flow splits, *Highway Res. Board, HRR* 230 (1968).
4. W. R. Blunden, Some Applications of Linear Programming to Transportation and Traffic Problems, *ITTE*, University of California, Berkeley, Calif., 1956.
5. S. Yagar, *Analysis of the Peak Period Travel in a Freeway-Arterial Corridor*, Ph.D. Thesis, University of California, Berkeley, Calif., 1970.
6. M. A. Steel, Capacity restraint–A new technique, *Traffic Eng. Control* 7, 381–384 (1965).

Probabilistic Aspects of Traffic Assignment

Heinz Beilner and Friedrich Jacobs
University of Stuttgart, Stuttgart, Germany

Introduction

The basic question in the field of traffic assignment is the following: Given a certain traffic demand between some origin and the respective destination, determine the percentages of usage for several alternate routes.

The earliest assignment methods did not consider the effect of traffic flow on traffic assignment; however, they generally considered the fact, that not all drivers choose the same route. In contrast, most studies in flow-dependent assignment take for granted that all drivers, who are subjected to identical road and traffic conditions, use the same route.

A realistic method, however, should consider the different behavior of drivers under the given conditions including the induced traffic volume as one of these conditions. This paper is concerned with the relationships between empirical, deterministic and probabilistic methods for describing the behavior-dependent choice of routes.

Empirical-deterministic approaches

The earliest assignment methods assigned the total flow for each origin-destination pair to a single route (all-or-nothing method). This method was, particularly in interesting individual cases, unusable: When constructing bypasses, it became obvious that the traffic distributed on both the old and the new route. The problem is to predict the percentage of the flow that is drawn to each of the two routes.

One tries to formulate this partition of the traffic in relation to the characteristics of the routes. The first studies are of empirical nature. Fitted curves, so-called diversion-curves, are obtained from observed actual partitions. A typical example for this method is the diversion curve of the State of Washington.[12] The percentage of freeway usage is plotted against the ratio of the driving times on the freeway and on the quickest alternate route.

In a similar way, Campbell and Rothrock[16] examine the percentage of usage as a function of the time ratio, as a function of the time difference, and as a function of the distance ratio.

A better agreement with the observations is attained with several refined, graphic methods, that consider the percentage of usage of one of two routes as a function of two route characteristics. Trueblood[18] considers the time ratio and the distance ratio simultaneously and suggests straight lines as curves of equal usage percentages. Another such method is the one by Campbell,[5] who plots the usage of one of two routes against the distance ratio and speed ratio. The increasing complexity of these studies and the necessity of using computers require mathematical forms that describe the results of the observations as accurately as possible. A few examples for this:

Moskowitz[13] uses a family of hyperbolas:

$$p = 50 + \frac{50(d + mt)}{\sqrt{(d - mt)^2 + 2b^2}},$$

where

p = percentage of usage,
d = distance saved in miles,
t = time saved in minutes,
m = a coefficient relating the value of a mile saved to a minute lost,
b = a coefficient determining how far the vertices on the 100% and 0% boundaries are from the origin.

Murray[14] formulates Campbell's suggestions in the equations

$$p_1 = \frac{1}{1 + (1.162 \cdot t_1/t_2)^{5.85}} \tag{1}$$

and

$$p_1 = \frac{1}{1 + (0.86 \cdot d_1/d_2)^{6.7}}, \tag{2}$$

where

p_1 = percentage of traffic using route 1,
t_1 = time via route 1,
t_2 = time via route 2,
d_1 = distance via route 1,
d_2 = distance via route 2,

Brown and Weaver[4] use the formula

$$p_1 = 0.5 + 2.5 \frac{t_2 - t_1}{t_2 + t_1}$$

but suggest that the following might be better

$$p_1 = \frac{1}{1 + (t_1/t_2)^6} \tag{3}$$

All these studies show that the difference in usage can only partially be explained in terms of times and distances. Not only one or two characteristics of the routes have to be considered, but as many as possible of the factors that affect the decision of the driver. Remarkably, this perception was already in 1952 the basis of the studies of May and Michael.[11] They contemplated diversion as a function of a general cost index, that included time, distance, and speed.

In the first studies only the diversion between a route via roads of a higher standard and the best alternate route of the remaining network was considered. The next approximation to reality is gained through the consideration of more than two alternate routes. Graphic methods are impracticable from the start here. One tries to obtain a generalization of those formulas that have been acquired by the consideration of two routes. The most frequently used solution is called "Kirchhoff's Rule" because of its similarity to a formula in electrical engineering. Its form is

$$p_s = \frac{a_s}{\sum_r a_r} \, , \tag{4}$$

where p_s is the percentage of the drivers that choose the sth route, and a_s represents the "attractivity" of the sth route.

The formulas (1), (2), and (3) can be transformed to

$$p_1 = \frac{(1/t_1)^{5.85}}{(1/t_1)^{5.85} + (1.162/t_2)^{5.85}} \tag{1*}$$

$$p_1 = \frac{(1/d_1)^{6.7}}{(1/d_1)^{6.7} + (0.86/d_2)^{6.7}} \tag{2*}$$

$$p_1 = \frac{(1/t_1)^6}{(1/t_1)^6 + (1/t_2)^6} \tag{3*}$$

They already have then, in principle, the form of Eq. (4).

Such formulas for more than two routes are found, e.g., in the works of Broberg[3] and Nordquist[15] with

$$a_r = \frac{1}{(k_d d_r + t_r)^z} \, ,$$

Irwin[8] with

$$a_r = \frac{1}{(t_r)^z} \, ,$$

and Steierwald, Scholz and Haupt[17] with

$$a_r = \frac{1}{(k_t t_r + k_d d_r)^z} \, ,$$

where

d_r = distance via the rth route
t_r = time via the rth route
z, k_t, k_d = constants.

Probabilistic approaches

The methods and models that have been described so far all have in common that streams of vehicles are considered as a whole and therefore, in a manner of speaking, macroscopically. To the best of our knowledge Knödel[10] was the first, to make mention of another, as it were, microscopic way of thinking. The following reflections are the starting point:

On the one hand, each driver behaves in a way that is most advantageous for himself. If he has the possibility of choosing between different routes while driving, he chooses the "best" route.

On the other hand, drivers do not behave the same way. If several drivers have the possibility of choosing between different routes, then all drivers do not normally decide for the same route.

These differences in the drivers' judgments of the routes have two reasons:

(1) The characteristics of a route such as time required, distance, technical condition and paving of roads and intersections are judged differently by the drivers. Differences result in the overall judgment of a route.

(2) A driver normally judges a route according to his past driving experiences. The number of trips on a route and the course which these trips take are different. Consequently, the experiences are different.

The result of such a train of thought is that it is not possible to make certain statements about how an individual driver judges a route. Therefore, the inconvenience of a route can be mathematically described only by a random variable W_r. This variable has a distribution function $F_r(x)$ defining the probability $P(W_r < x)$ with which the inconvenience of the route r is estimated to be smaller than a value x.

An extended mathematical formulation is found in the works of Abraham,[1] and v. Falkenhausen,[6,7] who independently, to be sure, came to the same forms. Kato[9] and Beilner[2] carried the train of thought further. The common starting point of all these approaches is the idea that each driver considers simultaneously all the available routes r $(r = 1, 2, \ldots, m)$, and then estimates the inconvenience W_r of each of the alternate routes. The conditional probability of choosing the route s becomes if its inconvenience is estimated to be $W_s = x$, equal to the probability that for all other routes estimates of $W_r > x$ result. If the estimates are independent of each other, then this probability is equal to

$$\prod_{r=1, r \neq s}^{m} [1 - F_r(x)].$$

Hence it follows for the unconditional probability p_s of choosing the route s (Theorem of Total Probability):

$$p_s = \int_{-\infty}^{+\infty} f_s(x) \prod_{\substack{r=1 \\ r \neq s}}^{m} [1 - F_r(x)]\, dx$$

or

$$p_s = \int_{x=-\infty}^{+\infty} f_s(x) \prod_{\substack{r=1 \\ r \neq s}}^{m} \left\{ \int_{x_r=x}^{+\infty} f_r(x_r)\, dx_r \right\} dx, \qquad (5)$$

where $f_s(x)$ is the probability density belonging to $F_s(x)$. If the number of drivers is not too small, then the probability p_s of choosing the route s can be set equal to the percentage of drivers, who choose this route.

Relationships between empirical-deterministic and probabilistic approaches

The main purpose of this paper is to examine whether or not the probabilistic partition equations are in any way related to the empirical diversion curves and the deterministic partition rules, and if so, under which circumstances. In order to do this, we have to study Eq. (5) simultaneously with specific types and classes of distributions. In previous works on this subject there appeared, to our knowledge, only the two special cases of normal and lognormal distributed variables. Looking in general on distribution functions of random variables which shall be used to describe the inconvenience of routes, it seems reasonable to assume that:

Distributions should be continuous with density functions $f(x)$.

The density function $f(x)$ should be unimodal.

There is a fixed lower bound for the estimation of inconvenience, therefore $f(x) = 0$ for $x < x_1$.

On the other hand, a fixed upper bound of estimation does not exist necessarily.

Diversion curves and probabilistic partition equations

Let us consider the partition of traffic on two routes. We suppose that the distribution functions of estimated inconvenience $F_1(x; \mu_1, \sigma_1)$ and $F_2(x; \mu_2, \sigma_2)$, respectively, belong to a certain class of distribution functions and that they are determined by their means μ_1, μ_2, variances σ_1^2, σ_2^2 and, if necessary, several higher moments.

The probability, that for given $\mu_1, \sigma_1, \mu_2, \sigma_2$ route 1 will be chosen shall be denoted by

$$p_1(\mu_1, \sigma_1; \mu_2, \sigma_2) = \int_{-\infty}^{+\infty} f_1(x; \mu_1, \sigma_1)\, [1 - F_2(x; \mu_2, \sigma_2)]\, dx \qquad (6)$$

First we assume that for any a_1, a_2 the distribution functions satisfy the conditions

$$F_1(x; \mu_1, \sigma_1) = F_1(x + a_1; \mu_1 + a_1, \sigma_1)$$

and

$$F_2(x; \mu_2, \sigma_2) = F_2(x + a_2; \mu_2 + a_2, \sigma_2).$$

In this case, displacements of means only produce displacements of distribution functions on the abscissa. The central moments remain the same. Then we can write for the right side of Eq. (6):

$$\int_{-\infty}^{+\infty} f_1(x; \mu_1, \sigma_1) \left[1 - F_2(x; \mu_2, \sigma_2) \right] dx$$

$$= \int_{-\infty}^{+\infty} f_1(x; \mu_1 - \mu_2, \sigma_1) \left[1 - F_2(x; 0, \sigma_2) \right] dx.$$

Thus in this case the integral in Eq. (6) is invariant against displacements on the abscissa, and for fixed values of variances and higher central moments the probability p_1 only depends on the difference of the means

$$p_1(\mu_1 - \mu_2, \sigma_1; 0, \sigma_2) = \int_{-\infty}^{+\infty} f_1(x; \mu_1 - \mu_2, \sigma_1) \left[1 - F_2(x; 0, \sigma_2) \right] dx. \quad (7)$$

Equation (7) represents a function corresponding to that class of diversion curves, that describe the partition of traffic by functions of differences, such as time differences or distance differences.

Now we assume that for any a_1, a_2 the distribution functions satisfy the conditions

$$F_1(x; \mu_1, \sigma_1) = F_1(a_1 x; a_1 \mu_1, a_1 \sigma_1)$$

and

$$F_2(x; \mu_2, \sigma_2) = F_2(a_2 x; a_2 \mu_2, a_2 \sigma_2);$$

therefore,

$$f_1(x; \mu_1, \sigma_1) = a_1 \cdot f_1(a_1 x; a_1 \mu_1, a_1 \sigma_1)$$

and

$$f_2(x; \mu_2, \sigma_2) = a_2 \cdot f_2(a_2 x; a_2 \mu_2, a_2 \sigma_2).$$

In this case, variations in the means only produce proportional variations in the scale of the abscissa and particularly the square roots of the variances are always proportional to the means

$$\sigma_1 = c_1 \mu_1 \quad \text{and} \quad \sigma_2 = c_2 \mu_2.$$

Thus, we can write for the right side of Eq. (6):

$$\int_{-\infty}^{+\infty} f_1(x; \mu_1, c_1 \mu_1) \left[1 - F_2(x; \mu_2, c_2 \mu_2) \right] dx$$

$$= \int_{-\infty}^{+\infty} \frac{1}{\mu_2} \cdot f_1\left(\frac{x}{\mu_2}; \frac{\mu_1}{\mu_2}, c_1 \frac{\mu_1}{\mu_2} \right) \left[1 - F_2\left(\frac{x}{\mu_2}; 1, c_2 \right) \right] dx,$$

and substituting we obtain

$$\int_{-\infty}^{+\infty} f_1(x;\mu_1,c_1\mu_1)\,[1-F_2(x;\mu_2,c_2\mu_2)]\,dx$$

$$=\int_{-\infty}^{+\infty} f_1\left(x;\frac{\mu_1}{\mu_2},c_1\frac{\mu_1}{\mu_2}\right)[1-F_2(x;1,c_2)]\,dx$$

$$p_1\left(\frac{\mu_1}{\mu_2},c_1\frac{\mu_1}{\mu_2};1,c_2\right)=\int_{-\infty}^{+\infty} f_1\left(x;\frac{\mu_1}{\mu_2},c_1\frac{\mu_1}{\mu_2}\right)[1-F_2(x;1,c_2)]\,dx \quad (8)$$

Equation (8) represents a function corresponding to that class of diversion curves, which describe the partition of traffic by functions of some ratio, such as time ratio or distance ratio.

This shows, that diversion curves may be interpreted to be particular cases of the probabilistic partition equations. The assumption, that central moments, particularly the variance, of the distributions of estimated inconvenience are independent of the mean inconvenience, leads to difference diversion curves. On the other hand, under the assumption that the square roots of the variances are proportional to the mean inconvenience we obtain ratio diversion curves. The latter assumption seems to be more likely, and hence it is not surprising that it is often said, that ratio diversion curves give the better fit to empirical results.

Deterministic and probabilistic partition equations

In the course of examining a series of distributions according to the proposed conditions we found that the random variable with the following probability density is of particular interest

$$f(x)=\begin{cases} g'(x)\cdot\exp[-g(x)], & x\geq 0,\\ 0, & x<0, \end{cases} \quad (9)$$

where $g(x)$ is for $x>0$ a positive, monotonously increasing unique function with $g(0)=0$ and $g(\infty)=\infty$.

This random variable has the mean

$$\mu=\int_0^\infty x\cdot g'(x)\cdot\exp[-g(x)]\,dx,$$

$$\mu=\int_0^\infty h(x)\cdot\exp[-x]\,dx,$$

with $h(x)$ denoting the inverse function of $g(x)$. $h(x)$ is under the given conditions a unique function.

The variance is given by

$$\sigma^2 = \int_0^\infty h^2(x) \cdot \exp\left[-x\right] dx - \mu^2.$$

By inserting this density function into Eq. (5) we obtain

$$p_s = \int_0^\infty g'_s(x_s) \cdot \exp\left[-g_s(x)\right] \prod_{r \neq s} \left\{ \int_{x_r = r_s}^\infty g'_r(x_r) \cdot \exp\left[-g_r(x)\right] dx_r \right\} dx_s$$

and after a transformation

$$p_s = \int_0^\infty \exp\left[-\left(x + \sum_{r \neq s} g_r(h_s(x))\right)\right] dx$$

This integral is easy to handle if we let

$$g_r(x) = \frac{a_r}{a_s} \cdot g_s(x), \quad r \neq s$$

and, therefore,

$$h_r(x) = h_s\left(\frac{a_s}{a_r} \cdot x\right).$$

We then obtain the following relationships:
For the means

$$\mu_s = \int_0^\infty h_s(x) \cdot \exp\left[-x\right] dx \tag{10}$$

and

$$\mu_r = \int_0^\infty h_s\left(\frac{a_s}{a_r} \cdot x\right) \cdot \exp\left[-x\right] dx, \quad r \neq s. \tag{11}$$

For the Variances

$$\sigma_s^2 = \int_0^\infty h_s^2(x) \exp\left[-x\right] dx - \mu_s^2 \tag{12}$$

and

$$\sigma_r^2 = \int_0^\infty h_s^2\left(\frac{a_s}{a_r} x\right) \cdot \exp\left[-x\right] dx - \mu_r^2, \quad r \neq s. \tag{13}$$

For the percentages

$$p_s = \int_0^\infty \exp\left[-\left(x + \sum_{r \neq s} \frac{a_r}{a_s} x\right)\right] dx$$

$$= \int_0^\infty \exp\left[-\frac{\sum_r a_r}{a_s} \cdot x\right] dx$$

and

$$p_s = \frac{a_s}{\sum_r a_r} . \tag{14}$$

On the other hand, we obtain under the assumption of unbiased estimations for the inconveniences of the routes, Kirchhoff Rules in the form:

$$p_s = \frac{(1/\mu_s)^n}{\sum_r (1/\mu_r)^n} \tag{15}$$

In order that Eqs. (14) and (15) conform with each other we have to write

$$\mu_r = \frac{k}{a_r^{1/n}} .$$

This means, we have to find a function $h_s(x)$ so that

$$\int_0^\infty h_s(x) . \exp[-x] \, dx = \frac{k}{a_s^{1/n}} .$$

It can be easily shown that

$$h_s(x) = \left(\frac{x}{a_s}\right)^{1/n}$$

is a solution of the integral equation. We obtain

$$k = \Gamma(1 + 1/n), \tag{16}$$

$$\mu_s = \frac{\Gamma(1 + 1/n)}{a_s^{1/n}}, \tag{17}$$

and

$$\sigma_s^2 = \frac{\Gamma(1 + 2/n) - [\Gamma(1 + 1/n)]^2}{a_s^{2/n}} . \tag{18}$$

The inverse function to $h_s(x)$ is

$$g_s(x) = a_s \cdot x^n$$

Analogous results may be found for μ_r, μ_r^2, g_r. Hence, the probability density in Eq. (9) becomes

$$f_r(x) = \begin{cases} a_r \cdot n \cdot x^{n-1} \cdot \exp\left[-a_r x^n\right], & x \geqslant 0 \\ 0, & x < 0. \end{cases} \quad \text{(See Figs. 1 and 2.)}$$

This density belongs to the so-called Weibull variate and has its maximum at $[(1 - 1/n)/a_r]^{1/n}$, near the mean for not too small values of n. As can be seen from Eqs. (16–18) there is a constant ratio between mean and standard deviation if n is fixed. Hence the Kirchhoff Rules include the assumption of this proportionality between standard deviation and mean in the same manner as the ratio diversion curves do.

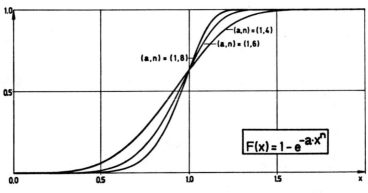

Figure 1

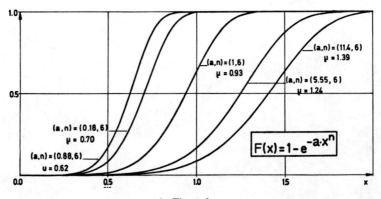

Figure 2

Equations of the form (14), which are always Kirchhoff-similar ones, may be obtained by inserting other appropriate functions $g_r(x)$, e.g.,

$$g_r(x) = a_r(x + c)^n, \quad x \geqslant 0$$

leads to

$$\mu_r = \frac{\Gamma(1 + 1/n)}{a_r^{1/n}} - c$$

and

$$p_s = \frac{[1/(\mu_s + c)]^n}{\sum_r [1/(\mu_r + c)]^n}.$$

This shows that Kirchhoff Rules may be interpreted to be special cases of the probability partition equations. Where Weibull distributions are adequate to describe the distribution functions for the estimated inconvenience of routes, the Kirchhoff Rules possess the advantage of easiness.

Extensions

Flow-dependence

Although it was not explicitly mentioned, flow-dependence was not excluded in the probabilistic approach. The flow is certainly one of the factors, which determine the judgment of the routes and therefore their distribution functions; the parameters of the distribution functions are functions of the flow. Certainly, the effects of the flow differ on different routes. Nevertheless, roads with similar flow-dependence can be placed into categories. For every road of a certain category, the same functions are chosen to characterize the dependence of the parameters on the flow.[2]

Further problems

There were two assumptions for the derivation of the essential Eq. (5):
The first assumption concerned the decisionmaking pattern. Here it was assumed, that a driver, when choosing one of several possible routes, considered all of the routes at the same time. If there are more than two routes, then this is not the only conceivable decisionmaking pattern. One can imagine that the driver considers, in a certain sequence, only some of the routes at the same time. Mathematical formulations analogous to Equation 5 can be found in the same manner.

The second assumption was that the random variables for describing the routes could be regarded as being independent. However, if alternate routes have common stretches then independence of the random variables can in general not be expected. In this case, dependence of the random variables can be taken into account by regarding the judgment of a route as a function of the route's stretches.

The underlying question in both problems is the nature of the mechanism of judgment which, in turn, opens the door for a number of other unsolved questions.

References

1. C. Abraham, La répartition du traffic entre intinéraires concurrents, *Révue général des routes et des aerodromes*, 1961.
2. H. Beilner, *Über Modelle und Methoden zur Prognose der Aufteilung des Straßenverkehrs in Abhängigkeit von der Verkehrsmenge*, Dissertation, Stuttgart, 1969.
3. A. Broberg, A computer method for traffic assignment developed and used in Stockholm, *Int. Road Safety Traffic Rev.* **11,** 27 (1963).
4. R. M. Brown and H. H. Weaver, Traffic assignment using IBM computations and summation, *Highway Res. Board Bull.* **130,** 47–58 (1956).
5. E. W. Campbell, A mechanical method for assigning traffic to expressways, *Highway Res. Board Bull.* **130,** 27–46 (1956).
6. H. v. Falkenhausen, Ein Verfahren zur Prognose der Verkehrsverteilung in einem geplanten Stra ennetz, *Unternehmensforschung* **7.** 75–88 (1963).
7. H. v. Falkenhausen, *Ein stochastisches Modell zur Verkehrsumlegung,* Dissertation, Darmstadt, 1966.
8. N. A. Irwin and H. G. v. Cube, Capacity restraint in multi-travel mode assignment programs, *Highway Res. Board Bull.* **347,** 258–289 (1962).
9. A. Kato, Traffic Assignment and Highway Systems, *Annual Report of Roads*, pp. 20–40, Japan Road Association, 1965.
10. W. Knödel, Verkehrsplanung und Mathematik III: Netzmodelle, *Math. Tech. Wirtschaft* **7,** 147–154 (1960).
11. A. D. May, Jr. and H. L. Michael, Allocation of Traffic to Bypasses. *Highway Res. Board Bull.* **61,** 38–58 (1952).
12. J. K. Mladinov and R. J. Handen, A mechanized procedure for traffic assignment of traffic to a new route, *Highway Res. Board Bull.* **130,** 59–68 (1956).
13. K. Moskowitz, California method of assigning diverted traffic to proposed freeways, *Highway Res. Board Bull.* **130,** 1–26 (1956).
14. F. J. Murray, Comparative usage of Olentangy River Road and alternate arterials in Columbus, Ohio, *Highway Res. Board Bull.* **61,** 18–37 (1952).
15. S. Nordquist and B. Wilhelmson, Data processing in the analysis, forecast and assignment of traffic at Gavle, Sweden, *Int. Road Safety Traffic Rev.* **11,** 36 (1963).
16. C. A. Rothrock and W. E. Campbell, Traffic assignment, *Highway Res. Board Bull.* **130,** 497–506 (1962).
17. G. Steierwald, G. Scholz, and D. Haupt, Probleme der Verkehrsumlegung, *Straßenverkehrstechnik* **7,** 469–480 (1963).
18. D. L. Trueblood, Effect of travel time and distance on freeway usage, *Highway Res. Board Bull.* **61,** 38–58 (1952).

Theory of Traffic Assignment to a Road Network

Tetsuzo Hoshino
Tokyo Branch Office, Japan Highway Public Corporation, Tokyo, Japan

Abstract

The existing methods of traffic assignment to a road network may be roughly classified into two groups, one is the "All or Nothing" method based upon "minimum path" and the other is the "Assignment Rate" method.

My method is the assignment rate method by computation using simultaneous equations in consideration of capacity restraint. There are several unique characteristics in this method. They are that (1) it considers the hourly fluctuations of traffic volumes during a day, (2) it considers intrazonal trips, (3) more than ten routes can be taken for the assignment of each interzonal transfer, and (4) the evaluation value of each route which is related to the assignment rates is determined by the probability density functions of both travel time and toll fares.

Classification of the Theories of Traffic Assignment and Their Characteristics

The author classifies the theories of traffic assignment which have been proposed as follows:

1 optimum solution method
2 others
  2-1 traffic demands (with no capacity restraint)
    2-1-1 all or nothing method
    2-1-2 assignment rate method
  2-2 actual volumes (with capacity restraint)
    2-2-1 all or nothing method
    2-2-2 assignment rate method
      2-2-2-1 simulation method
      2-2-2-2 analytical method

Among the all or nothing methods (2-2-1) there are methods which are shown in the papers by J. C. Carrol,[8] R. Smock,[9] B. V. Martin and M. L. Manheim,[10] and Kitayama,[1] and in the reports of B. P. R.[7] and M. I. T.[10] The method of Kitayama[1] is that when it converts the toll fares to time values using time evaluation values, it assumes the distribution of evaluation values and uses the different values for each assignment by generation of random numbers.

The simulation methods (2-2-2-1) are shown in the papers by N. A. Irwin[11] and A. Kato[2] and the report of B. P. R.

Among the analytical methods (2-2-2-2) there is a method[3] proposed by T. Sasaki, Kyoto University, which assigns an O-D transfer to several routes by the principle of equal time arrival after the search for the minimum path, and

---

References pp. 213-214

the author's method[4] which at first selects several routes for each interzonal transfer that an O–D demand will choose according to traffic conditions and assigns the volume between them by using time-ratio diversion curves. In addition, there is a method proposed by W. W. Mosher.[12] The author's method[4] was applied to the third Keihin highway (link number 14, 15, 16 in Fig. 7), Haneda-Yokohama expressway (link number 40, 41 in Fig. 7), a road network of the southwest of Kanagawa prefecture, north Kyushu highway and Meihan highway (which connects Osaka and Nagoya).

When the assignment rate methods are used, the problem is how to determine the assignment rates. Among the papers in Japan concerning this problem, one[5] applies information theory and another[6] considers time evaluation values as random variables. The paper by C. Abraham[13] is also very unique. In the application example in section 8, Abraham's theory is used as the principle of the assignment rate.

Route Search

If the assignment process is based on the All or Nothing method in which demands are assigned to the minimum paths, computers are indispensable to select the minimum path for each O–D movement among various routes in a complicated road network. However, if the assignment is to be done using the assignment rate method, it is not necessary to find the minimum path exactly.

The appropriate number of feasible routes for a given O–D movement could be selected through engineer's judgment, then traffic can be assigned by applying an assignment rate to each route. In the real world, drivers decide their own travel routes before they start, and the routes selected by each driver differ from each other. Therefore the above method by which an engineer selects the routes for each O–D pair is considered realistic. Among the assignment rate methods, there is a systematic method of examining all the possible routes between a given O–D based on the theory of graphs by W. W. Mosher, although there is the possibility of examining unrealistic routes such as whirlpools and zigzags. Anyway, human judgments are indispensable for the route search even for a computerized route search process. Furthermore, there is no assurance that drivers follow the minimum path determined by the route search. Therefore, for this study, the routes for assignment are to be determined through engineer's judgment from the viewpoint that it is enough to select the appropriate number of routes which drivers would usually follow. The appropriate number described above, would be attained by only adopting main routes for O–D movements which are important to the area under study, or by limiting the number of routes for such O–D movements that have comparatively low demand but are significant to the area under study. A group of such routes is called a "route table."

Establishment of a Series of Equations (1) – Short Distance Method

Let $\overline{Q}_{ij}$ be the O–D traffic volume between i and j which we want to assign. The point in the network where $\overline{Q}_{ij}$ is originated, destined, joined, or branched

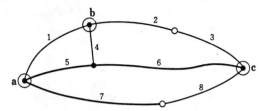

Fig. 1. a indicates a trip end; ● indicates an A-node; ○ indicates a B-node; thick line represents a toll road; thin line represents an ordinary road; a number signifies a link name.

off is called an "A-node." The point where the design standard of highways is changed, or there is a big change in local traffic, or there is an origin or destination of a toll road is called a "B-node." The highway section between A-nodes or B-nodes is called a "link" (see Fig. 1).

In this model, the possible routes which are chosen based on the concepts of the previous section are as follows.

$$\overline{Q}_{ab}: \quad \text{Route 1} \quad 1$$
$$\text{Route 2} \quad 5 - 4$$
$$\overline{Q}_{ac}: \quad \text{Route 1} \quad 1 - 2 - 3$$
$$\text{Route 2} \quad 5 - 6$$
$$\text{Route 3} \quad 7 - 8$$
$$\overline{Q}_{bc}: \quad \text{Route 1} \quad 2 - 3$$
$$\text{Route 2} \quad 4 - 6$$

If $\overline{Q}_{ij}$ is assigned to m routes, then

$$\overline{Q}_{ab} = {}_1Q_{ab} + {}_2Q_{ab},$$
$$\overline{Q}_{ac} = {}_1Q_{ac} + {}_2Q_{ac} + {}_3Q_{ac},$$
$$\overline{Q}_{bc} = {}_1Q_{bc} + {}_2Q_{bc},$$

and in general,

$$\overline{Q}_{ij} = \sum_{r=1}^{m} {}_rQ_{ij}$$

where $_rQ_{ij}$ = assigned volume in the rth route. This is named the *continuation equation*.

When a typical point of each link is chosen, the total traffic volume at the point, q_l, is the sum of the total assigned volume $(_rQ_{ij})$ and the local traffic, $_Kq_l$. Then, for

Link 1	$q_1 = {}_1Q_{ab} + {}_1Q_{ac} + {}_Kq_1,$
Link 2	$q_2 = {}_1Q_{ac} + {}_1Q_{bc} + {}_Kq_2,$
Link 3	$q_3 = {}_1Q_{ac} + {}_1Q_{bc} + {}_Kq_3,$
Link 4	$q_4 = {}_2Q_{ab} + {}_2Q_{bc} + {}_Kq_4,$
Link 5	$q_5 = {}_2Q_{ab} + {}_2Q_{ac} + {}_Kq_5,$
Link 6	$q_6 = {}_2Q_{ac} + {}_2Q_{bc} + {}_Kq_6,$
Link 7	$q_7 = {}_3Q_{ac} + {}_Kq_7,$
Link 8	$q_8 = {}_3Q_{ac} + {}_Kq_8.$

In general,

$$q_l = \sum {}_r\delta_{ij} \cdot {}_rQ_{ij} + \kappa q_l,$$

where ${}_r\delta_{ij} = 0$ or 1. This is named the *cumulation equation.*

Local traffic can be defined as the "traffic volume other than ${}_rQ_{ij}$ at a typical point of each link." It affects the running speed of ${}_rQ_{ij}$ when they pass the link. It usually consists of the "intrazonal traffic volume," but there is also a small volume of traffic other than the intrazonal volume. They are of course the volumes which are not considered as the object of assignment. At present the local traffic is determined by subtracting $\overline{Q}_{ij}$ which is thought to pass the link from the total traffic volume at a typical point of the highway section. Since it is uncertain what percentage of the volume use those links when $\overline{Q}_{ij}$ is using several routes, it is desirable to do an origin–destination survey in which all the routes and the distribution of traffic between them are found for the major O-D volumes. When the routes cannot be found by the origin–destination survey, a method to determine the local traffic is to subtract $\Sigma_r Q_{ij}$ (the total ${}_rQ_{ij}$) which is determined as follows from the total volume of each link. $\Sigma_r Q_{ij}$ of all the links are calculated using this assignment theory in the present road network which excludes the planned highway by giving the necessary factors such as the present O-D volume table, the route table when there is not the planned highway, speed-flow relations, the relations between hourly and daily volumes, toll rates and evaluation functions, and as known numbers the present total traffic volumes of all the links.

The method to assign $\overline{Q}_{ij}$ to each route will be shown in section 6. The assignment rate, ${}_r\beta_{ij}$, is defined as follows.

$$\begin{aligned}
{}_1Q_{ab}/\overline{Q}_{ab} &= {}_1\beta_{ab}, \\
{}_2Q_{ab}/\overline{Q}_{ab} &= {}_2\beta_{ab} = 1 - \beta_{ab}, \\
{}_1Q_{ac}/\overline{Q}_{ac} &= {}_1\beta_{ac}, \\
{}_2Q_{ac}/\overline{Q}_{ac} &= {}_2\beta_{ac}, \\
{}_3Q_{ac}/\overline{Q}_{ac} &= {}_3\beta_{ac} = 1 - {}_1\beta_{ac} - {}_2\beta_{ac}, \\
{}_1Q_{bc}/\overline{Q}_{bc} &= {}_1\beta_{bc}, \\
{}_2Q_{bc}/\overline{Q}_{bc} &= {}_2\beta_{bc} = 1 - {}_1\beta_{bc}.
\end{aligned}$$

In general, ${}_rQ_{ij}/\overline{Q}_{ij} = {}_r\beta_{ij}$. This is named the *assignment rate definition function.*

When a driver selects a route among many routes, he should make some evaluation of the route and judge it as the most attractive one. The grounds of the judgement are thought to be small travel time, cheap toll rate, comfort, small stress of driving, etc. If they can be quantified and called the "evaluation value" (${}_lE$ or ${}_rE_{ij}$) of road links or routes, the assignment rate of a route can be expressed as a function of the evaluation values of the route and the other possible routes. Then

$$_1\beta_{ab} = {_1f_{ab}}(_1E_{ab}, {_2E_{ab}}),$$
$$_1\beta_{ac} = {_1f_{ac}}(_1E_{ac}, {_2E_{ac}}, {_3E_{ac}}),$$
$$_2\beta_{ac} = {_2f_{ac}}(_1E_{ac}, {_2E_{ac}}, {_3E_{ac}}),$$
$$_1\beta_{bc} = {_1f_{bc}}(_1E_{bc}, {_2E_{bc}}).$$

In general, $_r\beta_{ij} = {_rf_{ij}}(_1E_{ij}, {_2E_{ij}},..._mE_{ij})$. This is named the *assignment equation*.

The evaluation value $(_rE_{ij})$ is considered to be a function of various factors such as travel time, $_rT_{ij}$, toll rate, $_rF_{ij}$, and comfort. Among these, comfort can be expressed as a function of running speed V, or travel time $_rT_{ij}$, and a constant, $_rK_{ij}$, which indicates the degree of comfort peculiar to each route, then,

$$_1E_{ab} = {_1g_{ab}}(_1T_{ab}, {_1K_{ab}}),$$
$$_2E_{ab} = {_2g_{ab}}(_2T_{ab}, {_2F_{ab}}, {_2K_{ab}}),$$
$$_1E_{ac} = {_1g_{ac}}(_1T_{ac}, {_1K_{ac}}),$$
$$_2E_{ac} = {_2g_{ac}}(_2T_{ac}, {_2F_{ac}}, {_2K_{ac}}),$$
$$_3E_{ac} = {_3g_{ac}}(_3T_{ac}, {_3F_{ac}}, {_3K_{ac}}),$$
$$_1E_{bc} = {_1g_{bc}}(_1T_{bc}, {_1K_{bc}}),$$
$$_2E_{bc} = {_2g_{bc}}(_2T_{bc}, {_2F_{bc}}, {_2K_{bc}}).$$

In general, $_rE_{ij} = {_rg_{ij}}(_rT_{ij}, {_r\sigma_{ij}} \cdot {_rF_{ij}}, {_rK_{ij}})$, where $_r\delta_{ij} = 0$ or 1. This is named the *evaluation equation*.

The total travel time of each route, $_rT_{ij}$, is the sum of the travel time of all the road links which constitute the route, and the travel time of each link is expressed as a function of the link distance, l_l, and the total traffic volume of a typical point of the link, q_l (which shows a different value in each link), then

$$_1T_{ab} = F_1(l_1, q_1),$$
$$_2T_{ab} = F_5(l_5, q_5) + F_4(l_4, q_4),$$
$$_1T_{ac} = F_1(l_1, q_1) + F_2(l_2, q_2) + F_3(l_3, q_3),$$
$$_2T_{ac} = F_5(l_5, q_5) + F_6(l_6, q_6),$$
$$_3T_{ac} = F_7(l_7, q_7) + F_8(l_8, q_8),$$
$$_1T_{bc} = F_2(l_2, q_2) + F_3(l_3, q_3),$$
$$_2T_{bc} = F_4(l_4, q_4) + F_6(l_6, q_6).$$

In general, $_rT_{ij} = \Sigma F_l(l_l, q_l)$, where the suffix l is taken for all the road sections which constitute the r route between i and j. This is named the *Q-T equation*.

When the assignment work is done, one must make the program such that $_rT_{ij}$ becomes infinite very rapidly when q_l reaches its capacity in order to execute capacity restrain[*].

By solving the above simultaneous equations, which consist of the continuation equations, the cumulation equations, the assignment equations, the evaluation equations and the Q-T equations, by giving $\bar{Q}_{ij}$ and $_Kq_l$ as known numbers and deciding the coefficients of $_rf_{ij}, {_rg_{ij}}$ and F_l functions, one can determine all the unknown numbers such as $_rQ_{ij}, q_l, {_r\beta_{ij}}$, and $_rE_{ij}$. In the above example the number of equations is:

continuation equations	3	cumulation equations	8
assignment rate definition functions	7	assignment equations	4
evaluation equations	7	Q-T equations	7
		total	36.

The number of unknown numbers is:

$_rQ_{ij}$	7	q_l	8
$_r\beta_{ij}$	7	$_rE_{ij}$	7
$_rT_{ij}$	7		
		total	36.

The number of known numbers is:

$\overline{Q}_{ij}$ 3 $_Kq_l$ 8
total 11.

In the above example, the total volumes of all kinds of vehicles are considered. But as these equations differ according to the vehicle kinds, one must make the equations for each vehicle kind.

In the example of Fig. 1 the number of equations is:

continuation equations	$3n$	cumulation equations	8
assignment rate definition functions	$7n$	assignment equations	$4n$
evaluation equation	$7n$	Q-T equations	$7n$
		total	$28n + 8.$

The number of unknown numbers is:

$_rQ_{ij}$	$7n$,	q_l	8
$_rE_{ij}$	$7n$,	$_rT_{ij}$	$7n$
		$_r\beta_{ij}$	$7n$
		total	$28n + 8.$

The number of known numbers is:

$\overline{Q}_{ij}$ $3n$, $_Kq_l$ 8.

Therefore, the simultaneous equations can be solved.

Establishment of a Series of Equations (2)–Long Distance Method

The series of equations which are discussed in the previous section are based on an assumption that the hourly volume stays constant during certain hours in a day and when a $\overline{Q}_{ij}$ is being assigned the other $\overline{Q}_{ij}$'s and local traffic volumes also stay constant. Although the assumption is reasonable when one makes the assignment in a relatively small area, it is not valid for the assignment of long

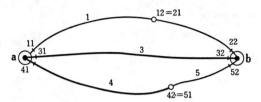

Fig. 2. Legend is the same as in Fig. 1.

distance trips like those between Tokyo and Osaka, because this traffic moves in both the daytime peak hours and the nighttime off-peak hours during the trips from the origins to the destinations. Also considering the peculiar phenomenon (although it may change in the future) that these long distance trips usually start at the origins late in the evening, the equations in the previous section cannot be used without changes. In this case the equations should be modified as follows (see Fig. 2).

For convenience, the explanation is made for one vehicle type. Suppose there are $\bar{Q}_{ab}$ long distance trips and they have three routes, 1- 2, 3, and 4–5, and $l1$ is the origin and $l2$ is the destination of each road section.

If $_{\tau(l)K}q_{l2}$ is meant as the local traffic at $l2$ after $\tau(l)$ hours from the standard time, the total volumes at the origins of Links 1, 3, and 4 are

$$q_{11} = {_1}Q_{ab} + {_{\tau(0)K}}q_{11},$$
$$q_{31} = {_2}Q_{ab} + {_{\tau(0)K}}q_{31},$$
$$q_{41} = {_3}Q_{ab} + {_{\tau(0)K}}q_{41}.$$

In the total volumes at the destinations of Links 1, 3 and 4, there are no changes in $\bar{Q}_{ij}$, but the local traffic volumes are the volumes after $\tau(l)$ hours in which $_rQ_{ij}$ takes to pass each road section, then

$$q_{12} = {_1}Q_{ab} + {_{\tau(1)K}}q_{12},$$
$$q_{32} = {_2}Q_{ab} + {_{\tau(3)K}}q_{32},$$
$$q_{42} = {_3}Q_{ab} + {_{\tau(4)K}}q_{42},$$

and

$$q_{22} = {_1}Q_{ab} + {_{\tau(1+2)K}}q_{22},$$
$$q_{52} = {_3}Q_{ab} + {_{\tau(4+5)K}}q_{52}.$$

These correspond to the *cumulation equations.*

The *continuation equation* is the same as in the previous section.

$$\bar{Q}_{ab} = {_1}Q_{ab} + {_2}Q_{ab} + {_3}Q_{ab}$$

The *assignment rate definition equations* are

$${_1}Q_{ab}/\bar{Q}_{ab} = {_1}\beta_{ab},$$
$${_2}Q_{ab}/\bar{Q}_{ab} = {_2}\beta_{ab},$$
$${_3}Q_{ab}/\bar{Q}_{ab} = {_3}\beta_{ab} = 1 - {_1}\beta_{ab} - {_2}\beta_{ab}.$$

The *assignment equations are*

$$_1\beta_{ab} = {}_1f_{ab}({}_1E_{ab}, {}_2E_{ab}, {}_3E_{ab}),$$
$$_2\beta_{ab} = {}_2f_{ab}({}_1E_{ab}, {}_2E_{ab}, {}_3E_{ab}).$$

The *evaluation equations* are

$$_1E_{ab} = {}_1g_{ab}({}_1T_{ab}, {}_1K_{ab}),$$
$$_2E_{ab} = {}_2g_{ab}({}_2T_{ab}, {}_2F_{ab}, {}_2K_{ab}),$$
$$_3E_{ab} = {}_3g_{ab}({}_3T_{ab}, {}_3F_{ab}, {}_3K_{ab}).$$

The *Q-T equations* are

$$_1T_{ab} = F_1(l_1, q_{11}, q_{12}) + F_2(l_2, q_{12}, q_{22}),$$
$$_2T_{ab} = F_3(l_3, q_{31}, q_{32}),$$
$$_3T_{ab} = F_4(l_4, q_{41}, q_{42}) + F_5(l_5, q_{42}, q_{52}).$$

In addition to these equations, the hourly fluctuation data of local traffic should be given.

$$_{\tau(1)K}q_{12} = f(\tau(1)),$$
$$_{\tau(3)K}q_{32} = f(\tau(3)),$$
$$_{\tau(4)K}q_{42} = f(\tau(4)),$$
$$_{\tau(1+2)K}q_{22} = f(\tau(1 + 2)),$$
$$_{\tau(4+5)K}q_{52} = f(\tau(4 + 5)).$$

These are named the *fluctuation equations.*

By solving the above simultaneous equations, the 25 unknown numbers ($_1Q_{ab}$, $_2Q_{ab}$, $_3Q_{ab}$, q_{11}, q_{31}, q_{41}, q_{12}, q_{32}, q_{42}, q_{22}, q_{52}, $_1\beta_{ab}$, $_2\beta_{ab}$, $_3\beta_{ab}$, $_{\tau(1)K}q_{12}$, $_{\tau(3)K}q_{32}$, $_{\tau(4)K}q_{42}$, $_{\tau(1+2)K}q_{22}$, $_{\tau(4+5)K}q_{52}$, $_1E_{ab}$, $_2E_{ab}$, $_3E_{ab}$, $_1T_{ab}$, $_2T_{ab}$, $_3T_{ab}$) can be determined.

When there are many kinds of vehicles, one can set up the equations as in the previous section.

The above equations are for the case of one $\bar{Q}_{ij}$. The equations for the case that more than two $\bar{Q}_{ij}$'s which have the wave type fluctuation mutually intervene are shown in the next section. It is very difficult and complicated to solve them. Therefore, the author thinks that, for a practical solution method in the case of long distance trips, one should divide the network involving long distance trips into several small networks with which the short distance method can deal. One first determines any one of long distance $\bar{Q}_{ij}$'s using the long distance method by regarding q_l, which is determined by the short distance method, as the local traffic in the long distance method. Then consider the sum of the previous q_l and the calculated $\bar{Q}_{ij}$ as the new q_l and solve for the next long distance $\bar{Q}_{ij}$. Another method is to assign each $\bar{Q}_{ij}$ only on the short distance q_l without adding the long distance q_l.

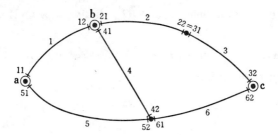

Fig. 3. Legend is the same as in Fig. 1.

Establishment of a Series of Equations (3)—The Strict Method

For convenience of explanation, consider traffic composed of only one kind of vehicle. Let $\overline{Q}_{ac}$ and $\overline{Q}_{bc}$ be O–D traffic volumes between the points a and c, and b and c, respectively (see Fig. 3). The routes for $\overline{Q}_{ac}$ are supposed to be (1 – 2 – 3) and (5 – 6), and the routes for $\overline{Q}_{bc}$ are (2 – 3) and (4 – 6).

Then $_rQ_{ij}$ would be $_1Q_{ac}, _2Q_{ac}, _1Q_{bc}$ and $_2Q_{bc}$. Let

$_{(t\alpha)}\tau_l$ = travel time for a vehicle over section l starting from point $l1$ at the time $(t\alpha)$ and reaching the end point $l2$;

$_{(t\alpha)}q_{lm}$ = total volume at the point lm ($m = 1$ or 2) at the time $(t\alpha)$;

$_{(t\alpha + {}_{t\alpha}\tau_l)r}Q_{ij}$ = O–D traffic volume from i to j at the time $_{(t\alpha)}\tau_l$ hours after the starting time $(t\alpha)$;

$_{(t\alpha)K}q_{lm}$ = local traffic at the point lm at the time $(t\alpha)$.

Since the total volume at the beginning point 11 of Link 1 at the time $(t\alpha)$ is the sum of the local traffic and the O–D traffic volume from a to b,

$$_{(t\alpha)}q_{11} = {}_{(t\alpha)k}q_{11} + {}_{(t\alpha)1}Q_{ac}.$$

The total volume at the end point 12 of link 1 at time $(t\alpha + {}_{t\alpha}\tau_1)$ is the sum of the local traffic $_{(t\alpha)}\tau_1$ hours after the time $(t\alpha)$, and the O–D traffic volume from a to b starting at point 11 at time $(t\alpha)$. Therefore,

$$_{(t\alpha + {}_{t\alpha}\tau_1)}q_{12} = {}_{(t\alpha + {}_{t\alpha}\tau_1)k}q_{12} + {}_{(t\alpha)1}Q_{ac}.$$

Points 51 and 52 on link 5 are derived in the same way as above,

$$_{(t\alpha)}q_{51} = {}_{(t\alpha)k}q_{51} + {}_{(t\alpha)2}Q_{ac},$$

$$_{(t\alpha + {}_{t\alpha}\tau_5)}q_{52} = {}_{(t\alpha + {}_{t\alpha}\tau_5)k}q_{52} + {}_{(t\alpha)2}Q_{ac}.$$

For link 4,

$$_{(t\alpha)}q_{41} = {}_{(t\alpha)k}q_{41} + {}_{(t\alpha)2}Q_{bc},$$

$$_{(t\alpha + {}_{t\alpha}\tau_4)}q_{42} = {}_{(t\alpha + {}_{t\alpha}\tau_4)k}q_{42} + {}_{(t\alpha)2}Q_{bc}.$$

The total volume at the beginning point 21 of link 2 at the time $_{(t\alpha)}\tau_1$ hours after the time $(t\alpha)$ is the sum of the local traffic at the time $(t\alpha + {}_{t\alpha}\tau_1)$, Q_{ac}

starting at point 11 at the time $(t\alpha)$ and $_1Q_{bc}$ starting at point 21 at the time $(t\alpha + {}_{t\alpha}\tau_1)$.

$$(t\alpha+{}_{t\alpha}\tau_1)q_{21} = (t\alpha+{}_{t\alpha}\tau_1)kq_{21} + (t\alpha)_1Q_{ac} + (t\alpha+{}_{t\alpha}\tau_1)_1Q_{bc}$$

Point 22 is derived in the same way as above.

$$(t\alpha+{}_{t\alpha}\tau_1+[t\alpha+{}_{t\alpha}\tau_1]\tau_2)q_{22} = (t\alpha+{}_{t\alpha}\tau_1+[t\alpha+{}_{t\alpha}\tau_1]\tau_2)kq_{22}$$
$$+ (t\alpha)_1Q_{ac} + (t\alpha+{}_{t\alpha}\tau_1)_1Q_{bc}$$

While point 61 and 62 give

$$(t\alpha+{}_{t\alpha}\tau_5)q_{61} = (t\alpha+{}_{t\alpha}\tau_5)kq_{61} + (t\alpha)_2Q_{ac}$$
$$+ (t\alpha+{}_{t\alpha}\tau_5-{}_{t\alpha}\tau_4)_2Q_{bc},$$

$$(t\alpha+{}_{t\alpha}\tau_5+[t\alpha+{}_{t\alpha}\tau_5]\tau_6)q_{62} = (t\alpha+{}_{t\alpha}\tau_5+[t\alpha+{}_{t\alpha}\tau_5]\tau_6)kq_{62}$$
$$+ (t\alpha)_2Q_{ac} + (t\alpha+{}_{t\alpha}\tau_5-{}_{t\alpha}\tau_4)_2Q_{bc}.$$

The total volume at point 31 is equal to that at point 22. The total volume at 32 is

$$(t\alpha+{}_{t\alpha}\tau_1+[t\alpha+{}_{t\alpha}\tau_1]\tau_2+[t\alpha+{}_{t\alpha}\tau_1+\{t\alpha+{}_{t\alpha}\tau_1\}\tau_2]\tau_3)q_{32}$$
$$= (t\alpha+{}_{t\alpha}\tau_1+[t\alpha+{}_{t\alpha}\tau_1]\tau_2+[t\alpha+{}_{t\alpha}\tau_1+\{t\alpha+{}_{t\alpha}\tau_1\}\tau_2]\tau_3)kq_{32}$$
$$+ (t\alpha)_1Q_{ac} + (t\alpha+{}_{t\alpha}\tau_1)_1Q_{bc}.$$

The $_2Q_{bc}$ starting at b at time $(t\alpha + {}_{t\alpha}\tau_5 - {}_{t\alpha}\tau_4)$ is

$$(t\alpha+{}_{t\alpha}\tau_5-{}_{t\alpha}\tau_4)q_{41} = (t\alpha+{}_{t\alpha}\tau_5-{}_{t\alpha}\tau_4)kq_{41} + (t\alpha+{}_{t\alpha}\tau_5-{}_{t\alpha}\tau_4)_2Q_{bc},$$
$$(t\alpha+{}_{t\alpha}\tau_5)q_{42} = (t\alpha+{}_{t\alpha}\tau_5)kq_{42} + (t\alpha+{}_{t\alpha}\tau_5-{}_{t\alpha}\tau_4)_2Q_{bc}.$$

These 13 equations form the *cumulation equations*. The following four equations are the *continuation equations*.

$$(t\alpha)\overline{Q}_{ac} = (t\alpha)_1Q_{ac} + (t\alpha)_2Q_{ac},$$
$$(t\alpha)\overline{Q}_{bc} = (t\alpha)_1Q_{bc} + (t\alpha)_2Q_{bc},$$
$$(t\alpha+{}_{t\alpha}\tau_1)\overline{Q}_{bc} = (t\alpha+{}_{t\alpha}\tau_1)_1Q_{bc} + (t\alpha+{}_{t\alpha}\tau_1)_2Q_{bc},$$
$$(t\alpha+{}_{t\alpha}\tau_5-{}_{t\alpha}\tau_4)\overline{Q}_{bc} = (t\alpha+{}_{t\alpha}\tau_5-{}_{t\alpha}\tau_4)_1Q_{bc} + (t\alpha+{}_{t\alpha}\tau_5-{}_{t\alpha}\tau_4)_2Q_{bc}.$$

The first two equations show that the O-D demand at the time $t\alpha$ $(_{(t\alpha)}Q_{ij})$ is equal to the sum of the volumes assigned to each link $(_{(t\alpha)r}Q_{ij})$. The lower two show similar relations.

The *assignment rate definition functions* for the time $(t\alpha)$ are given by the following equations:

$$_{(t\alpha)1}Q_{ac}/_{(t\alpha)}\overline{Q}_{ac} = {}_{(t\alpha)1}\beta_{ac},$$

$$_{(t\alpha)2}Q_{ac}/_{(t\alpha)}\overline{Q}_{ac} = {}_{(t\alpha)2}\beta_{ac} = 1 - {}_{(t\alpha)1}\beta_{ac},$$

$$_{(t\alpha)1}Q_{bc}/_{(t\alpha)}\overline{Q}_{bc} = {}_{(t\alpha)1}\beta_{bc},$$

$$_{(t\alpha)2}Q_{bc}/_{(t\alpha)}\overline{Q}_{bc} = {}_{(t\alpha)2}\beta_{bc} = 1 - {}_{(t\alpha)1}\beta_{bc}.$$

Let $_{(t\alpha)r}E_{ij}$ be the evaluation value for the travel time and the toll fares for each route for given O–D traffic $_{(t\alpha)}\overline{Q}_{ij}$ at the time $t\alpha$. Then *assignment equations* are as follows.

$$_{(t\alpha)1}\beta_{ac} = {}_{(t\alpha)1}f_{ac}({}_{(t\alpha)1}E_{ac}, {}_{(t\alpha)2}E_{ac}),$$

$$_{(t\alpha)1}\beta_{bc} = {}_{(t\alpha)1}f_{bc}({}_{(t\alpha)1}E_{bc}, {}_{(t\alpha)2}E_{bc}).$$

The evaluation value for each O–D traffic at time $t\alpha$ can be considered as a function of travel time $_{(t\alpha)r}T_{ij}$, toll fare $_{(t\alpha)r}F_{ij}$ and a constant value, $_{(t\alpha)r}K_{ij}$, which shows the comfort of the route. Therefore, the *evaluation equations* are,

$$_{(t\alpha)1}E_{ac} = {}_{(t\alpha)1}g_{ac}({}_{(t\alpha)1}T_{ac}, {}_{(t\alpha)1}K_{ac}),$$

$$_{(t\alpha)2}E_{ac} = {}_{(t\alpha)2}g_{ac}({}_{(t\alpha)2}T_{ac}, {}_{(t\alpha)2}F_{ac}, {}_{(t\alpha)2}K_{ac}).$$

$_{(t\alpha)1}E_{bc}$ and $_{(t\alpha)2}E_{bc}$ are also described as above.

Concerning *Q-T equations* $_{(t\alpha)r}T_{ij}$ is expressed as the functions of road link lengths and link volumes. Then

$$_{(t\alpha)1}T_{ac} = F_1(l_1, {}_{(t\alpha)}q_{11}, {}_{(t\alpha+t\alpha\tau_1)}q_{12})$$

$$+ F_2(l_2, {}_{(t\alpha+t\alpha\tau_1)}q_{21}, {}_{(t\alpha+t\alpha\tau_1+[t\alpha+t\alpha\tau_1]\tau_2)}q_{22}$$

$$+ F_3(l_3, {}_{(t\alpha+t\alpha\tau_1+[t\alpha+t\alpha\tau_1]\tau_2)}q_{22}, {}_{(t\alpha+t\alpha\tau_1}$$

$$+ {}_{[t\alpha+t\alpha\tau_1]\tau_2+[t\alpha+t\alpha\tau_1+\{t\alpha+t\alpha\tau_1\}\tau_2]\tau_3)}q_{32}).$$

$_{(t\alpha)2}T_{ac}$, $_{(t\alpha+t\alpha\tau_1)1}T_{bc}$, and $_{(t\alpha+t\alpha\tau_5-t\alpha\tau_4)2}T_{bc}$ are also determined as above.

Fluctuation equations are described by the variations of O–D traffic volumes $\overline{Q}_{ij}$, for the assignment and local traffic volumes, of road link during a day. Then

$$_{(t\alpha+t\alpha\tau_1)}\overline{Q}_{bc} = f_{bc}(t\alpha + {}_{t\alpha}\tau_1),$$

$$_{(t\alpha+t\alpha\tau_5-t\alpha\tau_4)}\overline{Q}_{bc} = f_{bc}(t\alpha + {}_{t\alpha}\tau_5 - {}_{t\alpha}\tau_4),$$

$$_{(t\alpha+t\alpha\tau_1)k}q_{12} = f_{12}(t\alpha + {}_{t\alpha}\tau_1).$$

$_{(t\alpha+t\alpha\tau_5)k}q_{52}$, $_{(t\alpha+t\alpha\tau_4)k}q_{42}$, $_{(t\alpha+t\alpha\tau_1)k}q_{21}$, $_{(t\alpha+t\alpha\tau_1+[t\alpha+t\alpha\tau_1]\tau_2)k}q_{22}$, $_{(t\alpha+t\alpha\tau_1+[t\alpha+t\alpha\tau_1]\tau_2+[t\alpha+t\alpha\tau_1+\{t\alpha+t\alpha\tau_1\}\tau_2]\tau_3)k}q_{32}$, $_{(t\alpha+t\alpha\tau_5)k}q_{61}$, $_{(t\alpha+t\alpha\tau_5+[t\alpha+t\alpha\tau_5]\tau_6)k}q_{62}$, $_{(t\alpha+t\alpha\tau_5-t\alpha\tau_4)k}q_{41}$, $_{(t\alpha+t\alpha\tau_5)k}q_{42}$, are also described as above.

In this example the number of equations and unknown numbers are both 43. Therefore, the simultaneous equations can be solved.

Assignment Rate

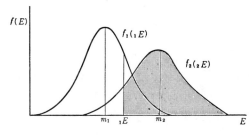

$f(E)$
$f_1({}_1E)$
$f_2({}_2E)$
m_1 ${}_1E$ m_2 E

Fig. 4

For the assignment rate, Abraham's theory is adopted as a basis, as mentioned in the first section of this paper.

The route evaluation value, ${}_rE_{ij}$, which was mentioned in the third section corresponds to the expected value which car drivers consider as an index for choosing their routes. All the drivers do not always give the same evaluation to a route. To simplify the explanation, let's assume that there are two routes, 1 and 2, and their evaluation values are ${}_1E$ and ${}_2E$. These evaluation values are considered to be distributed around the means, m_1 and m_2 with distribution functions $f_1({}_1E)$ and $f_2({}_2E)$ (see Fig. 4). If ${}_1E \leqslant {}_2E$, route 1 will be chosen, otherwise route 2 will be chosen. The probability that a driver evaluates route 1 as ${}_1E$ and routes 2 as ${}_1E \leqslant {}_2E$ is expressed by

$$f_1({}_1E) \int_{{}_1E}^{\infty} f_2({}_2E)\, d_2E\, d_1E.$$

Therefore, the probability to choose route 1 is

$$\int_{-\infty}^{\infty} f_1({}_1E) \int_{{}_1E}^{\infty} f_2({}_2E)\, d_2E\, d_1E$$

and the probability to choose the route 2 is,

$$\int_{-\infty}^{\infty} f_2({}_2E) \int_{{}_2E}^{\infty} f_1({}_1E)\, d_1E\, d_2E.$$

If one extends this argument to the case of *n* routes, the probability that route *k* is chosen, or the assignment rate to the route *k* is,

$$_k\beta_{ij} = \int_{-\infty}^{\infty} f_k({}_kE) \int_{{}_kE}^{\infty} f_1({}_1E) \cdots \int_{{}_kE}^{\infty} f_{k-1}({}_{k-1}E) \cdot \int_{{}_kE}^{\infty} f_{k+1}({}_{k+1}E)$$

$$\cdots \int_{{}_kE}^{\infty} f_n({}_nE) \cdot d_nE \cdots d_{k+1}E \cdot d_{k-1}E \cdots d_2E \cdot d_1E.$$

The above is considered the essence of a generalized Abraham's theory. Although his theory uses, as the route evaluation value, the amount of gasoline consumption or the converted money value of safety, comfort, or travel time, one could use the "time" itself as the route evaluation value. It is often said that the value of toll expressways is time, and it seems fairly reasonable to take "time" as an index of the route evaluation value.

It is not considered proper to obtain converted time values of toll rates by using a constant coefficient of conversion when there are toll roads in the routes, because there is a distribution in time evaluation values. In this method, therefore, the distribution of time evaluation values is considered.

"Time" is taken as an index of route evaluation, then,

$$_rE_{ij} \equiv {}_rE = \frac{{}_rF_{ij}}{\lambda} + {}_rT_{ij}$$

$$\equiv x \cdot {}_rF + t,$$

where

   $_rF_{ij}$ = toll rate of the route r between i and j,

   $_rT_{ij}$ = travel time of the route r between i and j,

   λ = time evaluation value,

  $_rF_{ij} = {}_rF, {}_rT_{ij} = t, \dfrac{1}{\lambda} = x.$

If we assume that the probability density functions of x and t are logarithmic normal distributions, then $\log x$ and $\log t$ have normal distributions, and $\log {}_rE$ also has a normal distribution (where $x > 0, t > 0, x, t$: independent) by the convolution of normal distribution. Therefore, when we express the means and variances of $\log x$ and $\log t$ as

$$\log x: \ N(m_x, S_x^2)$$

and

$$\log t: \ N(m_t, S_t^2);$$

those of $\log {}_rE$ as

$$\log {}_rE: \ N({}_rF \cdot m_x + m_t, {}_rF^2 \cdot S_x^2 + S_t^2).$$

Then the probability density function of $_rE_{ij}$ is

$$f_r({}_rE) = \frac{1}{S_E \sqrt{2\pi}} \cdot \frac{1}{{}_rE} \cdot \exp\left\{ -\frac{(\log {}_rE - m_E)^2}{2S_E^2} \right\}$$

where

$$m_E = {}_rF \cdot m_x + m_t$$
$$S_E^2 = {}_rF^2 \cdot S_x^2 + S_t^2$$

Therefore, when m_x, S_x, m_t, and S_t are determined, we can get $f_r(_rE)$. Where the mean and variance of logarithmic normal distribution are μ and σ,

$$\mu = e^{m + S^2/2},$$

$$\sigma = e^{2m + S^2} (e^{S^2} - 1).$$

Then if we can get (μ_x, σ_x) and (μ_t, σ_t) from the actual distributions of x and t, (m_x, S_x) and (m_t, S_t) can be determined from the above equations.

Application Examples of Short Distance Method

Description of outline

Examples where the method described earlier are applied to practical cases as follows:

Large scale application examples:

(i) a road network that involves the second Yokohama bypass diverging from the third Keihin highway located east of Kanagawa prefecture, extending to Fujisawa city parallel with the existing Yokohama bypass; South Yokohama bypass diverging from the second Yokohama bypass, running through the Miura peninsula to Yokosuka city; and main highways in Kanagawa prefecture,

(ii) Kan-etsu expressway from Tokyo to Shibukawa city,

(iii) the urban freeway and other related highways in Kawasaki city.

As the small scale examples, there are the cases of the Seisho bypass and Odawara-Atsugi highway in Kanagawa prefecture.

In this section, the outline of the case of the second Yokohama bypass and South Yokohama bypass is explained.

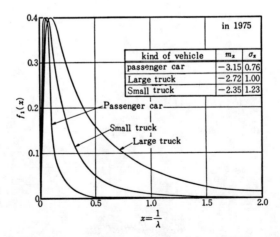

Fig. 5. Distribution of $1/\lambda$ (the reciprocal of time evaluation value).

TABLE I

Route Table

Origin 1		Destination 17 (14)
Route	1	38 – 39 – 40 – 41 – 42 – 43 – 44 – 46 – 81 – 82 – 83 – 84 – 85 – 86
"	2	38 – 39 – 40 – 41 – 42 – 43 – 44 – 45 – 34 – 35 – 36 – 37 – 96
"	3	38 – 39 – 40 – 41 – 42 – 43 – 44 – 45 – 34 – 35 – 36 – 108 – 95 – 96
"	4	48 – 49 – 50 – 51 – 52 – 53 – 54 – 76 – 77 – 78 – 79 – 80 – 81 – 82
"		83 – 84 – 85 – 86
"	5	48 – 49 – 50 – 51 – 52 – 53 – 54 – 76 – 97 – 98 – 34 – 35 – 36 – 37
		– 96
"	6	48 – 49 – 50 – 51 – 52 – 53 – 54 – 76 – 97 – 98 – 34 – 35 – 36 – 108
"		– 95 – 96
"	7	1 – 169 – 13 – 14 – 15 – 17 – 32 – 33 – 34 – 35 – 36 – 37 – 96
"	8	1 – 169 – 13 – 14 – 15 – 17 – 32 – 33 – 34 – 35 – 36 – 108 – 95 – 96
"	9	1 – 169 – 13 – 14 – 15 – 16 – 56 – 33 – 34 – 35 – 36 – 37 – 96
"	10	1 – 169 – 13 – 14 – 15 – 16 – 56 – 33 – 34 – 35 – 36 – 108 – 95 – 96
"	11	118 – 119 – 13 – 14 – 15 – 17 – 32 – 33 – 34 – 35 – 36 – 37 – 96 –
"	12	118 – 119 – 13 – 14 – 15 – 17 – 32 – 33 – 34 – 35 – 36 – 108 – 95 – 96
"	13	118 – 119 – 13 – 14 – 15 – 16 – 56 – 33 – 34 – 35 – 36 – 37 – 96
"	14	118 – 119 – 13 – 14 – 15 – 16 – 56 – 33 – 34 – 35 – 36 – 108 – 95 – 96

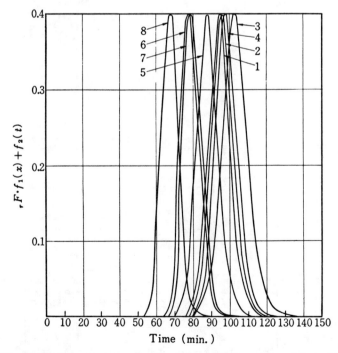

Fig. 6. The distribution of route evaluation values (the sum of travel time and time converted value of toll rate).

TABLE II

Travel Time and Traffic Volume of Road Links

Link number	Link distance	Link travel time (min)			Local traffic (vehicles/day)	Assigned volume (vehicles/day)						Total volume (vehicle/day)
		Passenger car	Small truck	Large truck		Passenger car	Bus	Small truck	Large truck	Light vehicle	Total	
1	9.1	11.2	12.5	12.7	68,338	6,307	129	1,624	1,069	627	9,756	78,094
2	7.6	6.6	7.3	7.7	48,055	2,646	177	938	539	489	4,789	52,844
3	12.1	10.6	11.8	12.5	51,355	2,981	202	1,137	639	672	5,631	56,986
4	15.3	12.5	13.9	14.9	31,942	3,919	182	1,216	598	508	6,423	38,365
5	22.9	17.8	19.7	21.2	21,472	136	15	49	0	46	255	21,727
6	45.3	38.3	42.4	45.7	20,199	0	0	0	0	0	0	20,199
7	5.0	5.0	5.5	6.0	5,799	0	0	0	0	0	0	5,799
8	9.1	8.4	9.3	9.9	11,730	2,413	160	723	390	388	4,074	15,804
9	5.4	4.6	5.0	5.3	12,361	1,618	160	633	354	275	3,040	15,401
10	3.5	2.9	3.2	3.4	12,361	1,618	160	633	354	275	3,040	15,401
11	7.1	6.0	6.6	7.0	12,782	1,322	128	489	255	235	2,429	15,211
12	6.6	5.6	6.1	6.5	13,843	707	78	258	119	131	1,293	15,136
13	2.0	2.4	2.8	2.8	88,998	24,661	360	6,450	3,226	2,868	37,565	126,563
14	2.2	2.1	2.3	2.4	61,798	24,661	360	6,450	3,226	2,868	37,565	99,363
15	11.1	10.4	11.6	11.9	40,109	31,906	439	9,062	5,023	4,643	51,073	91,182

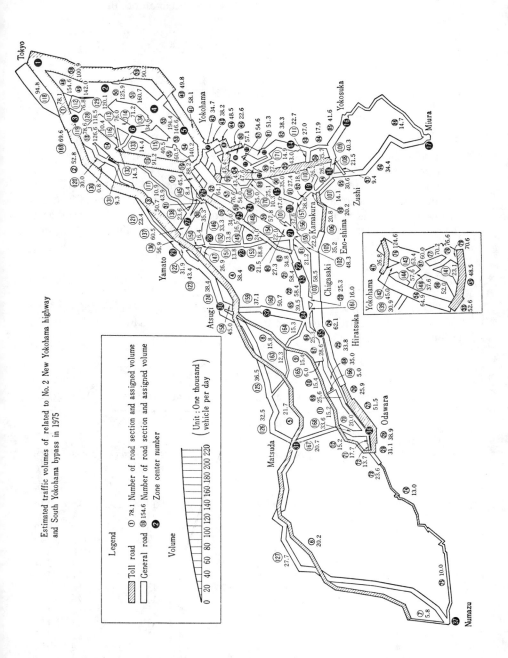

Estimated traffic volumes of related to No. 2 New Yokohama highway and South Yokohama bypass in 1975

Legend

Toll road

General road

① 78.1 Number of road section and assigned volume

⑲ 154.6 Number of road section and assigned volume

❷ Zone center number

Volume

0 20 40 60 80 100 120 140 160 180 200 220

(Unit : One thousand vehicle per day)

To do the traffic assignment using the short distance method, the following data must be prepared.

1. Future O–D (origin and destination) traffic demand table,
2. Road network data for assignment,
3. Route table,
4. Q-V (Volume and speed) functions,
5. Relationship between hourly and daily traffic volumes,
6. Toll rates,
7. Evaluation function,
8. Local traffic.

Among the above data items, the route table (3) and the evaluation functions (7) are shown in Fig. 5 and Table I, respectively.

Before computation, evaluation functions must be made for each O-D pair and route. These data are not necessarily needed as the output. However, Fig. 6 is shown for reference. This figure shows the combined distribution curves of assumed travel times and converted time values of toll rates for the pair of origin (1) and destination (12 and 15).

Results of computation

The computation was done by Facom 230-20 computer. The results are shown in Table II (travel time and traffic volume of each highway link), Table III (those of each O-D pair) and Fig. 7 (estimated traffic volumes).

TABLE III

Route travel time and Assigned volume of O–D pairs

			Route travel time (min)			Assigned volume (vehicles/day)					
O	D	Route number	Passenger car	Small truck	Large truck	Passenger car	Bus	Small truck	Large truck	Light vehicle	Total
1	17										
		1	115.5	125.4	129.1	13	1	8	12	3	37
		2	101.4	110.9	114.2	81	3	30	20	10	144
		3	106.6	116.6	119.8	56	2	23	18	8	107
		4	159.0	173.2	177.5	0	0	0	0	0	0
		5	106.9	116.5	119.6	16	1	9	11	3	40
		6	139.8	153.0	156.8	0	0	0	1	0	1
		7	93.8	104.4	107.8	119	5	41	25	14	204
		8	99.0	110.1	113.4	92	4	33	23	11	163
		9	90.2	100.5	103.6	124	5	43	28	15	215
		10	95.4	106.2	109.2	96	4	34	25	12	171
		11	98.3	108.8	112.4	82	3	31	23	11	150
		12	103.5	114.5	118.0	55	2	23	18	8	106
		13	94.7	104.9	108.2	84	4	31	25	11	155
		14	99.9	110.6	113.8	54	2	22	19	8	105
					SUM	872	36	328	248	114	1598

Closing Comment

The weak point of the assignment method by simultaneous equations has been that, when the number of O–D pairs are too many, its solution becomes almost impossible, because the memory capacity of computers are exceeded. But we can have the optimistic prospect that the rapid progress of the ability of computers will overcome the weak point. Also, the following method can be taken in order to avoid it, although the accuracy of the assignment will go down a little. Solve the simultaneous equations by giving large O–D traffic volumes and local traffic volumes, as the velocity of highway links is determined by link volumes. Then other small O–D traffic volumes can be assigned using assignment rate functions by fixing the travel time of each highway link as a known number.

In addition, there is a problem of the saturation of the link volumes, but I did not have enough pages to describe in detail how to deal with this problem. When a link volume reaches its capacity, the O–D demand will flow to another route which does not include the saturated link. When all the routes are saturated with the demands, it is considered that the generation of traffic itself is restricted. There is a method in which we consider a cut section in the highway network, which contains at least one $\overline{Q}_{ij}$ in which all $_rQ_{ij}$ are included and restricts the traffic generation.

Finally, the number of elements in the assignment computation of large scale application examples is shown in Table IV for reference.

TABLE IV

Number of Elements in Assignment Computation

	Name of Highway	Number of equation for each vehicle	Number of vehicle kind	Total number of equations	Number of O–D pairs	Number of links	Number of routes	Machine name
(1)	The 2nd Yokohama Bypass South Yokohama Bypass	441	3	1,323	128	178	569	Facom 230-20
(2)-1	Kan-etsu Expressway (1975)	842	3	2,526	287	157	1,129	Facom 230-50
(2)-2	Kan-etsu Expressway (1985)	1,343	3	4,029	287	247	1,631	Facom 230-50
(3)	Kawasaki Urban Expressway	1,664	3	4,992	917	386	2,581	Facom 230-35

References

1. T. Kitayama, K. Toyama, and Y. Horie, Traffic assignment in road network applied a new technique (in Japanese), *Expressways and Automobiles* 10 (8) (1967).
2. A. Kato, Application of traffic flow analysis in road network planning (in Japanese), *Highway* (Sept. & Oct. 1964).

3. T. Sasaki, *Theory of Traffic Flow* (in Japanese), Gijutsu-shoin, 1965.
4. T. Hoshino, Theory of Traffic Assignment in Road Network (in Japanese), Proceedings, The 5th Japan Road Congress, 1959; Theory of traffic assignment in road network (in Japanese), *Highway* (April & May, 1963).
5. K. Hirahara, On Estimation of the Rate of Highway Using (in Japanese), Proceedings, The 6th Japan Road Congress, 1961.
6. N. Sakashita, Microscopic theory of traffic assignment (in Japanese), *Expressways* 5 (8) (1962).
7. *Traffic Assignment Manual,* Bureau of Public Roads (June 1964), Office of Planning, Urban Planning Division.
8. J. D. Carrol, A Method of Assignment to an Urban Network, H. R. B. Bulletin 224 (1959).
9. Robert Smock, A Comparative Description of a Capacity Restrained Traffic Assignment, H.R.B. Record No. 6 (1963).
10. Final Report of the Highway Transportation Demand Research Project, Research Report R 65-24 MIT (1965); B. V. Martin and M. L. Manheim, A Research Program of Traffic Assignment Techniques, H.R.B. Record No. 88
11. N. A. Irwin, D. Norman, and H. G. Von Cube, Capacity Restraint Programs, H.R.B. Bulletin 297 (1961).
12. W. W. Mosher, A Capacity Restrained Algorithm for Assigning Flow to a Transport Network, H.R.B. Record No. 6 (1963).
13. C. Abraham, La Repartition du Traffic entre Itineraires Concurrents, Revue General des Routes et des Aerodromas, Oct. (1961).

Nine Estimators of Gap-Acceptance Parameters

Alan J. Miller
Transport Section, University of Melbourne, Melbourne, Australia

Abstract

There is a vast array of different methods available for analyzing gap-acceptance data. This paper contains a comparative study of a selection of nine of these methods.

For the purpose of comparison, the following simple model is used. Each person has a critical size of gap. These critical gaps vary from person to person. One person always rejects gaps less than his or her critical gap, and always accepts larger gaps.

Desirable properties of estimators are discussed. Each method in turn is described. To compare methods, 100 sets of artificial data have been generated. Each method has been applied to each set of data. Of the nine methods, four only use information on lags accepted or rejected, while the other five methods use all lags and gaps considered. Five of the methods give results which are seriously biassed and cannot be recommended unless corrections for the bias are made. The method suggested by Ashworth (Ref. 4) is fairly efficient and can be recommended provided that the offered gaps are not highly correlated. The method of maximum likelihood is shown to give satisfactory results.

A few ideas for handling more realistic models are given at the end of the paper.

Principal Notation

There appears to be no accepted notation in this field. The following notation is adapted from that used by McNeil and Morgan.[1] In keeping with standard statistical practice, upper case letters are used for the *names* of variables, lower case letters denote *values* of variables. Where a lower-case letter is used for a probability density function (p.d.f.), the upper-case letter denotes the cumulative frequency function.

X, x = critical gap of a person,
A, a = largest gap or lag rejected by a person,
B, b = gap or lag accepted by a person,
R, r = number of gaps, including the lag, rejected by a person,
$f( )$ = p.d.f. of critical gaps (or lags),
$g( )$ = p.d.f. of gaps in the traffic stream,
$g_0( )$ = p.d.f. of lags [used when this differs from $g( )$],
Q, q = traffic flow.

The Model

Drivers and pedestrians are required to decide whether to accept or reject gaps in many situations including the following:

(i) pedestrians waiting to cross a road;

(ii) unsignalized intersections at which drivers in one stream wait for a gap in another stream;

(iii) signalized intersections at which there is no separate phase for turning across oncoming traffic;

(iv) merging, weaving and lane changing on freeways;

(v) overtaking on two-lane, two-way roads.

While each of these situations gives rise to special problems of its own, there is much in common between them. In each case there is some time, t_0, at which the driver or pedestrian first wants a gap. Then consecutive vehicles in the stream in which the gap is sought, pass at times $t_1, t_2, \ldots$. The time interval from t_0 to t_1 is usually called a lag; the other time intervals are called gaps. For each person seeking a gap, a sequence of lag and gaps rejected (if any) is recorded together with the gap accepted (if any). This is sometimes augmented with ancillary information on speed and type of vehicle, sex and age of pedestrian or driver, etc. Occasionally distances (which are what people actually perceive) are measured instead of times.

For clarity, the following simple model is used throughout this paper. Each person has a critical size of gap. These critical gaps vary from person to person. One person always rejects gaps smaller than his or her critical gap and always accepts larger gaps. Lags are treated in the same way as gaps in this paper. That is it is assumed that the critical lag for a person is identical with that person's critical gap.

This model is of course a gross simplification of any gap-acceptance situation. This is deliberate. Only in this way can the basic advantages and disadvantages of the many different estimators be seen. This paper is concerned with the efficient estimation of parameters not with the comparison of models. Efficient and understood methods of parameter estimation are of course necessary before hypotheses can be tested and models compared. At the end of this paper are a few ideas for handling data analysis using more realistic models than that just described.

The Importance of Accuracy

The parameters of gap-acceptance distributions are required in formulae for delays to persons waiting for gaps. Probably the simplest of such formulae is that derived by Adams[2] for the average waiting time for a gap $> x_0$ in a random traffic stream. If the traffic flow is q vehicles per unit time, then the average waiting time is

$$\{\exp(qx_0) - 1 - qx_0\}/q. \tag{1}$$

If (qx_0) is fairly small, say less than one, the exponential can be expanded to show that the average waiting time is approximately

$$\tfrac{1}{2} qx_0^2, \tag{2}$$

i.e., the average waiting time increases approximately as the square of the minimum gap required.

When the critical gap, X, varies from person to person, the average waiting time for a gap is obtained by averaging (1) over x_0. If the flow is not too high, then the same type of expansion can be used. This gives a similar result to (2) but with the x_0^2 replaced by the second moment of the distribution of X. Thus the average waiting time depends strongly upon the mean and variance of the gap-acceptance distribution, but is fairly insensitive to the shape of this distribution except for heavy traffic.

A Common Bias

Before proceeding to look at the properties of the nine estimators considered here, it is necessary to point out a common source of bias. We are interested in the proportion of people who will accept a given size of gap. In general this proportion is greater than the proportion of gaps of that size which are accepted. This important difference was probably first pointed out by Raff (Ref. 3, footnote to pages 27, 28), and the difference has since been quantified by Ashworth.[4]

The reason for this difference is that the person who wants a short gap will often accept the lag, while the person who wants a long gap will often reject the lag and many gaps before obtaining an acceptable gap. If we look at all gaps which are considered, the cautious person is found to be very much over-represented compared with the risk-taker.

This difference is the main reason why many of those who have analyzed gap-acceptance data have only used the lag for each driver and have neglected information on any subsequent gaps considered by people.

The Estimators

Strictly speaking, more than nine estimators are compared in this paper. Nine basically different *methods* have been investigated. One of these methods yields just one estimator of some kind of central value, the others yield a mean and a standard deviation, and one method estimates the complete frequency function of the distribution of critical gaps between drivers. As all the methods yield an estimate of a central value, most of the comparisons are based upon these estimators.

The nine methods are each described below. In most cases, it is extremely tedious to try to obtain exact results, or there are approximate analytical results, the accuracy of which is unknown. To compare methods, a large amount of artificial data has been generated and each method in turn has been used on this data. This simulation and its results are given after the description of the methods of estimation.

In order to be able to pick a best estimator, we need to decide what properties are desirable. In this case, they seem to be:

   (i) Estimates should be fairly close to the true (but usually unknown) value.
   (ii) Estimates should be insensitive to the distribution of gaps offered.
   (iii) Estimates should be insensitive to the distribution of critical gaps.

Closeness to the true value can be measured in many ways. The simulation results give the distribution of estimators about the true value. These are summarized in two statistics, the mean and the standard deviation for each estimator. The difference between the mean value of an estimator and the true value is called the bias. Most methods give estimates which are biassed. This is not a serious disadvantage provided that we can estimate the bias and correct for it. A method suggested by Ashworth[4] is known to give biassed results but incorporates an approximate correction for the bias.

Several of the methods are sensitive to the distribution of gaps offered. This sensitivity is particularly important in testing, for instance, whether people are prepared to accept shorter gaps in heavy traffic than in light traffic.

A disadvantage of the method of maximum likelihood is that a specific distribution must be assumed for the critical gaps. Quite often though, estimators obtained by using the method of maximum likelihood are still good estimators even though the distribution which was assumed in order to obtain them was unrealistic. This is not always the case, but it applies here.

Method 1 (Raff)

The method of analyzing gap-acceptance data which was proposed by Raff[3] is probably still the most commonly used. In this method only lags are considered. Only one parameter is estimated, and this was called the "critical lag" by Raff. This was defined by Raff as follows:

> "This critical lag x_0 is the size lag which has the property that the number of accepted lags shorter than x_0 is the same as the number of rejected lags longer than x_0."

(N.B. The notation has been changed from that used by Raff.)

What is the "critical lag"? Is it the lag which 50% of people will accept, or is it the mean, or mode of the gap-acceptance distribution? Nobody seems to have examined this question previously.

Consider a large number of people, N, who each consider one lag. If $f(x)$ is the probability density function of critical lags, x, between people, and $g_0(l)$ is the probability density function of the lags offered to the N people, the expected number of accepted lags which are shorter than some lag x_0 is

$$N \int_0^{x_0} f(x) \{G_0(x_0) - G_0(x)\} dx \qquad (3)$$

since $f(x) dx$ is the probability that any person's critical gap is in the range $(x, x + dx)$, and $\{G_0(x_0) - G_0(x)\}$ is the probability that the gap offered to that person is greater than x but less than x_0. Similarly the expected number of rejected lags greater than x_0 is

$$N \int_{x_0}^{\infty} f(x) \{G_0(x) - G_0(x_0)\} dx \qquad (4)$$

Raff's critical lag is obtained by equating actual numbers in a sample, not expected numbers, but as N tends to infinity, we would anticipate that the sample critical lag will tend towards that value which equates formulas (3) and (4) above. (N.B. For the rigorous mathematician, this does require assumptions of continuity of both f and g_0.) Equating (3) and (4) and rearranging shows that as N tends to infinity, Raff's critical lag x_0 tends to the solution of

$$G_0(x_0) = \int_0^{\infty} f(x) G_0(x) dx. \qquad (5)$$

For any given pair of distributions f and g_0, numerical solutions of (5) can easily be found. Approximate solutions can be obtained by expanding $G_0(x)$ about the mean, μ, of the gap-acceptance distribution.

$$G_0(x) \simeq G_0(\mu) + (x - \mu) g_0(\mu) + \tfrac{1}{2} (x - \mu)^2 g_0'(\mu)$$

Substituting this in the right side of (5) and integrating gives

$$\text{R.H.S.} \simeq G_0(\mu) + \tfrac{1}{2} \sigma^2 g_0'(\mu).$$

If we expand the left side of (5) as

$$\text{L.H.S.} \simeq G_0(\mu) + (x_0 - \mu) g_0(\mu),$$

then we have that Raff's critical lag is approximately

$$x_0 = \mu + \tfrac{1}{2} \sigma^2 g_0'(\mu)/g_0(\mu). \qquad (6)$$

If the distribution of lags is exponential, this reduces simply to

$$x_0 = \mu - \tfrac{1}{2} q \sigma^2, \qquad (7)$$

where q is the traffic flow.

At an unsignalized intersection, the values of μ and σ for drivers on the minor road waiting for gaps in the major road traffic may be say 5 and 2 sec, respectively. The bias in using Raff's critical lag as an estimator of the mean of the drivers gap-acceptance distribution is -0.11 sec if $q = 200$ veh./h, or -0.33 sec if $q = 600$ veh./h. These biasses are fairly small so that the method is a reasonably satisfactory one when the flow is fairly light in the traffic in which the lag is being sought.

It is desirable to have confidence limits for any estimator. The distribution of Raff's critical lag can be found, but it is very cumbersome mathematically. For the purpose of this paper, the distribution of x_0 has been obtained by simulation. Details are given later.

Method 2 (Probit Analysis)

If we again look only at lags, and the size of lag offered to a person is independent of that person's critical lag, then the probability that a lag of size x is accepted is just $F(x)$, that is the probability that the lag is offered to someone whose critical lag is less than x. The proportion of lags x (or usually within a small range of x) which are accepted gives an estimate of $F(x)$.

Cohen et al.[5] have argued that the distribution of critical lags between people can be expected to approximate a log-normal distribution. If this is so, when the proportion of lags of each size (grouped in say $\frac{1}{2}$ or 1-sec intervals) which are accepted is plotted on normal-probability paper against the logarithm of the lag size, the points should lie roughly on a straight line. Fairly good straight-line fits have been obtained by Robinson,[20] Cohen et al.,[5] Bissell,[6] and Solberg and Oppenlander.[7] Alternatively, Herman and Weiss[21] fitted a displaced exponential distribution to their experiment data, while Blunden, Clissold and Fisher fitted gamma distributions to their data (Ref. 10).

Probit analysis [8,9,16] is a standard statistical technique for fitting weighted linear regression lines to such data. The proportions are transformed to probits. The probit corresponding to any proportion is the number of standard deviations away from the mean such that the cumulative normal probability at that point equals the proportion. Five is usually added to the number of standard deviations to ensure that probits are nearly always positive; this practice dates from the days when such analyses were usually carried out on desk calculating machines. For instance, for the normal distribution, 10% of values are more than 1.28 standard deviations below average. The probit corresponding to a proportion of 10% is then 5 − 1.28 = 3.72.

Provided that the class interval is fairly small, probit analysis yields an unbiassed estimate of the lag which 50% of people will accept. The gradient of the regression line is asymptotically unbiassed, but for finite sample sizes there is a nonzero probability of an infinite gradient. This occurs if all the lags up to a certain size are rejected, and all of those greater are accepted. This probability rapidly tends to zero as the sample size increases. The reciprocal of this gradient gives an estimate of the standard deviation of the gap-acceptance distribution. Such estimates are biassed slightly, but the bias tends to zero with increasing sample size.

Method 3 (Ashworth)

A serious disadvantage of methods which utilize information on lags only is that a great deal of information is discarded. Also it is usual that gaps can be measured more reliably than lags, so that methods using only lags are using the less reliable data. Ashworth[4] has shown one way in which the information on gaps accepted or rejected can be used. He has shown that if consecutive gaps (and lags) are independent and have an exponential distribution, the average number of gaps, including the lag, considered by someone who wants a gap x, is exp (qx), where q is the flow. The probability, $P(t)$, that a lag or gap t is accepted is then

$$P(t) = C \int_0^t f(x) \, e^{qx} \, dx \qquad (8)$$

where

$$C = \left[\int_0^\infty f(x) \, e^{qx} \, dx \right]^{-1}.$$

Ashworth shows that if $f(x)$ is a normal distribution, then $P(t)$ is the cumulative frequency function for a normal distribution with the same standard deviation σ as $f(x)$ but with the mean increased by $q\sigma^2$. The probabilities, $P(t)$, can be estimated by finding the proportions of lags and gaps which are accepted for small ranges of t. Probit analysis can then be used to fit a cumulative normal distribution function to the proportions. This distribution can then be shifted to the left by $q\sigma^2$ (using appropriate estimates of q and σ) to obtain estimates of the parameters of $f(x)$.

If $f(x)$ is not a normal distribution, $P(t)$ can still be obtained from (8). For instance, if the critical gaps have the gamma distribution:

$$f(x) = \frac{\beta^{n+1}}{n!} \, x^n \, e^{-\beta x}, \qquad (9)$$

then

$$p(t) = \frac{(\beta - q)^{n+1}}{n!} \, t^n \, e^{-(\beta - q)t}, \qquad (10)$$

provided $q < \beta$. If the means and standard deviations of the distributions are μ_f, μ_p, σ_f, σ_p, then

$$\mu_p = \mu_f/(1 - q/\beta). \qquad (11)$$

and

$$\sigma_p = \sigma_f/(1 - q/\beta). \qquad (12)$$

That is, the mean and standard deviation of $p(t)$ are in the same ratio as those of $f(x)$. Notice also that

$$\mu_p = \mu_f/(1 - q\sigma_f^2/\mu_f)$$
$$\simeq \mu_f + q\sigma_f^2. \qquad (13)$$

That is, the shift of the mean is approximately the same as for the normal distribution.

Unfortunately if $f(x)$ is a log-normal distribution, $p(t)$ is not, though for prac-

tical purposes it is very similar to a log-normal. By appropriate expansions and
approximations, including truncating the range, it can be shown that the result
(13) is still a fairly good approximation, and that the coefficient of variation for
$p(t)$ is very close to that for $f(x)$.

Method 4 (Blunden, Clissold & Fisher)

Blunder, Clissold and Fisher[10] attempted to estimate the distribution of critical gaps, though in fact they were looking at the proportion of *gaps* of a given
size which are accepted rather than the proportion of *people* (drivers in their
case) who would accept that size of gap. Looking at proportions of gaps and
lags accepted in $\frac{1}{2}$-sec class intervals, they found that the proportions fluctuated
violently from one class to the next as a result of small sample sizes. To overcome these fluctuations, they decided to "build up the sample size." This was
done by assuming that everyone who accepted a lag or gap in the range $(t, t + \frac{1}{2})$
would accept lags or gaps in all larger classes, and that anyone who rejected lags
or gaps in the range $(t, t + \frac{1}{2})$ would reject any smaller lag or gap. The proportion of drivers who would accept a gap t was then estimated as $n_1(t)/\{n_1(t) +
n_2(t)\}$ where $n_1(t)$ was the number of drivers who accepted gaps less than or
equal to t, and $n_2(t)$ was the number who rejected gaps greater than or equal to
t. These proportions then increase monotonically for any sample size. Cumulative Erlang distributions were then fitted to these proportions.

In this paper, this method will be applied to lags only, so that each person is
represented once only, and a cumulative log-normal distribution will be fitted by
probit analysis to the proportions.

Method 5 (Drew)

Drew[11] has used three different methods to estimate parameters of gap-acceptance distributions for freeway merging. The first was Raff's method, the second
was the Blunden, Clissold and Fisher method, but fitting a cumulative log-normal
distribution by probit analysis, and the third will be called Drew's method here.

If someone rejects a 3-sec gap, then accepts a 5-sec gap, we know that that person's critical gap is between 3 and 5 sec provided we accept the basic model used
throughout most of this paper. If this person also rejects a 2-sec gap, it gives us
no further information on his or her critical gap. Drew considered all drivers
who rejected at least a lag. He then formed a "histogram" in which, if a driver's
largest rejected gap or lag was a and his accepted gap was b, the frequency in
each histogram class between a and b was increased by one. He then treated this
as a standard histogram and found its mean and standard deviation.

In this paper, the class intervals will be made infinitesimally small. Someone
whose largest rejected lag or gap is a, and whose accepted gap is b, then contrib-
utes an area $(b - a)$ to the histogram.

Method 6 (Dawson)

In the discussion of Drew's paper,[11] Dawson has suggested two further methods. One disadvantage of Drew's method is that someone who rejects a 1-sec lag then accepts a 21-sec gap is given 20 times as much weight as someone who rejects a 4-sec lag then accepts a 5-sec gap.

Dawson's first suggestion was that weight one should be given to a value midway between a and b in the "histogram." His second suggestion was that weight $1/(b - a)$ should be added to the "histogram" over the range from a to b. Clearly both methods will give the same mean for the "histogram" but the second method gives a larger standard deviation. The second method is the one referred to as Dawson's method in this paper.

Method 7 (Miller)

This method was put forward in a letter to E. M. Holroyd,[12] who promptly replied[13] that it does not give correct answers!

As with methods 3, 5, and 6, it was an attempt to use more information than just the value of the lag and whether it was accepted or rejected. If we look at the proportion of all lags and gaps of a particular size which are accepted then we find the cautious person who rejects many gaps is over-represented. The idea behind this method was that if a person considers say 4 gaps (including the lag), then a weight equal to $\frac{1}{4}$ should be attached to each gap or lag whether it was accepted or rejected. This means that the total weight given to each person is the same. To estimate the proportion of people who will accept a gap in a small range, say 5 to 5½ sec, the weights associated with all gaps and lags in this range are summed, firstly for just those lags and gaps which are accepted and then for all of those considered. The ratio of these two sums gives a proportion which it was hoped was an estimate of the proportion of people who would accept gaps in the range. A smooth curve can then be fitted through the set of proportions by regression (e.g., by probit analysis).

It can readily be shown that this method leads to very biassed results. Further, both Dr. G. F. Yeo and Dr. M. C. Dunne have shown in private correspondence that it is not possible to find weights which are functions of R only which lead to unbiassed estimators.

This method has only been included here, as in the simulation results the estimate of the mean critical gap had the smallest variance of all the methods. If a method of bias correction can be found, we may obtain a good estimator.

Method 8 (McNeil & Morgan)

McNeil and Morgan[1] have given two methods for estimating parameters of gap-acceptance distributions. In each method the distribution of critical gaps is first estimated, though this estimation is implicit rather than explicit, and the moments are then estimated from the distribution. The first of their methods, which is method 8 in this paper, uses only information relating to lags. The second method uses the largest rejected gap (or lag) and the accepted gap in cases where the lag is rejected. The second method gives an infinite bias in estimating the mean critical gap.

The argument behind McNeil and Morgan's first method is as follows. The probability that a lag is in a small range $(t, t + \delta t)$ *and* is rejected is

$$g_0(t) \, \delta t \, [1 - F(t)] + o(\delta t). \tag{14}$$

Now this probability can be estimated by using the proportion of all lags which are in the range $(t, t + \delta t)$ *and* are rejected. Let this proportion be denoted by $h(t)$, then from (14)

$$\frac{h(t)}{g_0(t) \, \delta t} \tag{15}$$

can be used as an estimate of $[1 - F(t)]$, that is of the porportion of people whose critical gaps are greater than t. Then the mean of the critical gap distribution is given by

$$\mu = \int_0^\infty t f(t) \, dt$$

$$= \int_0^\infty [1 - F(t)] \, dt. \tag{16}$$

If we replace $[1 - F(t)]$ by its estimate (15), and let $\delta t \to 0$, we then have the following estimator for μ,

$$\frac{1}{n} \sum_{i=1}^n u_i \, [g_0(t_i)]^{-1}, \tag{17}$$

where $u_i = 0$ if the lag is accepted and 1 if it is rejected, and $t_i, i = 1, 2, \ldots, n$ are the lags offered to n persons. This is obtained by making δt so small that only one lag is within each range. Then $h(t_i) =$ either $0/n$ or $1/n$.

To use this method, the distribution of lags must be known or estimated. McNeil and Morgan show how the lag distribution is related to the gap distribution, and use this to obtain good estimates of the lag distribution. They also show how higher moments of the gap-acceptance distribution can be obtained using similar formulas to (16) and the same estimates of $[1 - F(t)]$.

Formula (16) can be used in conjunction with any method which gives estimates of $[1 - F(t)]$. Table I gives the data from Raff's intersection *C*. The proportions p_i are estimates of $[1 - F(t_i)]$ where t_i is the midpoint of the ith range. Since these proportions are from binomial trials, the variance of each estimate is $F(t_i) \, [1 - F(t_i)]/n_i$. The last column of Table I gives estimates of these vari-

TABLE I

Lags Offered and Rejected at Raff's Intersection *C*, Plus the Calculation of the Mean
Critical Lag and Its Variance

Range of lags (sec)	Number of lags (n_i)	Number rejected	Proportion rejected (p_i)	$\dfrac{p_i(1-p_i)}{n_i}$
0–0.9	224	220	0.98	0.0001
1–1.9	265	259	0.98	0.0001
2–2.9	254	232	0.91	0.0003
3–3.9	251	214	0.85	0.0005
4–4.9	231	172	0.74	0.0008
5–5.9	220	139	0.64	0.0010
6–6.9	174	63	0.36	0.0013
7–7.9	182	56	0.31	0.0012
8–8.9	177	35	0.20	0.0009
9–9.9	161	17	0.10	0.0006
10–10.9	125	15	0.12	0.0008
11–11.9	123	1	0.01	0.0001
12–12.9	95	2	0.02	0.0002
13–13.9	95	1	0.01	0.0001
14–14.9	103	2	0.02	0.0002
Over 15	961	4	–	–
Totals	3,641	1,432	6.25	0.0082

ances. Since these proportions are independent, the variance of their sum is
equal to the sum of their variances. The estimated mean critical lag is then 6.25
sec with a variance = 0.0082 (sec)^2. Approximate 95% confidence limits for the
mean are then 6.25 ± 0.18 sec. Raff's method gave a "critical lag" of 5.89 sec
for this data; a value which is well outside the confidence limits for the mean.

Method 9 (Maximum Likelihood)

Maximum likelihood is a standard statistical technique for estimating parameters of distributions, yet there appear to be only two instances of its use to estimate the parameters of gap-acceptance distributions. The first of these was by
Moran[14] who looked only at the lags accepted and rejected. He showed that if
both the distribution of offered lags and of peoples' critical lags are normal, then
the method of maximum likelihood leads to probit analysis. In fact the method
of maximum likelihood estimation applied to the lags only is probit analysis irrespective of the distribution of offered gaps.

The other use of maximum likelihood was made by Miller and Pretty.[15] In
this case, gap-acceptance in overtaking on rural two-lane roads was being studied.
It was found that the log-normal distribution for critical gaps did not fit well as a
small percentage of drivers did not want to overtake, and would not accept any
gap. An extra parameter had to be introduced into the model and estimated for
such drivers. In that study, the gaps were to the next oncoming vehicle on

straight level roads. Subsequent studies (unpublished) have shown that when sight distance is the factor limiting overtaking, it is necessary to add another parameter for the proportion of drivers who will overtake with almost zero sight distance.

For the simple model of the present paper, the probability that the largest lag or gap rejected by a person will be in the small range $(a \pm \frac{1}{2} \delta a)$ and the gap accepted will be in a small range $(b \pm \frac{1}{2} \delta b)$ can be shown to be

$$\delta a \, \delta b \int_a^b f(x) \sum_{r=1}^\infty r g(a) \left[G(a) \right]^{r-1} g(b) \, dx, \qquad \text{if } a \neq 0, \tag{18}$$

$$\delta b \int_0^b f(x) g(b) \, dx, \qquad \text{if } a = 0. \tag{19}$$

In deriving (18) we have to consider all persons with critical gaps, x, between a and b, and then find the probability that of the first r gaps (including the lag), one is in the range $(a \pm \frac{1}{2} \delta a)$ and the other $(r - 1)$ are all less than a. This has to be multiplied by the probability that the next gap is in the range $(b \pm \frac{1}{2} \delta b)$, and finally all values of r have to be considered. Performing the summation and integration in (18) gives

$$\delta a \, \delta b \, \{ F(b) - F(a) \} \frac{g(a) \, g(b)}{\left[1 - G(a) \right]^2} . \tag{20}$$

(N.B. It has been assumed that $g_0(a) = g(a)$ and that consecutive lags and gaps are independent. Most of the results which follow do not require either assumption.)

Similarly integrating (19) gives

$$\delta b \, F(b) \, g(b). \tag{21}$$

If the δa and δb are dropped, we have the "likelihood" of the data pair (a, b).

If we have a sample of n persons with values of (a_i, b_i) associated with each, where $a_i = 0$ if the lag was accepted and $b_i = \infty$ if no gap was accepted, then the likelihood of this sample of data is

$$\prod_{i=1}^n \{ F(b_i) - F(a_i) \} \prod_{i=1}^n \frac{g(a_i) \, g(b_i)}{\left[1 - G(a_i) \right]^2} , \tag{22}$$

provided that we define $g(0) = 1$.

The method of maximum likelihood finds the values of the parameters which maximize the likelihood, or for practical reasons, those which maximize the logarithm of the likelihood. We are interested in the parameters of the gap-acceptance distribution which appear in only the left product in (22), and so the second product is a constant for any given set of data. Taking logarithms of (22), the log-likelihood L is

$$L = \text{constant} + \sum_{i=1}^{n} \log_e \{F(b_i) - F(a_i)\}. \tag{23}$$

To maximize L it is necessary to assume a distribution for the critical gaps. Let us assume that the logarithms of peoples' critical gaps have a normal distribution with mean μ and standard deviation σ. Then it can be shown (para. 59 of Ref. 15), that

$$\frac{\partial L}{\partial \mu} = - \sum_{i=1}^{n} \frac{f(b_i) - f(a_i)}{F(b_i) - F(a_i)}, \tag{24}$$

$$\frac{\partial L}{\partial \sigma^2} = - \frac{1}{2\sigma^2} \sum_{i=1}^{n} \frac{(b_i' - \mu) f(b_i) - (a_i' - \mu) f(a_i)}{F(b_i) - F(a_i)}, \tag{25}$$

where $a_i' = \log_e a_i$ and $b_i' = \log_e b_i$. The maximum likelihood estimates for μ and σ are obtained by setting these two derivatives equal to zero and solving iteratively.

The estimates of μ and σ obtained by the use of maximum likelihood are biassed, but the bias appears to be extremely small except for unnatural gap distributions (e.g., with no offered gaps between say 4 and 6 sec), or for very small sample sizes (say less than 10). Under conditions which certainly apply here (see, e.g., Chapter 5 of Ref. 18), as the sample size increases, the bias in maximum likelihood estimators tends to zero. It is hoped to be able to explore the properties of these estimates more fully.

Comparison of Estimators Using Simulation

To compare the nine methods just described, 100 sets of artificial data have been generated. In each set of data generated there were 100 persons. These people each had a critical gap, and these critical gaps had an overall distribution which very closely approximated a log-normal distribution. The critical gaps were selected so that there was one in each percentile of the log-normal distribution; the position within the percentile was selected using tables of random numbers. The same set of critical gaps was used for all the data sets.

Each person was offered gaps sampled from a displaced exponential distribution until they accepted one. The displaced exponential distribution had a minimum headway of 1 sec and a mean of 5 sec. No distinction was made between lags and gaps; that is they were both sampled from the same distribution.

Each method in turn was then applied to each data set. For those methods which require that gaps accepted or rejected be grouped into classes, the following classes were used: $<2, 2 - 2.5, 2.5 - 3, 3 - 3.5, 3.5 - 4, 4 - 5, 5 - 6, 6 - 7, 7 - 8, 8 - 9, 9 - 10, 10 - 12, 12 - 15,$ and > 15 sec. In using probit analysis, the mid-points of these classes were taken as 1.5, 2.25, 2.75, 3.25, 3.75, 4.48, 5.48, 6.48, 7.48, 8.48, 9.48, 10.92, 13.31, and 19.0 sec. These values are the averages of the gaps offered in each of the classes.

In those methods in which the analysis was performed on logarithmic values, the mean critical gap was estimated as

$$\exp \{\bar{x} + \tfrac{1}{2}s^2\}, \tag{26}$$

where $\bar{x}, s^2$ were the estimates of the mean and variance of the logarithmic values. Aitchison and Brown (formula 5.38, p. 45, Ref. 16) give a slightly better transformation, but as $\tfrac{1}{2}s^2$ is small compared with $\bar{x}$, the benefit from using the more elaborate transformation did not seem justified.

In method 3, the bias correction used was $q\sigma_0^2$ where σ_0^2 was the variance of critical gaps, even though the gap distribution was not exponential, and the critical gap distribution was lognormal not normal. q was estimated as the reciprocal of the average of all gaps generated, and the estimate used for σ_0^2 was

$$\exp (2\bar{x} + s^2) \{\exp (s^2) - 1\}. \tag{27}$$

The justification for this is that if the logarithm of critical gaps have mean μ and standard deviation σ, then the mean and standard deviation of the critical gaps are $\exp (\mu + \tfrac{1}{2}\sigma^2)$ and $\exp (\mu + \tfrac{1}{2}\sigma^2) \sqrt{\{\exp (\sigma^2) - 1\}}$ respectively (see, e.g., Ref. 16, p. 8 or Ref. 17, p. 63). The bias corrections for the 100 sets of data ranged from 0.24 to 1.22 sec, with a mean of 0.61 sec.

In method 8, the distribution of offered lags was taken as the displaced exponential distribution with an average equal to the average of all gaps generated.

Simulation Results

The parameter which can be compared most readily is the estimate of the mean critical gap. Figure 1 shows the histograms obtained for the estimates of the mean critical gap for the nine methods. The actual value being estimated is 5.1667 sec. From this figure it is clear that methods 1, 4, 5, 6, and 7 give biassed results. The methods which give least variation in the estimates are 3, 7, and 9. The means and standard deviations of the estimates are given in Table II.

TABLE II

Means and Standard Deviations of the 100 Estimates of the Mean Critical Gap

Method	1	2	3	4	5	6	7	8	9
Mean (sec)	4.956	5.196	5.144	5.029	7.887	6.580	4.623	5.148	5.156
St. dev. (sec)	0.321	0.308	0.197	0.287	0.638	0.315	0.168	0.332	0.173

Methods 1, 2, 4, and 8 use only lag information and so would be expected to give more variation than methods using all gaps. The standard deviations of the estimates given by these methods are all about 0.3 sec. The standard deviations of the estimates from method 3, 7, and 9 are only about 60% of this. The relative advantage is using estimators based upon all gaps is directly related to the

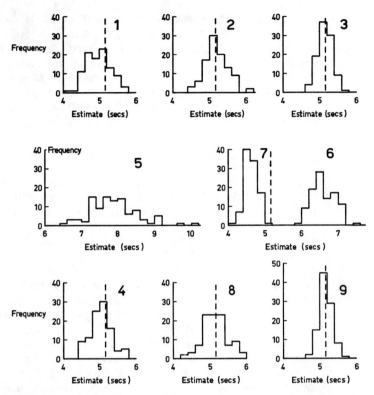

Fig. 1. Histograms of nine estimators of the mean obtained from 100 sets of artificial data. Broken line denotes the true mean.

average number of gaps per person considered, and this in turn depends upon the traffic flow. The average number of gaps per person in this simulation was 3.08.

For all methods except Raff's, an estimate of the standard deviation of critical gaps can be obtained. In the methods which make use of probit analysis, the reciprocal of the gradient of the regression line gives an estimate of the standard deviation of the logarithms of the critical gaps. Table III shows the means and standard deviations of these estimates. Method 8 can give negative variance es-

TABLE III

Means and Deviations of the 100 Estimates of the Standard Deviation of Critical Gaps

Method	2	3	4	5	6	7	8	9
Mean (sec)	1.59	1.61	1.11	4.21	3.20	1.39	1.51	1.54
St. dev. (sec)	0.40	0.19	0.24	0.84	0.34	0.20	0.60	0.16

timates. In the five cases in which this occurred, the estimate of the standard deviation has been taken as zero. The actual value being estimated is 1.5507 sec.

In the case of estimated standard deviations, we see that methods 4, 5, 6, and 7 obviously give very biassed results. Compared with these four methods, the estimates for method 3 look reasonable, but this mean is significantly large at the 1% level. Methods 2, 8, and 9 give estimates which are either unbiassed or have very small biasses. Of these, clearly the estimates given by method 9 (maximum likelihood) are much more reliable than those of methods 2 and 8.

Sensitivity of Maximum Likelihood Estimates

From the above results it appears that maximum likelihood applied to the largest rejected gap and the accepted gap (method 9) is the most satisfactory of these methods. Ashworth's method (no. 3) is a good runner-up. In the 100 sets of data generated, Ashworth's estimate of the mean was closer to the true value in 36 sets, the maximum likelihood estimate was closer in 63 sets, and there was one tie.

One of the disadvantages of using maximum likelihood to estimate parameters is that distributional assumptions must be made. Sometimes the estimators obtained are fairly insensitive to the true distribution(s)—they are "robust" in statistical jargon—in other cases they are very sensitive. For instance, in estimating the proportion of drivers who would not overtake in Ref. 15, this proportion was probably very sensitive to the assumed distribution of critical gaps.

To test the sensitivity of the estimators in the present application of maximum likelihood, it was necessary either to change the actual distribution of critical gaps, or to change the assumed distribution and hence the estimation equations. The second alternative was chosen as it could be carried out very simply by assuming a normal rather than a log-normal distribution for the critical gaps. In the simulation program, the only change necessary was that the maximum likelihood procedure was carried out first without taking logarithms of the largest rejected gap and the accepted gap. Figure 2 shows these two distributions, each with mean = 5.167 sec and standard deviation = 1.551 sec.

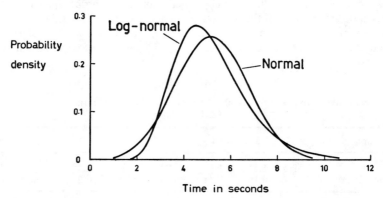

Fig. 2. Normal and log-normal distributions with the same mean and standard deviation.

For 97 out of the 100 data sets, the mean critical gap estimated assuming a normal distribution was larger than that obtained assuming a log-normal distribution. The differences between estimates for the same data set ranged from −0.02 sec to +0.11 sec, with an average of +0.040 sec. The differences between estimates of the standard deviation ranged from −0.18 sec to +0.05 sec, with an average of −0.044 sec. For 73 of the data sets, the log-normal gave the higher standard deviation estimate.

The effect of changing the assumed distribution clearly made a difference in the estimates, but the order of magnitude of the changes is so small for practical purposes that the estimators can be declared robust.

Extensions to the Model

Firstly let us look at the problem presented by the person who rejects a 6-sec gap then accepts a 4-sec gap. There are several ways of introducing such possibilities into the model according to the type of explanation we want to give for this behavior. If we feel that impatience is the reason, then we could make the critical gap a function of the number of gaps, r, already rejected. Such functions as $\{c + d/(r + 1)\}$ and $c \exp(-dr)$ are possibilities. Maximum likelihood could then be used to estimate the parameters c and d. Unless a moderate number of people accept shorter gaps than they have previously rejected, the estimates of d will be very poor. The equations to be solved will be similar to (24) and (25), and care will need to be taken to restrict the domain of values of c and d to those for which $\{F(b_i) - F(a_i)\} > 0$ for all the data points.

An alternative explanation for this behavior is that people misjudge the sizes of gaps. In most situations, gaps are perceived in terms of distance. The person wanting a gap makes a rough estimate of speed and hence of the size of a gap in time units. If the actual gap is t (in time units), then a person may judge it to be t' say. It seems reasonable to expect that the deviations of t' from t are proportional to t. That is that if people make errors of up to 1 sec in judging a 3-sec gap, they will make errors of up to 3 sec in judging a 9-sec gap. If we take logarithms of the gaps then these errors will be additive rather than multiplicative. Let us assume then that the logarithm of the size of a gap t as perceived by a driver is

$$\log_e t + \epsilon, \tag{28}$$

where the ϵ's are independently sampled from a normal distribution, and that the same normal distribution applies for all persons.

If a person's critical gap is x, he accepts a gap t if

$$\log_e t + \epsilon > \log_e x, \tag{29}$$

i.e., if

$$\epsilon > \log_e x - \log_e t. \tag{30}$$

If we assume that the ϵ's have a normal distribution with mean zero and standard deviation σ, then the probability that a person with critical gap x accepts an offered gap t is the probability that ϵ is greater than the right side of (30). This is then an integral over that part of the normal distribution to the right of the value given by (30).

Let $z(\epsilon)$ be the probability density function of ϵ. We can now write down the likelihood function. This time we have to consider all the gaps offered to a person, not just the largest rejected and the one accepted. If a person rejected gaps of say 3 and 4 sec we do not know that his judged value of the second gap was larger than his judged value of the first.

To simplify the derivation, the gaps offered will be treated as given values, not as random variables as in the earlier derivation. As was seen in formula (22) and had earlier been pointed out by McNeil and Morgan,[1] the probability that a person will be offered any particular size of gap, and the probability that they will accept a gap of that size, are independent. Since we are not interested in the first of these probabilities, it can be omitted from the likelihood formulation.

Let $t_1, t_2 \ldots t_r$ be the gaps, including the lag, rejected by a person, and let t_{r+1} be the lag or gap accepted, where r may be zero. If the person's critical gap

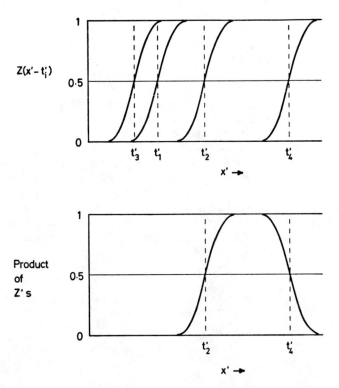

Fig. 3. (a) Probabilities of rejecting each of the gaps t_i', $i = 1, 2, 3, 4$, as a function of critical gap size, x'. (b) Joint probability of rejecting gaps t_1, t_2, t_3, and accepting gap t_4, as a function of x'.

is x, then the probability that he rejects the first r gaps and accepts the next is

$$Z(x' - t'_1) Z(x' - t'_2) \ldots Z(x' - t'_r) \{1 - Z(x' - t'_{r+1})\}, \qquad (31)$$

where $Z(\ )$ denotes a cumulative normal probability and the dashes ($'$) denote logarithms of the values. The likelihood of a person who will reject the above sequence of r gaps and then accept a gap t_{r+1} is then found by considering all possible values of x and is thus

$$\int_0^\infty f(x) Z(x' - t'_1) Z(x' - t'_2) \cdots Z(x' - t'_r) \{1 - Z(x' - t'_{r+1})\} \, dx. \qquad (32)$$

This is the likelihood for one person. Similar quantities can be calculated for each person in a set of data, and multiplied to give the overall likelihood. This can then be maximized numerically with respect to the parameters of the distributions of x and ϵ.

Formula (32) looks very formidable but in fact is little different from what we had before. Figure 3 (a) shows the cumulative normal integrals Z for the case of

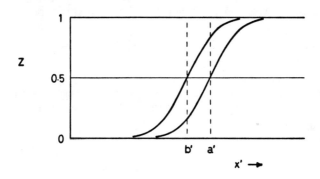

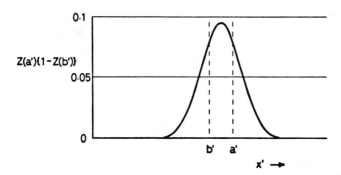

Fig. 4. (a) Probabilities of rejecting each of the gaps a', b', as a function of critical gap size, x'. (b) Joint probability of rejecting a gap a' and accepting a smaller gap b' as a function of x'.

a person who has rejected gaps t_1, t_2, t_3, and accepted t_4, as functions of the critical gap x. The case illustrated is one in which the gap accepted is larger than the largest rejected. Figure 3 (b) shows the product (32) for this case. Notice that in this case the distributions associated with t_1 and t_3 make very little difference to the product. In the previous formulation, the equivalent of the product shown in Fig. 3 (b) was a step function with an up step at t_2 and a down step at t_4. Formula (32) then involves the integration of $f(x)$ weighted by the product shown in Fig. 3 (b). Clearly the integral in this case will be fairly close to $\{F(t_4) - F(t_2)\}$.

When the largest gap rejected (a) exceeds the gap accepted (b), and the other gaps rejected are sufficiently smaller than a or b to have a negligible effect, the situation is as illustrated in Fig. 4 (a) and (b). The vertical scale in Fig. 4 (b) is different from other scales as the product is now very small. In maximizing the likelihood, the absolute magnitude of these products does not matter; what is important is the relative magnitude. The shape of Fig. 4 (b) is very similar to that which is obtained if a and b are interchanged. For practical purposes, if we are only interested in the parameters of the critical gap distribution and not in those of the distribution of ϵ's, method 9 can be used with a and b reversed wherever $a > b$. It is hoped to investigate this approximate method further.

Conclusions

Methods of analysis of gap-acceptance data which use only lag information are inferior to those which make use of all gaps, for two reasons: (i) They use less information and are consequently less precise; (ii) the measurement of lags is usually less precise than that of gaps.

Of the methods using all gaps, only Ashworth's and maximum likelihood give results with a satisfactorily small bias. Method 7 would be a good method if a practical formula for bias correction could be found.

Ashworth's method (method 3) is slightly less precise than maximum likelihood (method 9) but is suitable for use with desk calculating machines. Quick, rough estimates can be obtained by plotting on log-probability paper, fitting straight lines by eye, and then using Ashworth's correction. Another quick variation is to use the method illustrated in Table I using all lags *and* gaps, and then applying Ashworth's correction. The method requires that offered gaps are independent, and may be fairly sensitive to this.

Maximum likelihood estimation is much more laborious, though the author has carried through the iterations for a sample of 69 drivers, using a slide rule and tables of the normal distribution. Standard formulae exist for finding confidence limits for the parameters (e.g., Chapter 5, Ref. 18). This method can be extended to cover other models than that used in this paper.

Since the first draft of this paper was prepared in October 1970, Ashworth's paper[19] has been published. He has derived the result given in formula (7) which is exact with his set of assumptions. Blunden, Clissold, and Fisher [10] describe several methods of analyzing gap-acceptance data. The method which is called method 4 here differs from that by the same authors which was chosen by Ashworth, in that the numbers of rejected gaps have not been adjusted. Ashworth's simulation results were for a range of flows from 600 to 1200 veh./h and show similar results to those obtained here for his method (No. 3) and for methods 5 (Drew) and 6 (Dawson).

References

1. D. R. McNeil and J. H. T. Morgan, Estimating minimum gap acceptances for merging motorists, *Trans. Sci.* **2**, 265–277 (1968).
2. W. F. Adams, Road traffic considered as a random series, *J. Inst. Civ. Eng.* **4**, 121–130 (1936).
3. M. S. Raff, A Volume Warrant for Urban Stop Signs, Eno Foundation, Saugatuck, U.S.A., 1950.
4. R. Ashworth, A note on the selection of gap acceptance criteria for traffic simulation studies, *Trans. Res.* **2**, 171–175 (1968).
5. J. Cohen, E. J. Dearnaley, and C. E. M. Hansel, The risk taken in crossing a road, *Opnal. Res. Qty.* **6**, 120–128 (1955).
6. H. H. Bissell, *Traffic Gap Acceptance from a Stop Sign,* unpubl. res. report, I.T.T.E., University of California, 1960.
7. P. Solberg and J. C. Oppenlander, Lag and gap acceptances at stop-controlled intersections, Highway Res. Record 118, H.R.B., Washington, D.C., 1966.
8. D. J. Finney, *Probit Analysis,* Cambridge University Press, 1947.
9. R. A. Fisher and F. Yates, *Statistical Tables for Biological, Agricultural and Medical Research,* Oliver and Boyd, Edinburgh, 1948.
10. W. R. Blunden, C. M. Clissold, and R. B. Fisher, Distribution of acceptance gaps for crossing and turning manoeuvres, *Proc. Aust. Rd. Res. Board* **1**, (1), 188–205 (1962).
11. D. R. Drew, Gap acceptance characteristics for ramp-freeway surveillance and control, Highway Res. Record 157, H.R.B., Washington, D.C., 1967.
12. A. J. Miller, Letter to E. M. Holroyd, Sept. 1966.
13. E. M. Holroyd, Letter in reply to Ref. 12, Oct. 1966.
14. P. A. P. Moran, Estimation from inequalities, *Aust. J. Stats.* **8** (1), 1–8 (1966).
15. A. J. Miller and R. L. Pretty, Overtaking on two-lane rural roads, Proc. *Aust. Rd. Res. Board* **4** (1), 582–591 (1968).
16. J. Aitchison and J. A. C. Brown, *The Lognormal Distribution,* Monograph no. 5, Dept. of Appl. Economics, University of Cambridge, Cambridge University Press, 3rd ed., 1966.
17. J. C. Tanner and J. R. Scott, Sample Survey of the Roads and Traffic of Great Britain, Road Res. Tech. Paper 62, H.M.S.O., 1962.
18. C. R. Rao, *Linear Statistical Inference and its Applications,* John Wiley, New York, 1965.
19. R. Ashworth, The analysis and interpretation of gap acceptance data, *Trans. Sci.* **4**, 270–280 (1970).
20. C. C. Robinson, Pedestrian interval acceptance, Proc. Inst. *Traffic Eng. (U.S.A.),* 144–150 (1951).
21. R. Herman and G. H. Weiss, Comments on the highway crossing problem, *Operations Res.* **9**, 828–840 (1961).

Theoretical Analysis of Expressway Ramp Merge Controls as Single-Server and Tandem Queues[*][†]

Allan Marcus
Department of Statistics, The Johns Hopkins University, Baltimore, Maryland

Abstract

The formation of a stationary or slow-moving queue at an expressway entrance behind a slow-moving vehicle forces *all* the vehicles in queue to merge at low speed. The large velocity differential increases merging delays and reduces ramp capacity, in addition to creating an accident hazard. This can be improved by adding a "stop", "yield" or "stop here until ramp is clear" (SHURIC) sign at sufficient distance from the ramp nose to allow high speed merging whenever visibility permits. We analyze the performance of ramps with such signs as single-server tandem queues with Poisson input and exceptional service for the first vehicle in a "busy period" at the sign. The ramp travel time is taken as the service time of counter 1, and the delay in merging at the ramp nose as service time at counter 2. For the SHURIC sign these simply add, hence we can use a single-server model; but a tandem queue model is needed for the set-back "stop" sign. It is assumed that a queue can form at the sign, but not on the ramp.

Realistic model functions for ramp delay and merging delay distributions can be constructed from empirical studies and from theoretical analyses of delays for different merging strategies. Prabhu and Neuts have obtained the most general results for bilateral transforms of the waiting time and queue size distributions. Even for the simplest realistic models, it was not possible to invert these transforms to obtain explicit expressions for delay time and queue size distributions in the transient case. Useful asymptotic time dependent expressions for the mean and variance are obtained for traffic intensity $\rho > 1$. Some useful explicit results can be obtained for stationary distributions.

The model functions include generally different ramp travel time distributions with a bonus for jumping the sign and a penalty for stopping at the ramp nose, and an arbitrary probability of always rejecting the initial gap and stopping at the ramp nose. The travel time distributions are assumed to be displaced Gamma (Erlang), and the merging delay distribution for a stopped vehicle assumed to be Gamma (Erlang). Numerical calculations of stationary average waiting and queue size are carried out for displaced exponential and exponential distributions and a plausible range of parameter values.

These expressions can be used to compare the trade-off between increased delay on the ramp and (possible) decreased delay on merging due to use of the SHURIC sign. Numerical calculations for the "stop" and SHURIC sign suggest that for a plausible range of model values, the additional delay at the SHURIC sign in moderate traffic is relatively small and may possibly be offset by safety considerations. In heavy traffic, however, the additional delay at the SHURIC sign may cause the ramp to become saturated ($\rho > 1$).

The model considerably extends and generalizes some aspects of a Markov renewal model of a gap-acceptance entrance ramp controller proposed by Wiener et al. (1970)

[*]Research supported by the United States Department of Transportation, Federal Highway Administration, awarded to the Department of Statistics, The Johns Hopkins University.

[†]The opinions, findings, and conclusions expressed in this publication are those of the author and not necessarily those of the Federal Highway Administration.

References pp. 251-252

237

Introduction

A major difficulty in the use of expressway entrance ramps during periods of moderate and heavy traffic is that vehicles entering the expressway often do so at very low speeds as a result of slowing nearly to a stop at the entrance. One possible remedy, which we will not consider in this paper, is to control traffic flow on the outside lane of the expressway itself. A usually more useful approach is to change the mode of usage of the ramp. A "stop" sign placed on the ramp at a sufficient distance from the expressway entrance to allow acceleration to a safe merging speed is a better solution, but forcing *all* drivers to stop reduces the capacity of the ramp. A "yield right of way" sign avoids this difficulty, but not the problem that arises when a driver approaching the entrance revises his decision about the adequacy of the next headway and decides instead to stop; not only does this driver eventually merge at a dangerously low speed, but drivers in line behind him, *incorrectly anticipating* his initial decision to merge and thus following too closely, are forced to stop at the expressway entrance and merge eventually also at low speeds.

Several kinds of signs may be employed to control the mode of expressway entrance:

(a) "Stop" or "yield" sign at ramp nose;

(b) "Stop" or "yield" sign sat back from ramp nose at a sufficient distance to allow high-speed merging for a vehicle accelerating from the sign.

(c) A "stop here until ramp is clear" (SHURIC) sign set back from ramp nose. It is seen that there are two sources of delay:

(1) Ramp travel time (b and c)

(2) Possible delay at ramp nose until an acceptable gap appears in traffic on the outside lane of the expressway.

One might consider the problem in a queue-theoretic context as that of two servers in series, i.e., as a *tandem queue*, with the ramp travel time as the service time in the first server, and the merging delay as the service time in the second server. This is not strictly correct. There is, of course, no ramp travel time delay in (a). In (c) however, if the sign instructions are obeyed, there is at most one car on the ramp at any time; thus the delay is ramp travel time plus no merging delay if the driver uses the gap he initially accepts when he departs from the SHURIC sign, and ramp travel time plus merging delay if he stops at the ramp nose. Thus the use of a SHURIC sign as well as the use of "stop" or "yield" signs at the ramp nose can be modelled as single-server queues with exceptional service for the first customer in the busy period (Yeo and Weesakul,[1] McNeil and Smith[2]).

On the other hand, for the set-back "stop" or "yield" sign, the service at the first server can begin even though a car (customer) blocks the second server. There is, of course, only a finite space on the ramp (finite waiting room), and when this is filled one can say that service at the first server is blocked, and that the queue which forms at the sign (in front of the first server) is just an extension of the queue at the ramp nose (in the waiting room in front of the second server). For mathematical convenience we will assume that if a car is waiting at the ramp nose, then at most one other car will begin to move up the ramp from the sign (no intermediate waiting room). The transient solution for a tandem queue with no intermediate waiting room has been obtained by Prabhu,[3] and the most general results for a tandem queue with finite waiting room by Neuts.[4] Neither of these deals with the generalization required here— exceptional service for the first customer in a busy period. The generalization is easily obtained, however, so that comparisons can be made of delay times and queue sizes among the various modes of sign usage.

Formulation of the problem

Let $V_n^{(1)}$ be the ramp travel time of the nth vehicle to use the ramp after time $t = 0$. Let $V_n^{(2)}$ be merging delay of the nth vehicle. Define

$U_n^{(1)}$ = ramp travel time from a standing start to moving merge
$U_n^{(2)}$ = ramp travel time from a standing start with stop at ramp nose
$U_n^{(3)}$ = ramp travel time from a moving start without stopping at ramp nose
$U_n^{(4)}$ = ramp travel time from a moving start with stop at ramp nose
$V_n'^{(2)}$ = remaining merger delay of nth vehicle at time when n + 1st vehicle
  arrives at the sign.

For a given driver and vehicle, we usually have the following (stochastic) in-equalities.

$$U_n^{(1)} < U_n^{(2)}, U_n^{(3)} < U_n^{(4)}, U_n^{(1)} > U_n^{(3)}.$$

The ramp operation can be dichotomized in the following ways:
  (A) vehicle n is waiting at ramp nose when vehicle n + 1 arrives at sign,
  (B) vehicle n is not waiting at ramp nose when vehicle n + 1 arrives,
or
  (a) merging vehicle accepts initial gap at ramp nose,
  (b) merging vehicle rejects initial gap at ramp nose, and stops until accept-able gap is available. Further define
  L_n = merging delay for vehicle stopped at ramp nose
  H_n = merging delay for vehicle which merges at high speed, without stopping
    at nose.
We may usually take $H_n = 0$.
  Convenient models of service for a "stop" or "yield" sign, and for a "SHURIC" sign, are as follows:

CONTROL. "Stop" or "Yield" sign set back from ramp nose.
MODEL. Tandem queue with exceptional service for first customer.
OPERATION. (Realistic Case).
  (A) Vehicle n is stopped at ramp nose when vehicle n + 1 arrives at control. Vehicle n + 1 immediately moves up ramp, after stopping.
If $V_n'^{(2)} > U_{n+1}^{(1)}$,

$$V_{n+1}^{(1)} = U_{n+1}^{(2)}, V_{n+1}^{(2)} = L_{n+1}. \qquad (1)$$

If $V_n'^{(2)} < U_{n+1}^{(1)}$,

$$V_{n+1}^{(1)} = U_{n+1}^{(1)}, \quad V_{n+1}^{(2)} = H_{n+1} \quad \text{with probability } h_{n+1},$$

$$V_{n+1}^{(1)} = U_{n+1}^{(2)}, \quad V_{n+1}^{(2)} = L_{n+1} \quad \text{with probability } 1 - h_{n+1}. \tag{2}$$

(B) No vehicle at ramp nose when vehicle $n + 1$ arrives at sign.

$$V_{n+1}^{(1)} = U_{n+1}^{(3)}, \quad V_{n+1}^{(2)} = H_{n+1} \quad \text{with probability } g_{n+1} \tag{3}$$

$$V_{n+1}^{(1)} = U_{n+1}^{(4)}, \quad V_{n+1}^{(2)} = L_{n+1} \quad \text{with probability } 1 - g_{n+1} \tag{4}$$

CONTROL. "Stop Here Until Ramp is Clear" sign.
MODEL. Single server queue with exceptional service for first customer.
OPERATION. V_n is total service time.

(A) Vehicle $n + 1$ stays at control until vehicle n departs ramp nose

$$V_{n+1} = U_{n+1}^{(1)} + H_{n+1} \quad \text{with probability } h_{n+1}, \tag{5}$$

$$V_{n+1} = U_{n+1}^{(2)} + L_{n+1} \quad \text{with probability } 1 - h_{n+1}. \tag{6}$$

(B) No vehicle at ramp nose when vehicle $n + 1$ arrives at sign.

$$V_{n+1} = U_{n+1}^{(3)} + H_{n+1} \quad \text{with probability } g_{n+1}, \tag{7}$$

$$V_{n+1} = U_{n+1}^{(4)} + L_{n+1} \quad \text{with probability } 1 - g_{n+1}. \tag{8}$$

These equations are useful for simulation. However, in order to simplify the analysis for the setback "stop" sign we must assume that the service times $V_{n+1}^{(1)}$ and $V_{n+1}^{(2)}$ do not depend on $V_n^{(1)}$ or $V_n^{(2)}$—in particular, that there is no dependence on $V_n'^{(2)}$. It is clear that if vehicle $n + 1$ starts up the ramp even though vehicle n is waiting at the ramp nose, it may have to stop (blockage). Let us suppose:

CONTROL. "Stop" or "Yield" sign set back from ramp nose.
OPERATION. (Simplified case).

(A)

$$V_{n+1}^{(1)} = U_{n+1}^{(1)}, \quad V_{n+1}^{(2)} = H_{n+1} \quad \text{with probability } k_{n+1}, \tag{9}$$

$$V_{n+1}^{(1)} = U_{n+1}^{(2)}, \quad V_{n+1}^{(2)} = L_{n+1} \quad \text{with probability } 1 - k_{n+1}. \tag{10}$$

(B) As before.

The probability g_n that vehicle n makes a high-speed merge from a moving start at the sign, and the probability h_n that vehicle n makes a high-speed merge from a standing start at the sign, reflect only the variable gap acceptance criteria of driver n, the (stationary) properties of the expressway traffic flow, and the possibility that driver n will reject the gap he finds available as he approaches the ramp nose. The probability k_{n+1} including not only the event that vehicle $n + 1$ makes a high-speed merge from a standing stop, but also that vehicle n will clear the ramp nose and merge before vehicle $n + 1$ arrives, i.e., $V_n'^{(2)} < U_{n+1}^{(1)}$. In this paper we will not usually be concerned with transient results. Hence we assume the existence of stationary limits

$$\lim_{n \to \infty} g_n = g, \lim_{n \to \infty} h_n = h, \lim_{n \to \infty} k_n = k. \tag{11}$$

In this paper we will not attempt to evaluate g, h, k, or to derive the distribution of L_n from basic considerations of major road flow; work relevant to this problem has appeared elsewhere (see Ref. 2 for additional references). The (asymptotic) independence or otherwise of the sequence of events $\{U_{n+1}^{(1)} > V_n'^{(2)}\}$ is an interesting problem in renewal theory. If, as in the section on model functions, we choose a negative exponential distribution for L_n and set $H_n \equiv 0$, then $V_n'^{(2)}$ is also negative exponentially distributed and successive events are clearly independent. Hence, in order to prevent contradictions in the analysis of the simplified model a stop sign set back from the ramp nose, we must set

$$k \leq \lim_{n \to \infty} P\{U_{n+1}^{(1)} > L_n\} = k_{\max}. \tag{11a}$$

It is thus assumed that a driver will remain at the stop sign so long as the preceding vehicle (if any) is on the ramp; but once the preceding vehicle reaches the ramp nose, the waiting vehicle will move up the ramp whether or not the preceding vehicle actually merges (slowing and stopping at the ramp nose if the preceding vehicle still blocks it a few seconds later.) In a similar way we also require

$$g \leq g_{\max} = \int_{v=0}^{\infty} \int_{u=0}^{v} [1 - \exp(-\lambda(v - u))] \, dP\{U_{n+1}^{(3)} \leq u\} dP\{L_n \leq v\} \tag{11b}$$

Theoretical results

The stationary distributions of waiting time and queue size for both single-server and tandem queues with exceptional service can be characterized by means of appropriate Laplace transforms $\psi(\theta)$ and $\psi_0(\theta)$, which correspond to service time increments to waiting time for busy and empty queues respectively (the mathematical details will be published elsewhere.[5] For our particular problem, we have:

CONTROL. "Stop" or "Yield" sign set back from ramp nose (simplified case). Let

$$B(v) = kP\{U_n^{(1)} \leq v\} P\{H_n \leq v\} + (1 - k)P\{U_n^{(2)} \leq v\}P\{L_n \leq v\} \tag{12}$$

and

$$D(v,t) = gP\{U_n^{(3)} \leq v\}P\{H_n \leq v + t\} + (1 - g)P\{U_n^{(4)} \leq v\}P\{L_n \leq v + t\} \tag{13}$$

be the same for each n, and

$$\psi(\theta) = \int_0^\infty \exp(-\theta v)\,dB(v), \tag{14}$$

$$\psi_0(\theta,\mu) = \int_0^\infty \mu \exp(-\mu t)\,dt \int_0^\infty \exp(-\theta v)\,d_v D(v,t). \tag{15}$$

CONTROL. "Stop Here Until Ramp is Clear" sign.
Let

$$B(v) = h\,P\{U_n^{(1)} + H_n \leq v\} + (1-h)P\{U_n^{(2)} + L_n \leq v\} \tag{16}$$

and

$$D(v) = g\,P\{U_n^{(3)} + H_n \leq v\} + (1-g)P\{U_n^{(4)} + L_n \leq v\} \tag{17}$$

be the same for each n, and

$$\psi(\theta) = \int_0^\infty \exp(-\theta v)\,dB(v) \tag{18}$$

and

$$\psi_0(\theta) = \int_0^\infty \exp(-\theta v)\,dD(v). \tag{19}$$

We suppose that vehicles arrive Poissonwise (negative exponential interarrival time distribution) at rate λ. Then define for the tandem server queue

$$\psi_0(\theta) = \psi_0(\theta, \lambda). \tag{20}$$

Let t be an arbitrary instant of time,

$W_1(t)$ = virtual waiting time in front of sign (server 1) at time t;
$W_2(t)$ = time spent on ramp (service at first server plus time blocked by vehicle at second server) at time t (tandem server);
$Q(t)$ = number of vehicles in queue in front of sign, plus the vehicle (if any) moving up ramp or blocked by a vehicle at server two, at time t.

Let

$$\rho = -\lambda\,\psi'(0), \quad \nu = -\psi_0'(0). \tag{21}$$

If $\rho < 1$, there exist honest limiting distributions. Let

$$P\{W_1(t) \leq x,\, W_2(t) \leq y\} = F(x,y,t). \tag{22}$$

Then[6,7,3]

$$\lim_{t\to\infty} F(x,y,t) = F(x,y), \quad 0 < F(x,y) < 1 \text{ if } x,y < \infty \tag{23}$$

and

$$\lim_{t\to\infty} P\{Q(t) = j\} = u_j, \quad 0 < u_j < 1, j = 0, 1, 2, \ldots \tag{24}$$

with

$$\int_0^\infty \int_0^\infty \exp(-\theta x - \varphi y) dx\, dy\, F(x,y)$$

$$= \frac{1-\rho}{1-\rho+\lambda\nu} \frac{(\theta-\lambda)\psi_0(\varphi) + \lambda\psi(\varphi) + \lambda[\psi_0(\varphi)\psi(\theta) - \psi_0(\theta)\psi(\varphi)]}{\theta + \lambda - \lambda\psi(\theta)} \quad (25)$$

and

$$u_0 = \frac{1-\rho}{1-\rho+\lambda\nu}$$

and

$$\sum_{j=0}^\infty z^j u_j = \frac{1-\rho}{1-\rho+\lambda\nu} \frac{\psi(\lambda - \lambda z) - z\psi_0(\lambda - \lambda z)}{\psi(\lambda - \lambda z) - z} \quad (26)$$

(we have changed the notation slightly).

Moments are easily derived:

$$E\{W_1(\infty)\} = \frac{\lambda}{2(1-\rho)(1-\rho+\lambda\nu)} [1-\rho)\psi_0''(0) + \lambda\nu\psi''(0)], \quad (27)$$

$$E\{W_2(\infty)\} = \frac{\nu}{1-\rho+\lambda\nu}, \quad (28)$$

$$E\{Q(\infty)\} = \lambda[E\{W_1(\infty)\} + E\{W_2(\infty)\}], \quad (29)$$

$$\text{var}\{W_1(\infty)\} = \frac{\lambda^2}{4(1-\rho)^2(1-\rho+\lambda\nu)^2} \{\lambda\nu[2(1-\rho)+\lambda\nu][\psi''(0)]^2$$
$$- 2(1-\rho)^2\psi''(0)\psi_0''(0) + (1-\rho)^2[\psi_0''(0)]^2\}$$
$$- \frac{\lambda}{3(1-\rho)(1-\rho+\lambda\nu)} [(1-\rho)\psi_0'''(0) + \lambda\nu\psi'''(0)], \quad (30)$$

$$\text{var}\{W_2(\infty)\} = \frac{(1-\rho+\lambda\nu)[(1-\rho)\psi_0''(0) + \lambda\nu\psi''(0)] - \nu^2}{(1-\rho+\lambda\nu)^2}, \quad (31)$$

$$E\{Q^2(\infty)\} = \frac{\lambda^2}{2(1-\rho)^2(1-\rho+\lambda\nu)} [3(1-\rho) + \lambda^2\psi''(0)][1-\rho)\psi_0''(0)$$
$$+ \lambda\nu\psi''(0)]$$
$$+ \frac{\lambda\nu}{1-\rho+\lambda\nu} - \frac{\lambda^3[(1-\rho)\psi_0'''(0) + \lambda\nu\psi'''(0)]}{3(1-\rho)(1-\rho+\lambda\nu)}, \quad (32)$$

$$\text{covar}\{W_1(\infty), W_2(\infty)\} = \{\psi''(0)[1-\rho+\lambda\nu-\lambda^2\nu^2] - (1-\rho)^2\psi_0''(0)\}/2(1-\rho)$$
$$\cdot (1-\rho+\lambda\nu)^2 \quad (33)$$

Equation (29) is Little's theorem.[8]

Explicit formulas can be obtained for the stationary probabilities $F(x,\infty)$ and u_j in the case of the nonexceptional tandem queue [(Eq. (34)] if both $B_1(v)$ and $B_2(v)$ are negative exponential distributions; $F(x,\infty)$ is the sum of three exponential terms and u_j the sum of three geometric terms (two terms if B_1 and B_2 have the same mean values). These results are well known,[9,10] but will not be of great use here.

Most of the above results are well known, but some (Eqs. 12–15, 30–33) are believed to be new.

Transient results are much more difficult to obtain, particularly in the tandem-server case; the reasons are clearly explained by Prabhu.[3] In the case of a tandem server queue without exceptional first service, i.e., with

$$B_1(v) = P\{V_n^{(1)} \le v\}, \; B_2(v) = P\{V_n^{(2)} \le v\},$$
$$B(v) = B_1(v)B_2(v),$$
$$D(v,t) = B_1(v)B_2(t+v). \tag{34}$$

and $t = 0$ the epoch when server 1 is free and a customer just commences service at server 2, we have for Re $(\theta) > 0$, Re$(\varphi) > 0$, Re$(\omega) > 0$, $|z| < 1$,

$$\int_0^\infty \int_0^\infty \int_0^\infty \exp(-\omega t - \theta - \varphi y)\, dt\, d_x\, d_y\; F(x,y;t) = \frac{1}{\omega + \lambda - \lambda\psi_0(\eta(\omega),\omega+\lambda)}$$

$$\cdot \left\{ \psi_0(\varphi,\omega+\lambda) + \lambda\psi(\varphi) \frac{\psi_0(\theta,\omega+\lambda) - \psi_0(\eta(\omega)\,\omega+\lambda)}{\omega - \theta + \lambda - \lambda\psi(\theta)} \right. \tag{35}$$

and

$$\int_0^\infty \exp(-\omega t) \left[\sum_{n=0}^\infty P\{Q(t)=n\}z^n \right] dt$$

$$= \frac{1}{\omega + \lambda - \lambda\psi_0(\eta(\omega),\omega+\lambda)} \left\{ \frac{\omega + \lambda - \lambda z\psi_0(\omega + \lambda - \lambda z, \omega + \lambda}{\omega + \lambda - \lambda z} \right.$$

$$+ \frac{\lambda z[1 - \psi(\omega+\lambda-\lambda z][\psi_0(\omega+\lambda-\lambda z\omega+\lambda) - \psi_0(\eta(\omega),\omega+\lambda)]}{[\omega+\lambda-\lambda z][z - \psi(\omega+\lambda-\lambda z]} \right\}, \tag{36}$$

where $\eta(\omega)$ is the largest non-negative root of the equation

$$\eta(\omega) = \omega + \lambda - \lambda\psi(\eta(\omega)). \tag{37}$$

If $\rho < 1$, we have the easy limits (25) and (26). If $\rho > 1$, $F(x,y) = 0$ and $u_j = 0$ for all finite x, y, j. However, for large t, we can obtain from (35) and (36) by means of a Tauberian theorem the following useful asymptotic results:

$$E\{W_1(t)\} = (\rho - 1)t - \frac{\rho - 1 + \lambda v}{\lambda[1 - \psi_0(\eta_0)]} + o(1), \tag{37}$$

$$E\{W_2(t)\} = \rho/\lambda + o(1), \tag{38}$$

$$\text{var}\{W_1(t)\} = \left[\lambda\psi''(0) + \frac{2\nu(\rho - 1)}{1 - \psi_0(\eta_0)}\right] t + 0(1). \tag{39}$$

$$\text{var}\{W_2(t)\} = [\psi''(0) - \rho^2/\lambda^2] + o(1), \tag{40}$$

$$\text{covar}\{W_1(t), W_2(t)\} = \rho[\nu + (\rho - 1)D\psi_0(\eta)/\lambda[1 - \psi_0(\eta_0)] + o(1),$$

$$E\{Q(t)\} = \lambda t\left(1 - \frac{1}{\rho}\right) + \frac{1}{\rho[1 - \psi_0(\eta_0)]}$$

$$\cdot\{1 - \lambda D\psi_0(\omega) - \rho\psi_0(\eta_0) - \frac{\lambda^2}{2}[1 - \psi_0(\eta_0)]\psi''(0)\} + o(1), \tag{42}$$

$$\text{var}\{Q(t)\} = \lambda t\left(1 - \frac{1}{\rho}\right) + \frac{2\lambda t}{\rho[1 - \psi_0(\eta_0)]}$$

$$\cdot\{-\lambda D\psi_0(\omega) - \lambda\nu + \frac{[1 - \psi_0(\eta_0)]}{2\rho}\left(1 - \frac{1}{\rho}\right)\psi''(0)\} + 0(1), \tag{43}$$

where η_0 is the largest positive root of

$$\eta_0 = \eta(0) = \lambda - \lambda\psi(\eta_0)$$

and for any differentiable function $f(\omega)$,

$$D\psi_0(f) = \frac{d}{d\omega}\psi_0(f(\omega), \omega + \lambda)|_{\omega=0} \tag{44}$$

[The results of (39)–(43) are believed to be new]. We note that the leading terms in (37)–(40) and (42) are the same in the other models considered here. These asymptotic results are useful in describing the rate of growth of a queue during heavy (e.g., rush hour) traffic.

Model Functions

Ramp travel times are likely to be relatively concentrated around their mean value, with a definite positive lower bound (e.g., the time required to traverse the ramp at, say, 100 m.p.h.). We suppose thus that the distributions of $U_n^{(j)}$ ($j = 1, 2, 3, 4$) are *displaced* Gamma:

$$P\{U_n^{(j)} \le v\} = 0 \quad \text{if } v < a_j$$

$$= \int_0^{v-a_j} \frac{\alpha_j^{m_j}}{(m_j - 1)!} x^{m_j-1} \exp(-\alpha_j x)\,dx \quad \text{if } v \ge a_j, \tag{45}$$

where $a_j \ge 0$, $\alpha_j > 0$, m_j a positive integer ($j = 1, 2, 3, 4$) are possibly different parameters.
   For simplicity assume

$$H_n \equiv 0 \tag{46}$$

The low-speed merging delay L_n is likely to reflect the exponential tail of the headway distribution between expressway platoons. Assume then that L_n has a Gamma distribution:

$$P\{L_n \leq v\} = \int_0^v \frac{\beta^n}{(n-1)!} \, x^{n-1} \exp(-\beta x)\,dx \qquad (47)$$

where $\beta > 0$, n a positive integer.

After tedious but straightforward calculations, we obtain:

CONTROL. "Stop" sign set back from ramp nose (simplified).

$$\psi(\theta) = k\Psi(\theta; m_1, \alpha_1, a_1, 1, \infty) + (1-k)\Psi(\theta; m_2, \alpha_2, a_2, n, \beta) \qquad (48)$$

and

$$\psi_0(\theta, \mu) = g\Psi_0(\theta, \mu; m_3, \alpha_3, a_3, 1, \infty) + (1-g)\Psi_0(\theta, \mu; m_4, \alpha_4, a_4, n, \beta) \qquad (49)$$

where

$$\Psi(\theta; m, \alpha, a, n, \beta) = \exp(-\theta a)\left\{ \alpha^m/(\alpha+\theta)^m \right.$$
$$- \exp(-\beta a)\sum_{j=0}^{n-1} (\beta a)^j \, [1 - (\beta/\beta+\theta)^{n-j}]\, /j!$$
$$\left. + \theta \exp(-\beta a)\sum_{i=0}^{m-1}\sum_{j=0}^{n-1}\sum_{h=0}^{j}\binom{i+h}{h}\frac{\alpha^i \beta^j a^{j-h}}{(j-h)!}\frac{1}{(\alpha+\beta+\theta)^{i+h+1}} \right\}$$
$$(50)$$

and

$$\Psi_0(\theta, \mu; m, \alpha, a, n, \beta) = \exp(-\theta a)\left\{ \alpha^m/(\alpha+\theta)^m \right.$$
$$- \mu \exp(-\beta a)\sum_{j=0}^{n-1}\sum_{r=0}^{j}\frac{\beta^j a^{j-r}}{(\beta+\mu)^{r+1}}\,[1-(\beta/\beta+\theta)^{n-j}]/(j-r)!$$
$$+ \mu\theta \exp(-\beta a)\sum_{i=0}^{m-1}\sum_{j=0}^{n-1}\sum_{h=0}^{j}\sum_{r=0}^{j-h}\binom{i+h}{h}$$
$$\left. \cdot \frac{\alpha^i \beta^j a^{j-h-r}}{(j-h-r)!(\beta+\mu)^{r+1}(\alpha+\beta+\theta)^{i+h+1}} \right\} \qquad (51)$$

Of course, $\Psi(\theta; -) = \Psi_0(\theta; \infty; -)$.

CONTROL. "Stop here until ramp is clear" sign.

$$\psi(\theta) = h.\exp(-\theta a_1)\alpha_1^{m1}/(\alpha_1+\theta)^{m1}$$
$$+ (1-h)\exp(-\theta a_2)\alpha_2^{m2}\beta^n/(\alpha_2+\theta)^{m2}(\beta+\theta)^n, \qquad (52)$$

$$\psi_0(\theta) = g.\ \exp\left(-\theta a_3\right)\alpha_3^{m\,3}/(\alpha_3 + \theta)^{m\,3}$$
$$+ (1 - g)\exp\left(-\theta a_4\right)\alpha_4^{m\,4}\beta^n/(\alpha_4 + \theta)^{m\,4}(\beta + \theta)^n \qquad (53)$$

The derivatives required in the previous section are easily calculated from the above.

As noted earlier, some vexing problems of independence are removed by assuming that $n = 1$ when $V_n^{(2)}$ has the same distribution as L_n, the simplified model is the same as the realistic model, and so

$$k_{\max} = P\{U_{n+1}^{(1)} > L_n\} = 1 - \exp\left(-a_1\beta\right)\alpha_1^{m\,1}/(\alpha_1 + \beta)^{m\,1}$$

and

$$g_{\max} = 1 - \exp\left(-a_{3\beta}\right)\alpha_3^{m\,3}\lambda/(\alpha_3 + \beta)^{m\,3}(\lambda + \beta). \qquad (54)$$

The device $n = 1$ is also useful in other applications.[2] We would expect (if driver behavior upon leaving a "stop" sign or a SHURIC sign were completely consistent) that when $n = 1$,

$$k = h\,k_{\max}, \qquad g \ge h\,g_{\max} \qquad (55)$$

Note that $k_{\max}$ depends on the parameter β which decreases with increasing expressway traffic density. It seems likely that g, h, and k will also decrease with decreasing β, i.e., the probability of high-speed merging decreases with increasing expressway traffic density. Of course, consistency with (55) is only hypothetical.

Numerical Results

Although much interesting data is available on ramp merging (e.g., Drew et al.[11]) it is still not possible to obtain empirical estimates for all of the parameters which characterize this model. Accordingly, we present some instructive numerical calculations for plausible hypothetical values. We first assume: (1) There is a two-second penalty in ramp travel time for stopping at the ramp nose. (2) There is a one-second ramp travel time bonus for jumping the "stop" or SHURIC sign. For simplicity take

$$m_1 = m_2 = m_3 = m_4 = n = 1,$$
$$1/\alpha_1 = 1/\alpha_2 = 1/\alpha_3 = 1/\alpha_4 = 2\ \text{sec.}$$
$$a_1 = 6\ \text{sec}, a_2 = 8\ \text{sec}, a_3 = 5\ \text{sec}, a_4 = 7\ \text{sec.}$$
$$1/\lambda = 32, 20, 12.5, 10\ \text{sec.}$$
$$1/\beta = 2, 4, 8, 16, 32\ \text{sec.}$$

The values of g, h, k depend on β through $k_{\max}$. Using (54) we obtain:

$1/\beta$ sec	2	4	8	16	32
$k_{\max}$	0.975106	0.851314	0.623106	0.389076	0.219737

We assume set of values of g, h, and k which satisfy (55) and decrease with decreasing β except for the last value. With $1/\lambda = 32$ sec. and $1/\beta = 32$ sec, in order to have $\rho < 1$ we require

$$k > 0.057448$$

$$h = k/k_{max} > 0.260971$$

An increase in g, h, k for very small β would, perhaps, correspond to driver impatience in a long queue, hence to reduced gap or increased velocity differential acceptance criteria. For the chosen parameters g_{max} is large, $g_{max} \geq 0.95$. We thus choose fixed $g > h$:

$1/\beta$ sec	2	4	8	16	32
g	0.75	0.60	0.45	0.30	0.30
n	0.625	0.50	0.375	0.25	0.30
$k = hk_{max}$	0.609441	0.425657	0.222665	0.097269	0.065921

Numerical results are given in Tables I–VII. In addition to the moments we note also the ramp capacity

$$c = -3600/\psi'(0) \text{ veh./h.} \tag{56}$$

The results are easily interpreted and quite consistent. Everything else being the same, the user of the SHURIC sign suffers an additional delay before he begins to move up the ramp, a delay which the "stop" sign user may ignore. This difference is not too important for light and moderate traffic ($\rho < 0.5$, say) but can be very important for heavy traffic, possibly pushing the SHURIC sign to oversaturation ($\rho > 1$). Except for safety considerations the use of the SHURIC sign can be recommended only if there is reason to believe that it will greatly encourage high-speed merging, i.e., increase the value of h considerably.

TABLE I

Stop Sign Performance
Capacity c in vehicles per hour and traffic intensity ρ as a function
of arrival rate λ and average low-speed merging delay $1/\beta$.

$1/\beta$ sec	2	4	8	16	32
$1/\lambda$ sec (ρ)					
32	0.27462	0.29241	0.35659	0.48165	0.993156
20	0.43940	0.46785	0.57055	0.82935	>1
12.5	0.70304	0.74856	0.91288	>1	>1
10	0.87880	0.93570	>1	>1	>1
c(veh./h)	409.66	384.74	315.48	217.04	113.28

TABLE II

Stop Sign Performance
Stationary average delay in seconds at stop sign, $E\{w_1(\infty)\}$, as a function
of arrival rate λ and average low-speed merging delay $1/\beta$.

$1/\beta$ sec	2	4	16	16	32
$1/\lambda$ ($E\{w_1(\infty)\}$)					
32	1.426	1.623	2.525	10.312	4392.022
20	3.100	3.412	8.132	64.177	–
12.5	10.130	13.748	69.614	–	–
10	32.466	71.665	–	–	–

TABLE III

Stop Sign Performance
Stationary average time spent on ramp, $E\{w_2(\infty)\}$, as a function
of arrival rate λ and average low-speed merging delay $1/\beta$.

$1/\beta$ sec	2	4	8	16	32
$1/\lambda$ sec ($E\{w_2(\infty)\}$)					
32	7.814	8.201	9.267	12.998	31.590
20	8.016	8.460	9.959	15.351	–
12.5	8.362	8.911	11.105	–	–
10	8.609	9.239	–	–	–

TABLE IV

Stop Sign Performance
Stationary average queue size at stop sign, $E\{Q(\infty)\}$, as a function
of arrival rate λ and average low-speed merging delay $1/\beta$.

$1/\beta$ sec	2	4	8	16	32
$1/\lambda$ sec ($E\{Q(\infty)\}$)					
32	0.289	0.307	0.368	0.728	138.238
20	0.556	0.596	0.855	3.976	–
12.5	1.479	1.813	6.458	–	–
10	4.108	8.089	–	–	–

Markov Renewal Model for an Expressway Entrance Ramp Controller

The model of a gap acceptance entrance ramp controller proposed by Wiener, Yagoda and Pignataro[12] is a special case of the single-server queue with excep-

TABLE V

Stop Sign Performance
Capacity c in vehicles per hour and traffic intensity ρ as a function of arrival
rate λ and average low-speed merging delay $1/\beta$ at SHURIC sign.

$1/\beta$ sec	2	4	8	16	32
$1/\lambda$ sec (ρ)					
32	0.29688	0.34375	0.44531	0.67188	0.99375
20	0.47500	0.55000	0.71250	1.075	>1
12.5	0.76000	0.88000	1.16	>1	>1
10	0.95000	1.10	>1	>1	>1
c(veh./h)	375.95	327.27	252.63	167.44	113.21

TABLE VI

Stop Sign Performance
Stationary average delay in seconds at SHURIC sign, $E\{w_1(\infty)\}$, as a function
of arrival rate λ and average low-speed merging delay $1/\beta$.

$1/\beta$ sec	2	4	8	16	32
$1/\lambda$ sec ($E\{w_1(\infty)\}$)					
32	1.600	2.510	6.203	30.257	4967.890
20	3.429	5.856	19.148	—	—
12.5	12.000	35.133	—	—	—
10	72.000	—	—	—	—

TABLE VII

Stop Sign Performance
Stationary average queue size at SHURIC sign, $E\{Q(\infty)\}$, as a function
of arrival rate λ and average low-speed merging delay $1/\beta$.

$1/\beta$ sec	2	4	8	16	32
$1/\lambda$ sec ($E\{Q(\infty)\}$)					
32	0.312	0.388	0.607	1.597	156.240
20	0.604	0.804	1.642	—	—
12.5	1.687	3.673	—	—	—
10	8.141	—	—	—	—

tional service at the start of a busy period—but *only* in the case that merging
delays are exponentially distributed, i.e., the major road flow is Markovian
(otherwise vehicle dispatching by the controller must reflect the vehicle or
platoon spacing on the expressway). We then have:

Wiener et al.	Present paper
$A(z)$	$\psi(\lambda - \lambda z)$
$B(z)$	$\psi_0(\lambda - \lambda z)$
$\pi(z)$	$\sum_{j=0}^{\infty} z^j u_j$
π_0	u_0
F	$\lambda \nu$
G	ρ

and so on. In the case that the controller has a fixed, constant reset time,[11] the displaced Gamma density for ramp travel time can include the reset time in the displacement.

Future Work

As with any realistic model, there are many parameters to be estimated and it is not clear which are most important in determining system delay. Additional numerical investigations are underway for the model given in the section on Model Functions.

The possibility of altering the service mechanism when the queue becomes too long, e.g., by means of additional control devices on the outside lane of the expressway as well as on the ramp, can be studied as a tandem queue with a finite waiting room, but it was not possible to obtain simple analytical results for realistic model functions, so that further studies in this area will probably have to be done numerically (see Ref. 4).

References

1. G. F. Yeo, and B. Weesakul, Delays to road traffic at an intersection, *J. Appl. Probability* **2**, 297–310 (1964).
2. D. R. McNeil and J. T. Smith, A comparison of motorist delays for different merging strategies, *Transportation Sci.* **3**, 239–254 (1969).
3. N. U. Prabhu, Transient behavior of a tandem queue, *Management Sci.* **13**, 631–639 (1967).
4. M. Neuts, Two servers in series, studied in terms of a Markov renewal branching process, *Advan. Appl. Probability* **2**, 110–149 (1970).
5. A. H. Marcus, A note on tandem queues with exceptional service, Johns Hopkins Univ. Statistics Dept., Tech. Rept. 149 Dec. 1970 (submitted for publication).
6. G. F. Yeo, Single-server queues with modified service mechanisms, J. Australian Math. Soc. **2**, 499–507 (1961-1962).
7. P. D. Welch, On a generalized M/G/1 queueing process in which the first customer of each busy period receives exceptional service, *Operations Res.* **12**, 736–752 (1964).
8. J. D. C. Little, "A proof for the queueing formula: $L = \lambda W$", *Operations Res.* **9**, 383–387 (1961).
9. T. Suzuki, On a tandem queue with blocking, *J. Operations Res. Soc. Japan* **6**, 137–157 (1964).
10. B. Avi-Itzhak and M. Yadin, A sequence of two servers with no intermediate queue, *Management Sci.* **11**, 553–564 (1965).

11. D. R. Drew et al., Gap Acceptance and Traffic Interaction in the Freeway Merging Process, Texas Transportation Institute, Texas A and M University, College Station, Texas, 1970.

12. R. Wiener, L. J. Pignataro, and H. N. Yogoda, A discrete Markov renewal model for a gap-acceptance entrance ramp controller for expressways, *Transportation Res.* 4, 151–161 (1970).

Sensitivity of Delay at a Fixed Time Traffic Signal to Small Errors in the Observations Used for Calculating the Signal Settings

Richard E. Allsop

Research Group in Traffic Studies, University College London, London, England

Abstract

The cycle time and green times at a road junction controlled by fixed time traffic signals can be calculated so as to minimize, subject to certain constraints, Webster's simplified estimate of the delay per unit time to all traffic passing through the junction. The signal settings are determined by a vector whose components are the proportion of the cycle for which each group of approaching traffic streams has right of way and the proportion of the cycle taken up by lost time. The observed data required for the calculation of this vector are the average arrival rate and the saturation flow of traffic on each approach to the junction.

This paper shows how the vector determining the signal settings is affected by small errors in the observed arrival rates and saturation flows. It goes on to obtain an estimate of the difference between the delay per unit time at the junction with signal settings calculated from erroneous observations and the delay per unit time that would occur if the signal settings were calculated from the true arrival rates and saturation flows. This estimate is a quadratic form in the observational errors, and an expression is given for the matrix of coefficients in terms of known quantities. This matrix is shown to be symmetric.

Several numerical examples for a simple crossroads are given and the results are discussed. The delay per unit time at the example junction is found not to be very sensitive to errors in just one arrival rate or saturation flow, except when the junction is heavily loaded with traffic, but it is shown that the combined effect of several errors can be quite large.

The method developed in this paper should enable engineers to identify situations in which particularly accurate measurements of arrival rates and saturation flows are necessary, and to estimate the range of conditions for which particular signal settings are likely to be suitable.

Introduction

The author has previously shown (Allsop 1971)[1] that, if the delay at a road junction controlled by fixed time traffic signals is estimated by the simpler form of Webster's (1958)[2] expression, then the calculation of delay-minimizing signal settings can be expressed as a mathematical optimization problem having a unique solution that can be found. The observed data required for this calculation are the average arrival rate and the saturation flow[2] of traffic on each approach to the junction. The present paper shows how to estimate the effect of small errors in these observations on the average delay per unit time at the junction when the calculated settings are in force, and gives some numerical examples.

References p. 267

Definitions and Notation

It will be assumed (cf. Ref. 2) that for each approach to the junction the signal cycle is divided into an *effective red period* during which no traffic departs, and an *effective green period* during which traffic departs at a steady rate (the *saturation flow*) while there is a queue and is undelayed if there is no queue. Suppose that there are n approaches; for the jth approach, let

q_j = average rate at which traffic arrives
s_j = saturation flow
Λ_j = proportion of signal cycle that is effectively green $\qquad (j = 1, 2, \ldots, n)$.
d_j = average delay per vehicle

A part of the signal cycle in which one particular set of approaches has right of way will be called a *stage*. For each stage, the part of the cycle that is effectively green for every approach having right of way in that stage will be called the *effective green time for that stage*. The time between the end of the effective green time for one stage and the beginning of that for the next stage will be called the *lost time following the former stage*. It should be noted that some or all of this lost time may be effectively green for one or more approaches.

Suppose that there are m stages in the signal cycle. Let

λ_i = proportion of cycle that is effectively green for stage i $\qquad (i = 1, 2, \ldots, m)$,
L_i = lost time following stage i

$L = \sum_{i=1}^{m} L_i$ be the *total lost time* per cycle,

$a_{ij} = \begin{cases} 1 \text{ if approach } j \text{ has right of way in stage } i \\ 0 \text{ if not} \quad (i = 1, 2, \ldots, m; j = 1, 2, \ldots, n), \end{cases}$

$\mathbf{A} = (a_{ij})$ be the stage matrix (an $m \times n$ matrix)

a_{0j} = proportion of the total lost time that is effectively green for approach $j (j = 1, 2, \ldots, n)$

c = the *cycle time*, i.e., the duration of the signal cycle,

and $\quad \lambda_0 = L/c$.

The stage matrix and the lost time following each stage are usually determined by the layout of the junction, the requirements of pedestrians, and considerations of safety. Subject to certain practical constraints, the engineer has to choose the effective green time for each stage, and the cycle time. These times are determined by the vector $\lambda = (\lambda_0, \lambda_1, \lambda_2, \ldots, \lambda_m)$.

Expression for Delay

Webster[2] showed that the average delay per vehicle on approach j could be approximately estimated by

$$d_j = \frac{9}{10} \left\{ \frac{c(1 - \Lambda_j)^2}{2(1 - \Lambda_j x_j)} + \frac{x_j^2}{2q_j(1 - x_j)} \right\},$$

where $x_j = q_j/\Lambda_j s_j$

The estimated average delay per unit time on all approaches to the junction is $\sum_{j=1}^{n} q_j d_j$. This will be called the *rate of delay* for the junction, and can be written[1] as $9D(\lambda, \mathbf{q}, \mathbf{s})/10$, where

$$D(\lambda, \mathbf{q}, \mathbf{s}) = \sum_{j=1}^{n} \left\{ \frac{1}{\lambda_0} f_j(\Lambda_j) + g_j(\Lambda_j) \right\},$$

$$\left.\begin{array}{l} f_j(\Lambda_j) = \dfrac{L q_j s_j (1 - \Lambda_j)^2}{2(s_j - q_j)} \\[3mm] g_j(\Lambda_j) = \dfrac{q_j^2}{2 s_j \Lambda_j (\Lambda_j s_j - q_j)} \\[3mm] \Lambda_j = \displaystyle\sum_{i=0}^{m} a_{ij} \lambda_i \end{array}\right\} \quad (j = 1, 2, \ldots, n),$$

$$\mathbf{q} = (q_1, q_2, \ldots, q_n),$$

and
$$\mathbf{s} = (s_1, s_2, \ldots, s_n).$$

Practical Constraints and Choice of Signal Settings

The engineer's objective in choosing signal settings is often to minimize the rate of delay at the junction. This is achieved by choosing a value λ^*, say, of λ so as to minimize $D(\lambda, \mathbf{q}, \mathbf{s})$. The choice of λ^* is limited by some or all of the following constraints.

(a) By the definitions of the components of λ,
$$S(\lambda) = 1,$$

where
$$S(\lambda) = \sum_{i=0}^{m} \lambda_i.$$

(b) The cycle time may be subject to a maximum c_0, say, or may be required to take the value c_0. Then, respectively,

or
$$\left.\begin{array}{l} \lambda_0 \geqslant k_0 \\[2mm] \lambda_0 = k_0 \end{array}\right\} \quad \text{where } k_0 = L/c_0.$$

This will be called the *cycle time constraint*.

(c) The effective green times for one or more of the stages may be subject to minima g_{iM} ($i = 1, 2, \ldots, m$). Where no minimum is specified, let g_{iM} be zero. Then

$$\left.\begin{array}{l} \lambda_i c \geqslant g_{iM} \\[4pt] \lambda_i \geqslant k_i \lambda_0, \quad \text{where } k_i = g_{iM}/L \end{array}\right\} \quad (i = 1, 2, \ldots, m).$$

i.e.,

These will be called the *minimum green constraints*.

(d) The effective green time on each approach must be more than great enough to allow all the traffic arriving on that approach to pass through the junction in the long run. This will be true if

$$\left.\begin{array}{l} \Lambda_j > q_j/s_j \\[4pt] \displaystyle\sum_{i=0}^{m} a_{ij}\lambda_i > q_j/s_j \end{array}\right\} \quad (j = 1, 2, \ldots, n).$$

i.e.,

These will be called the *capacity constraints*.

The author has shown[1] that, subject to all these constraints, the value λ^* of λ that minimizes $D(\lambda, \mathbf{q}, \mathbf{s})$ exists, is unique, and can be found.

Effect of Small Errors in Observed Data on the Vector that Determines the Signal Settings

Suppose that observed arrival rates and saturation flows q_j and $s_j (j = 1, 2, \ldots, n)$ are used to calculate λ^*, while the corresponding true arrival rates and saturation flows are $q_j + \delta q_j$ and $s_j + \delta s_j (j = 1, 2, \ldots, n)$, so that the δq_j and δs_j represent errors in observation. Then λ^* will have been chosen so as to minimize $D(\lambda, \mathbf{q}, \mathbf{s})$, whereas the real aim is to minimize $D(\lambda, \mathbf{q}, + \delta\mathbf{q}, \mathbf{s} + \delta\mathbf{s})$, where $\delta\mathbf{q} = (\delta q_1, \delta q_2, \ldots, \delta q_n)$ and $\delta\mathbf{s} = (\delta s_1, \delta s_2, \ldots, \delta s_n)$. Let $\lambda^* + \delta\lambda^*$, where $\delta\lambda^* = (\delta\lambda_0^*, \delta\lambda_1^*, \ldots, \delta\lambda_m^*)$, be the value of λ that minimizes $D(\lambda, \mathbf{q} + \delta\mathbf{q}, \mathbf{s} + \delta\mathbf{s})$.

On the assumption that the errors in observation are small, an expression will now be derived for the vector $\delta\lambda^*$ in terms of the vectors $\lambda^*, \mathbf{q}, \mathbf{s}, \delta\mathbf{q}$, and $\delta\mathbf{s}$, correct to the first order in the components of $\delta\mathbf{q}$ and $\delta\mathbf{s}$. The resulting expression will be used in the next section to obtain a second-order estimate of the difference between the rate of delay corresponding to the signal settings given by λ^* and that corresponding to those given by $\lambda^* + \delta\lambda^*$.

The required expression for $\delta\lambda^*$ depends on the way in which the constraints set out in the previous section are satisfied by λ^*. $S(\lambda) = 1$ is an equality constraint, and the capacity constraints are strict inequalities, but each of the minimum green constraints can be satisfied either as an equality or as a strict inequality, and the same is true of the cycle time constraint.

If any of the minimum green constraints are satisfied as equalities by λ^*, let them be

$$\lambda_i^* = k_i \lambda_0^* \quad (i = 1, 2, \ldots, p),$$

where $1 \leqslant p \leqslant m$. This involves no loss of generality because no assumption has previously been made in this paper about the way in which the stages in the signal cycle are numbered. If $\lambda_i^* > k_i \lambda_0^*$ for $i = 1, 2, \ldots, m$, then let $p = 0$. Again, let

$$\Delta_0 = \begin{cases} 1 & \text{if } \lambda_0^* = k_0 \\ 0 & \text{if } \lambda_0^* \neq k_0, \end{cases}$$

so that $\Delta_0(\lambda_0^* - k_0) = 0$.

The vector λ^* is thus the value of λ that minimizes $D(\lambda, q, s)$ subject to the following equality constraints.

$$\left.\begin{aligned} \Delta_0(\lambda_0 - k_0) &= 0, \\ \lambda_i - k_i \lambda_0 &= 0 \quad (i = 1, 2, \ldots, p), \\ \sum_{i=0}^{m} \lambda_i &= 1. \end{aligned}\right\} \tag{1}$$

and

Hence, if $\mu_0, \mu_i \ (i = 1, 2, \ldots, p)$ and μ_s, respectively, are the Lagrange multipliers associated with these constraints,

$$\left.\begin{aligned} D_0(\lambda^*, q, s) + \Delta_0 \mu_0 - \sum_{i=1}^{p} k_i \mu_i + \mu_s &= 0, \\ D_i(\lambda^*, q, s) + \mu_i + \mu_s &= 0 \quad (i = 1, 2, \ldots, p) \\ D_i(\lambda^*, q, s) + \mu_s &= 0 \quad (i = p+1, p+2, \ldots, m), \end{aligned}\right\} \tag{2}$$

and

where $D_i = \partial D/\partial \lambda_i \ (i = 0, 1, 2, \ldots, m)$. It should be noted that λ has only $m + 1$ components, so that if there are more than m equality constraints and associated Lagrange multipliers, then either the settings are determined entirely by the constraints, regardless of the arrival rates and saturation flows, or there are no settings satisfying the constraints.

Now suppose that the errors in observation are small enough for the same equality constraints to be satisfied by $\lambda^* + \delta\lambda^*$ as are satisfied by λ^*, and let the corresponding Lagrange multipliers be $\mu_i + \delta\mu_i \ (i = 0, 1, 2, \ldots, p)$ and $\mu_s + \delta\mu_s$. Then

$$\left.\begin{aligned} \Delta_0(\lambda_0^* + \delta\lambda_0^* - k_0) &= 0, \\ \lambda_i^* + \delta\lambda_i^* - k_i(\lambda_0^* + \delta\lambda_0^*) &= 0 \quad (i = 1, 2, \ldots, p), \\ \sum_{i=0}^{m} (\lambda_i^* + \delta\lambda_i^*) &= 1; \end{aligned}\right\} \tag{3}$$

$$D_0(\lambda^* + \delta\lambda^*, q + \delta q, s + \delta s) + \Delta_0(\mu_0 + \delta\mu_0) - \sum_{i=1}^{p} k_i(\mu_i + \delta\mu_i) + \mu_s + \delta\mu_s = 0,$$

$$D_i(\lambda^* + \delta\lambda^*, q + \delta q, s + \delta s) + \mu_i + \delta\mu_i + \mu_s + \delta\mu_s = 0 \qquad (i = 1, 2, \ldots, p)$$

and

$$D_i(\lambda^* + \delta\lambda^*, q + \delta q, s + \delta s) + \mu_s + \delta\mu_s = 0 \qquad (i = p + 1, \ldots, m).$$

$$(4)$$

It will now be convenient to define a number of matrices: for $i = 0, 1, 2,$ $\ldots, m; j = 1, 2, \ldots, n; k = 0, 1, 2, \ldots, m$, let

B be the $(m + 1) \times (m + 1)$ matrix whose (i, k) th element is $\dfrac{\partial^2 D}{\partial\lambda_i \partial\lambda_k}$,

Q be the $n \times (m + 1)$ matrix whose (j, k) th element is $\dfrac{\partial^2 D}{\partial q_j \partial\lambda_k}$ and

S be the $n \times (m + 1)$ matrix whose (j, k) th element is $\dfrac{\partial^2 D}{\partial s_j \partial\lambda_k}$,

with all partial derivatives evaluated at (λ^*, q, s).
Let $\mathbf{K}_1$ be the $(m + 1) \times (m + 1)$ matrix

$$
\begin{bmatrix}
1 & -k_1 & -k_2 & \cdots & -k_p & \Delta_0 & 0 & \cdots & 0 \\
1 & 1 & 0 & \cdots & 0 & 0 & 0 & \cdots & 0 \\
1 & 0 & 1 & \cdots & 0 & 0 & 0 & \cdots & 0 \\
\cdot & \cdot & \cdot & & \cdot & \cdot & \cdot & & \cdot \\
\cdot & \cdot & \cdot & & \cdot & \cdot & \cdot & & \cdot \\
\cdot & \cdot & \cdot & & \cdot & \cdot & \cdot & & \cdot \\
1 & 0 & 0 & \cdots & 1 & 0 & 0 & \cdots & 0 \\
1 & 0 & 0 & \cdots & 0 & 0 & 0 & \cdots & 0 \\
\cdot & \cdot & \cdot & & \cdot & \cdot & \cdot & & \cdot \\
\cdot & \cdot & \cdot & & \cdot & \cdot & \cdot & & \cdot \\
\cdot & \cdot & \cdot & & \cdot & \cdot & \cdot & & \cdot \\
1 & 0 & 0 & \cdots & 0 & 0 & 0 & \cdots & 0
\end{bmatrix}
$$

and K_2 be the $(m + 1) \times (m + 1)$ matrix

$$\begin{bmatrix} 0 & \cdots & 0 & (1 - \Delta_0) & 0 & 0 & \cdots & 0 \\ 0 & \cdots & 0 & (1 - \Delta_0)k_1 & 0 & 0 & \cdots & 0 \\ \cdot & & \cdot & \cdot & \cdot & \cdot & & \cdot \\ \cdot & & \cdot & \cdot & \cdot & \cdot & & \cdot \\ \cdot & & \cdot & \cdot & \cdot & \cdot & & \cdot \\ 0 & \cdots & 0 & (1 - \Delta_0)k_p & 0 & 0 & \cdots & 0 \\ \hline 0 & \cdots & 0 & 0 & 1 & 0 & \cdots & 0 \\ 0 & \cdots & 0 & 0 & 0 & 1 & \cdots & 0 \\ \cdot & & \cdot & \cdot & \cdot & \cdot & & \cdot \\ \cdot & & \cdot & \cdot & \cdot & \cdot & & \cdot \\ \cdot & & \cdot & \cdot & \cdot & \cdot & & \cdot \\ 0 & \cdots & 0 & 0 & 0 & 0 & \cdots & 1 \\ 0 & \cdots & 0 & -(1 - \Delta_0)\left(1 + \sum_{i=1}^{p} k_i\right) & -1 & -1 & \cdots & -1 \end{bmatrix},$$

where in each case the partitions appear after the $(p + 1)$ th row and column. Let μ and $\delta\mu$ be the vectors

$$(\mu_s, \mu_1, \mu_2, \ldots, \mu_p, \mu_0, 0, \ldots, 0)$$

and

$$(\delta\mu_s, \delta\mu_1, \delta\mu_2, \ldots, \delta\mu_p, \delta\mu_0, 0, \ldots, 0)$$

of order $(m + 1)$. If the settings are affected by the arrival rates and saturation flows, then at least the last components of μ and $\delta\mu$ and at least the last column of K_2 must be nonzero because, as was previously noted, there can then be at most m equality constraints and Lagrange multipliers. If, on the other hand, there are $m + 1$ equality constraints, then $K_2 = 0$, the consequences of which will appear at the end of this section.

In the following matrix algebra, the superscript T will denote transposition, and all vectors will be regarded as row-vectors.

Putting $\lambda = \lambda^*$ in Eqs. (1) and subtracting them from the corresponding Eqs. (3) gives

$$\delta\lambda^* K_1 = 0. \tag{5}$$

Expanding the first terms of Eqs. (4) by Taylor's theorem and subtracting the corresponding Eqs. (2) gives, to first order in the error terms,

$$\delta\lambda^* B + \delta q Q + \delta s S + \delta\mu K_1^T = 0. \tag{6}$$

It follows from the definitions of K_1 and K_2, and the fact that $\Delta_0(1 - \Delta_0) = 0$, that $K_1^T K_2 = 0$, so that, on multiplying (6) on the right by K_2,

$$\delta\lambda^* BK_2 = -(\delta qQ + \delta sS)K_2. \tag{7}$$

Now the first $p + 1$ columns of K_1 and the last $m - p - 1$ columns of K_2 are nonzero, and the $(p + 2)$ th column is nonzero in K_1 or in K_2 according as $\Delta_0 = 1$ or 0. Hence (5) and (7) together give $m + 1$ nontrivial equations in the components of $\delta\lambda^*$. These equations can be written

$$\delta\lambda^*(K_1 + BK_2) = -(\delta qQ + \delta sS)K_2,$$

or, provided that $(K_1 + BK_2)$ is nonsingular,

$$\delta\lambda^* = -(\delta qQ + \delta sS)K_2 (K_1 + BK_2)^{-1}. \tag{8}$$

This is the required expression for $\delta\lambda^*$ in terms of λ^*, q, s, δq, and δs. It can be seen that if $K_2 = 0$, then $\delta\lambda^* = 0$. Since $K_2 = 0$ only if there are $m + 1$ equality constraints, the fact that the signal settings are then determined by the $m + 1$ equality constraints, and are therefore unaffected by errors in the observed arrival rates and saturation flows, is rightly reflected in the fact that $\delta\lambda^* = 0$.

Attempts to show analytically that $K_1 + BK_2$ is nonsingular have so far led to intractable algebra. It must, however, be nonsingular in normal cases because λ^* is uniquely determined by q and s, and $\lambda^* + \delta\lambda^*$ by $q + \delta q$ and $s + \delta s$, so that $\delta\lambda^*$ must be uniquely determined by q, s, δq, and δs.

Effect of the Errors on the Rate of Delay Corresponding to the Calculated Signal Settings

Since the true arrival rates and saturation flows are $q_j + \delta q_j$ and $s_j + \delta s_j$ $(j = 1, 2, \ldots, n)$, the rate of delay corresponding to signal settings determined by λ^* is $9D(\lambda^*, q + \delta q, s + \delta s)/10$, and that corresponding to the settings determined by $\lambda^* + \delta\lambda^*$ is $9D(\lambda^* + \delta\lambda^*, q + \delta q, s + \delta s)/10$. The use of the settings determined by λ^* therefore leads to an increase of $9\delta D/10$ in the rate of delay, where

$$\delta D = D(\lambda^*, q + \delta q, s + \delta s) - D(\lambda^* + \delta\lambda^*, q + \delta q, s + \delta s).$$

Hence, by Taylor's theorem, to second order in the error terms,

$$\delta D = -\nabla D(\lambda^* + \delta\lambda^*, q + \delta q, s + \delta s)\delta\lambda^{*T} + \tfrac{1}{2} \delta\lambda^* B\delta\lambda^{*T},$$

where ∇ denotes the gradient operator $\left(\dfrac{\partial}{\partial\lambda_0}, \dfrac{\partial}{\partial\lambda_1}, \cdots, \dfrac{\partial}{\partial\lambda_m}\right)$.

But Eqs. (4) can be written

$$\nabla D(\lambda^* + \delta\lambda^*, q + \delta q, s + \delta s) + (\mu + \delta\mu)K_1^T = 0,$$

so that, multiplying on the right by $\delta\lambda^{*T}$ and using the transpose of Eq. (5),

$$\nabla D(\lambda^* + \delta\lambda^*, q + \delta q, s + \delta s)\delta\lambda^{*T} = 0.$$

Hence, to second order in the error terms,

$$\delta D = \tfrac{1}{2} \delta\lambda^* B \delta\lambda^{*T}.$$

Again, multiplying Eq. (6) on the right by $\delta\lambda^{*T}$ and using the transpose of Eq. (5),

$$\delta\lambda^* B \delta\lambda^{*T} = -(\delta q Q + \delta s S)\,\delta\lambda^{*T}.$$

Thus on using expression (8) for $\delta\lambda^*$ and the fact that B is symmetric,

$$\delta D = \tfrac{1}{2} (\delta q Q + \delta s S)\,(K_1^T + K_2^T B)^{-1}\,K_2^T (Q^T \delta q^T + S^T \delta s^T).$$

This can be written

$$\delta D = (\delta q \ \delta s)\,E\,(\delta q \ \delta s)^T,$$

where E, which will be called the *error sensitivity matrix*, is the $2n \times 2n$ matrix given by

$$E = \frac{1}{2} \binom{Q}{S} (K_1^T + K_2^T B)^{-1}\,K_2^T (Q^T S^T). \tag{9}$$

B, Q, and S can be obtained analytically by partial differentiation of $D(\lambda, q, s)$ so that E can be evaluated easily in any given case with the aid of a matrix inversion routine.

The relative sizes of the elements of E give an indication of the relative sensitivity of the rate of delay at the junction to different errors in the observations from which the signal settings are calculated. This interpretation of E will be discussed further in the Discussion of Examples, with the aid of numerical examples. Before doing so, however, the symmetry of E will be established.

Symmetry of the Error Sensitivity Matrix

THEOREM. *If the error sensitivity matrix is well-defined, it is symmetric.*

Proof. The error sensitivity matrix E is well-defined only if $K_1^T + K_2^T B$ is non-singular. Equation (9) of the last section shows that it is then sufficient to prove that $(K_1^T + K_2^T B)^{-1}\,K_2^T$ is symmetric, i.e., to prove that $X = 0$, where

$$X = (K_1^T + K_2^T B)^{-1}\,K_2^T - K_2 (K_1 + B^T K_2)^{-1}.$$

Multiplying on the left by $K_1^T + K_2^T B$ and on the right by $K_1 + B^T K_2$,

$$(K_1^T + K_2^T B)\,X(K_1 + B^T K_2) = K_2^T (K_1 + B^T K_2) - (K_1^T + K_2^T B)\,K_2$$

$$= K_2^T K_1 + K_2^T B^T K_2 - K_1^T K_2 - K_2^T B K_2$$

$$= 0,$$

since B is symmetric and, as pointed out in a previous section, $K_1^T K_2 = 0$. Hence, multiplying on the left by $(K_1^T + K_2^T B)^{-1}$ and on the right by $(K_1 + B^T K_2)^{-1}$, $X = 0$ as required.

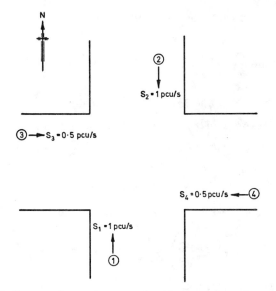

Fig. 1. Layout of example crossroads with approach numbers circled.

Numerical Examples

Consider a crossroads with four approaches (Fig. 1) controlled by signals with a two-stage cycle and total lost time of 10 sec/cycle. Suppose that the north-bound and southbound approaches, numbered 1 and 2, have right of way in stage 1 and have saturation flows of 1 pcu/second, that the eastbound and westbound approaches, numbered 3 and 4, have right of way in stage 2 and have saturation flows of 0.5 pcu/second, and that none of the lost time is effectively green for any approach. Then

$$A = \begin{pmatrix} 1 & 1 & 0 & 0 \\ 0 & 0 & 1 & 1 \end{pmatrix}$$

and $a_{01} = a_{02} = a_{03} = a_{04} = 0$. Suppose that the minimum green time is 8 sec for each stage and that the maximum cycle time is 100 sec, so that $k_0 = 0.1$ and $k_1 = k_2 = 0.8$.

The error sensitivity matrix has been calculated for seven cases, each corresponding to a different set of observed arrival rates. In each case, for simplicity, the arrival rates on approaches 1 and 2 have been made equal, and so have those on approaches 3 and 4. The arrival rates are given in the second and third columns of Table I. In cases 1–5 the ratio of the arrival rate to the saturation flow is the same on all approaches and increases by steps of 1/24 from $\frac{1}{4}$ in case 1 to 5/12 in case 5. Only in case 5 is it large enough to make the cycle time maximal. In case 6 the ratio of arrival rate to saturation flow is 1/18 on approaches 1 and

TABLE I

Effect on the Estimated Delay at the Example Junction of an Error of 10% in One Arrival Rate or Saturation Flow for Various Sets of Arrival Rates

Case	Assumed arrival rates in pcu./s.		Estimated rate of delay in pcu. with calculated settings in the absence of errors	Estimated increase in rate of delay (in pcu. and as a percentage of the rate of delay in the absence of errors) corresponding to an error of 10 per cent in observing							
	Approaches 1 & 2 (saturation flow 1 pcu/s)	Approaches 3 & 4 (saturation flow 0.5 pcu/s)		q_1		q_3		s_1	and	s_3	
				pcu.	%	pcu.	%	pcu.	%	pcu.	%
1 Neither cycle time	0.250	0.125	9.3	0.019	0.21	0.035	0.38	0.017	0.18	0.033	0.36
2 maximal nor either	0.292	0.146	13.1	0.047	0.36	0.085	0.65	0.046	0.35	0.084	0.64
3 green time minimal	0.333	0.167	18.9	0.14	0.73	0.22	1.14	0.14	0.74	0.21	1.14
4	0.375	0.188	28.8	0.48	1.68	0.66	2.29	0.50	1.73	0.67	2.31
5 Cycle time maximal	0.417	0.208	49.7	2.03	4.09	2.85	5.74	2.00	4.02	2.83	5.70
6 Green time on wider road minimal	0.056	0.306	9.2	0.0084	0.09	0.065	0.71	0.0040	0.04	0.056	0.61
7 Green time on narrower road minimal	0.611	0.028	9.7	0.029	0.30	0.024	0.25	0.021	0.21	0.020	0.20

2 and 11/18 on approaches 3 and 4; this makes the green time for the wider road minimal. In case 7 these ratios are interchanged to make the green time for the narrower road minimal: The estimated rates of delay corresponding to the delay-minimizing settings calculated from the assumed arrival rates are given in the fourth column of Table I.

Since there are 4 approaches, the error sensitivity matrix is an 8 × 8 matrix, but, because in each of the cases considered approaches 1 and 2 have the same arrival rate and saturation flow and have right of way in the same stage, and the same is true of approaches 3 and 4, each even-numbered row and column of the error sensitivity matrix is the same as the preceding odd-numbered row or column. It is therefore sufficient to give the matrix in an abbreviated form with the even-numbered rows and columns omitted, and the resulting 4 × 4 matrices for cases 1–7 are given in Table II.

TABLE II

Abbreviated Error Sensitivity Matrices for the Example Junction
with Elements in Units of $(\text{sec})^2/\text{pcu}$.

Case 1

$$\begin{bmatrix} 33.9 & -57.6 & -7.8 & 12.3 \\ -57.3 & 250.9 & 10.0 & -60.5 \\ -7.9 & 10.3 & 1.9 & -2.1 \\ 12.2 & -61.0 & -2.0 & 14.8 \end{bmatrix}$$

Case 2

$$\begin{bmatrix} 61.2 & -91.1 & -17.5 & 23.5 \\ -90.8 & 446.3 & 20.4 & -128.3 \\ -17.5 & 20.6 & 5.1 & -5.1 \\ 23.5 & -128.7 & -5.0 & 37.2 \end{bmatrix}$$

Case 3

$$\begin{bmatrix} 137.9 & -166.8 & -46.1 & 50.8 \\ -166.6 & 859.3 & 46.0 & 285.7 \\ -46.1 & 46.1 & 15.5 & -13.7 \\ 50.8 & -286.1 & -13.7 & 95.3 \end{bmatrix}$$

Case 4

$$\begin{bmatrix} 381.6 & -378.0 & -144.9 & 133.3 \\ -377.7 & 2090.0 & 124.5 & -786.5 \\ -144.9 & 124.7 & 55.2 & -43.4 \\ 133.1 & -786.8 & -43.3 & 296.3 \end{bmatrix}$$

Case 5

$$\begin{bmatrix} 1302.1 & -3084.2 & -537.8 & 1280.3 \\ -3084.2 & 7305.3 & 1273.9 & -3032.6 \\ -537.9 & 1274.0 & 222.2 & -528.9 \\ 1280.4 & -3032.7 & -528.8 & 1258.9 \end{bmatrix}$$

Case 6

$$\begin{bmatrix} 300.5 & -152.4 & -11.3 & 86.6 \\ -152.7 & 77.4 & 5.8 & -44.0 \\ -11.7 & 5.9 & 0.4 & -3.4 \\ 86.9 & -44.1 & -3.3 & 25.1 \end{bmatrix}$$

Case 7

$$\begin{bmatrix} 8.7 & -174.2 & -4.5 & 8.8 \\ -173.2 & 3469.7 & 89.2 & -174.5 \\ -4.5 & 89.7 & 2.3 & -4.5 \\ 8.7 & -174.9 & -4.5 & 8.8 \end{bmatrix}$$

The elements of the error sensitivity matrix are calculated in units of (seconds)^2/pcu, so that when the 8 × 8 matrix is multiplied on the left and right by vectors of errors in pcu/sec, and then by the factor 9/10 mentioned at the beginning of section 6 the result is the increase in rate of delay in pcu. (or pcu-sec/sec). The results of doing this for 4 simple error vectors are given in the last

8 columns of Table I, the increases being given both in pcu. and as a percentage of the rate of delay in the absence of errors. The 4 error vectors used are

$$(q_1/10, \quad 0, \quad 0, \quad 0, \quad 0, \quad 0, \quad 0, \quad 0),$$
$$(\quad 0, \quad 0, \; q_3/10, \quad 0, \quad 0, \quad 0, \quad 0, \quad 0),$$
$$(\quad 0, \quad 0, \quad 0, \quad 0, \; s_1/10, \quad 0, \quad 0, \quad 0),$$

and

$$(\quad 0, \quad 0, \quad 0, \quad 0, \quad 0, \quad 0, \; s_3/10, \quad 0),$$

representing respectively errors of 10% in the arrival rate on one of the wider and one of the narrower approaches, and in the corresponding saturation flows. In each case just one of the observations is assumed to be in error and all the other 7 to be correct.

The results given in Tables I and II are discussed in the next section.

Discussion of Examples

The first remark to be made about the increases in rate of delay shown in Table I is that, taken singly, they are quite small. Even in case 5, in which the junction is 92.6% saturated with traffic, the increase in delay resulting from an error of 10% in one observation is always less than 6%. The increases should not be neglected altogether, however, because errors in observation are unlikely to be confined to just one arrival rate or saturation flow. It is therefore necessary to consider how errors in different observations interact.

Examination of the signs of the elements in the error sensitivity matrices confirms that, as would be expected, the following pairs of errors reinforce each other's effect on delay at a crossroads of the kind considered in the examples.

(1) Errors in the same direction in arrival rates on approaches having right of way in the same stage.

(2) Errors in opposite directions in arrival rates on approaches having right of way in different stages.

(3) As (1) and (2) with arrival rates replaced by saturation flows.

(4) Errors in the same direction in the arrival rate for one approach and in the saturation flow for an approach having right of way in another stage.

(5) Errors in opposite directions in the arrival rate for one approach and in the saturation flow for an approach having right of way in the same stage.

Pairs of errors that wholly or partially counteract each other's effect on delay can be obtained from the pairs in (1)–(5) just listed by reversing the direction of one of the errors in each pair.

To take rather an extreme example, in case 3, where the largest increase in delay resulting from a single 10% error in observation is 1.14%, a 10% over-estimation of the arrival rates on approaches 3 and 4 and a 10% under-estimation of the saturation flows on the same approaches would lead to an estimated increase in delay of about 14% compared with the estimated delay in the absence of errors.

Consider next how the percentage increases in delay corresponding to a 10% error in one of the observations varies with the amount of traffic passing through the junction. In cases 1–5, the ratio of arrival rate to saturation flow has been made the same on all approaches and, since the lost time is one-tenth of the maximal cycle time, the junction is saturated with traffic when the ratio of arrival rate to saturation flow reaches 0.45. The estimated percentage increase in delay resulting from an error of 10% in the arrival rate on approach 1 is plotted against this ratio in Fig. 2, together with the corresponding curve for a 10% error in the arrival rate on approach 3. As would be expected, the increase in delay rises steeply as saturation is approached.

Another interesting feature of the results given in Table I is that in cases 1–5, where neither of the green times is minimal, the estimated effect on delay of a 10% error in the arrival rate on an approach is nearly equal to the effect of a similar error in the saturation flow on the same approach. The author has not yet been able to find a complete analytical explanation for this, but the equality is not surprising because, as shown by Webster[2] and confirmed by the author's more exact methods,[1] the delay-minimizing green times for the two stages at the example junction are approximately proportional to the ratios q_1/s_1 and q_3/s_3

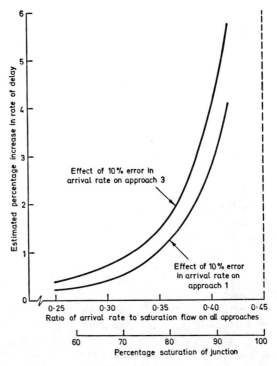

Fig. 2. Variation of the effects of given errors in observation with the amount of traffic passing through the example junction.

when neither of the green times is minimal. A small percentage change in the numerator of one of these ratios would therefore be expected to have an effect similar in magnitude to that of a similar change in the corresponding denominator.

It can also be seen from Table I and Fig. 2 that the effects on delay of 10% errors in the arrival rate and saturation flow on the wider approaches are smaller than those of similar errors on the narrower approaches. In case 6, where the green time on the wider road is minimal, the inequality is strongly reinforced, whereas in case 7, where the green time on the narrower road is minimal, the inequality is just reversed. These results suggest that, where the green time on an approach is minimal, the effect on delay of errors in observations on that approach is less than when the green time is not minimal.

Applications

The analysis described in this paper enables the traffic engineer to investigate the sensitivity of delay at traffic signals to errors in the observations of arrival rates and saturation flows that he makes in order to calculate the signal settings. As experience is gained in applying the methods in a wide variety of cases, it should be possible to distinguish cases in which particularly accurate observations are required from those in which quite rough estimates are adequate. It should also be possible to estimate the range of conditions for which particular signal settings are likely to be suitable.

References

1. R. E. Allsop, Delay-minimizing settings for fixed-time traffic signals at a single road junction, *J. Inst. Math. Appl.* 8 (2), 164–185 (1971).
2. F. V. Webster, Traffic Signal Settings, Road Research Technical Paper No. 39, H.M.S.O. London, 1958.

On-Line Feedback Control of Offsets for Area Control of Traffic

Masaki Koshi
Institute of Industrial Science, University of Tokyo, Tokyo, Japan

Abstract

The theory and the method of on-line feedback control of signal offsets for computerized area control of traffic are presented in this paper based on the data observed in the area traffic control system in downtown Tokyo. In this method, offsets are always gradually and continually shifted so that delay is reduced and, therefore, it is no longer necessary to switch offset patterns from one to another discontinuously which is inevitable in conventional offset pattern selection control.

The method is applicable without necessitating extra hardware, because the hardware necessary for the method is only a detector on an approach. Experiences in the Tokyo Area Control System indicate that the method works well and is feasible for the practical use.

Introduction

In most of the area traffic control systems, offsets have been controlled on the basis of the conventional pattern selection method where one out of the several predetermined offset patterns is selected. Switching of offset pattern from one to another usually requires several minutes during which traffic is often disturbed. Hence frequency of switching offset patterns must be limited in one way or another, causing the prevailing offsets to deviate from the optimum values. In addition, to design good offset patterns and to determine data processing procedures for selecting a particular one of them based on the detector informations are very important and at the same time very time consuming duties of traffic engineers.

The need for automatic control of offsets which can be expected to solve these problems encountered in the conventional offset pattern selection control was recognized in the course of operation of the experimental area traffic control system in Tokyo which was built in 1966. It controlled 36 signals with a small-scale digital computer using control methods very similar to those of the PR System.

The theory and the method presented here were first tested in the field pilot experiment in 1967 at three signalized intersections on National Highway 20, on which the cooperators of the experiment, T. Nakahara and others, reported at the First International Symposium on Traffic Control held in Versailles in June 1970. Immediately after the experiment, installation of the Tokyo Area

References p. 279

Control System, which started its operation in May 1970, was decided. The method mentioned here was selected together with such other control philosophies the author had proposed as multicriterion control and on-line formation of subareas.

A series of observations was carried out to test the actual behavior and characteristics of the offset control in the Tokyo Area Control System. Although the discussion in this paper will be based on the observed data, analysis in the light of theories of automatic control may be possible with some additional data on traffic characteristics.

Delay and Number of Stops of a Link as Related to the Relative Offset

Accumulated numbers of vehicles of input and output traffic at a stop line can generally be expressed in the form of a vehicles vs time diagram as shown in Fig. 1. Delay in a cycle at this stop line is represented by the shadowed area and the number of stopped vehicles in a cycle is equal to H in Fig. 1.

If the input flow is uniform, as illustrated in Fig. 2, then it is obvious that the delay and number of stops per cycle time at this stop line are unaffected by the offset of the signal. In Fig. 3, however, where the input flow is not uniform, the change of delay caused by shifting the start of green indication by dt is

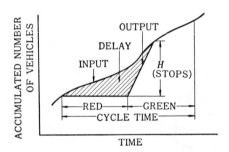

Fig. 1. Delay and number of stops in a form of accumulated number of vehicles vs time diagram.

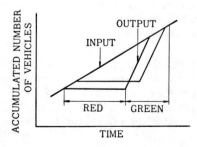

Fig. 2. Vehicle-time diagram for uniform input.

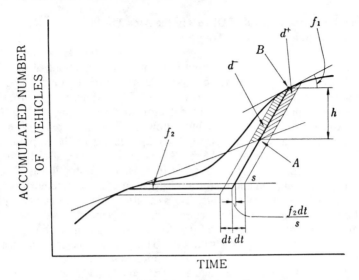

Fig. 3. Change in delay caused by offset shift.

d^+ $(dt > 0)$ or $d^-(dt < 0)$. It can be written as

$$d^- = d^+ = \left(1 - \frac{f_2}{s}\right) h\, dt$$

if dt is very small. Similarly, the change in number of stops caused by shifting the start of green is expressed as

$$\left(1 - \frac{f_2}{s}\right)(f_1 - f_2)\, dt. \quad \text{(See Fig. 3.)}$$

Let us define the relative offset of the signal for this stop line as the time from the beginning of green of the immediately upstream stop line to the beginning of green of the stop line in question. Then the differential coefficients of delay and stops with respect to the relative offset are $\left(1 - \frac{f_2}{s}\right) h$ and $\left(1 - \frac{f_2}{s}\right)(f_1 - f_2)$, respectively. Similar relationships are also true for traffic in the other direction on the same link, except that the signs are reverse.

The sum of the differential coefficients of both directions shows in which way the relative offset should be shifted to reduce the aggregate delay or the total number of stops in the link and the amount of reduction by offset shifting, and thus enables feedback control of relative offsets. This method can automatically take into consideration the effects of traffic volumes as well as sensitivity to offsets. The input traffic of a stop line is, for instance, usually less sensitive to offset when the immediately upstream intersection is a T intersection with a heavy stem.

On-Line Feedback Control of Offsets in the Area Traffic
Control System in Downtown Tokyo

It is necessary to know f_2, the input flow rate at the beginning of red; the saturation flow rate s; and B, the end point of the saturation output flow, in order to obtain the differential coefficient of delay with respect to the relative offset. These values are measured by detectors located as shown in Fig. 4 in the Tokyo Area Control System. The input flow rate f_2 is measured by the D2 detector which is located about 100 to 150 m upstream from the stop line in terms of its detection counts in 16 sec, 8 sec each before and after the beginning of red. The running time of the distance from D2 to the stop line is of course taken into consideration and therefore the computer always memorizes the counts in the latest several seconds. The saturation flow rate s can always be updated by computing the weighted sum of the latest value of s and counts by the D1 detector during the first 15 to 20 sec of the latest green. If, however, the fluctuation of s is small and the values can be assumed to be consistent, a certain constant value can be used, and the D1 detector can be eliminated.

The point A in Fig. 3 can be calculated from the f_2 and s, obtained as described above, and the length of red. The point B in Fig. 3 is obtained as the instant when the accumulated output which increases at the saturation flow rate s becomes equal to the accumulated input which is detected by the D2 detector.

If the differential coefficient of stops with respect to the relative offset is desired, f_1, the input flow rate at the end of saturation flow of output should be known, in addition to f_2. In the Tokyo Area Control System, however, the number of stops is not used as the criterion for feedback control of offsets.

In the current practice of the Tokyo System, the differential coefficients of delay are averaged over five minutes and the relative offsets are shifted once every 5 min by 2% of the cycle time based on the 5-min average of the coefficients. Offsets are not shifted, however, when the absolute value of the averaged differential coefficients are below 2 vehicle-sec/sec which means the threshold is ±2 vehicle-sec/sec. This practice is not theoretically derived but an empirically obtained results.

Multimode of control is the basic control philosophy of the Tokyo Area Control System. One of several control modes is selected according to the prevailing traffic conditions. On-line feedback control of offsets described here is used only for medium-heavy traffic conditions in usual daytime hours. For very light traffic at night and in the early morning, conventional program selection control similar to that in the PR System is used because random fluctuation of traffic would strongly affect the offsets if the method mentioned here were used. For very congested or oversaturated conditions, delay is not directly used

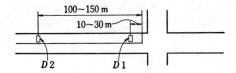

Fig. 4. Detector layout.

as the control criterion. Intersection capacity or queue length becomes the criterion and predetermined offsets are used, since the method and the theory of on-line feedback control are not valid for such traffic conditions.

The system in downtown Tokyo controls 123 traffic signals at present as the first stage of 360 traffic signals in the downtown area at the final stage. It has 196 vehicle detectors of which about 30 are D2 detectors. This means that on-line feedback control of offsets is possible at 30 stop lines, but, at present, it is applied only on several links of the major routes, because various constants for each of the approaches have not all been determined yet.

Observations in the Tokyo Area Control System

Signal indications as well as movements of vehicles on a link were recorded with a 10-pen recorder in order to test and to demonstrate how the method of offset control actually works.

Test Link

A link 240 m long as shown in Fig. 5 was selected as the test link. The link is quite heavily loaded with traffic and is near-saturated most of the day. It is located just in front of the Metropolitan Police Department Building on the roof

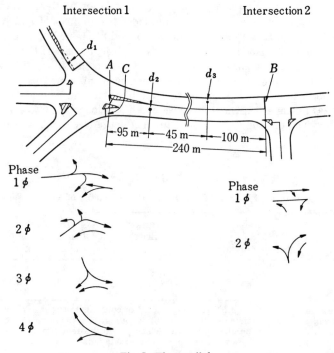

Fig. 5. The test link.

of which the observations were made. The major flows are those which pass stop lines A and B in Fig. 5 and offset optimization is applied for these two major flows. The flow which passes stop line C is fairly heavy but a large proportion of it is left turn traffic which is not stopped by the signal, hence there is no need of offset optimization for it. The third phase 3ϕ of the signal of Intersection 1 on the left side of Fig. 5 is an actuated phase and is skipped when no right turn traffic is detected by detector d_1. d_2 is the D2 detector for stop line A and d_3 is that for stop line B. No D1 detector is installed on this link for either of the stop lines. Approaches are all four lanes at stop lines A and B and also at detectors d_2 and d_3. Detectors d_2 and d_3 are both ultrasonic type and are set so as to detect vehicles on particular lanes (d_2 aims at the third lane and d_3 aims at the second lane from the curb), but they sometimes detect large vehicles on the adjacent lanes.

The relative offset is defined as the time lag between the beginning of the fourth phase 4ϕ at Intersection 1 to the beginning of the first phase 1ϕ at Intersection 2 expressed in terms of percent of the cycle length. A cycle length of around 140 sec, which may appear too long, is usually used for this link because of the need for coordination with the adjacent intersection which is extremely heavily loaded.

Accuracy of Detection

Table I shows the results of the regression calculation on the relationship between detector counts and manually counted actual number of vehicles in the same 15-sec periods. Although a sampling period of 16 sec is used in the actual control, Table I is made for 15 sec periods only for the sake of the conveniences of collecting and reducing data. It appears that detection of input flows by d_2 and d_3 is not very accurate but not too much deviated. Since offsets are shifted every 5 min, at least two cycles are sampled to obtain the average differential coefficients of delay even when the cycle length is as long as 150 sec. Hence the standard deviations shown in Table I are in effect reduced approximately $1/\sqrt{2}$ times.

TABLE I

Comparison of Detector Counts with Actual Number of Vehicles Passed

Detector	Number of data	Regression coefficient a^a	Standard deviation of detection counts from the regression line
$d_2{}^b$	685	0.345	1.62
d_3	679	0.259	1.23

[a]Regression equation: $Y = aX$. Where Y is detector count in 15 sec and X is number of vehicles which passed the detection point on all of four lanes in the same 15 sec.
[b]In the case of d_2, X is number of vehicles which passed the detection point and then went to stop line A. Those which went to stop line C are excluded.

Consistency of Saturation Flow Rates

Since no D1 detector is installed on the test link, it is assumed that the saturation flow rates at the stop lines are both consistent enough and can be assumed to be certain constant values. The actual saturation flow rates were observed by means of counting the number of vehicles passing the stop lines during saturated periods of 15 to 30 sec (excluding starting delay intervals at the beginning of green). Table II shows the average values and the standard deviations of the observed saturation flow rates.

Table II shows that the coefficients of variance are less than 10%; the saturation flow rates are quite consistent.

TABLE II

Averages and Standard Deviations of the Observed Saturation Flow Rates

Stop line	Number of data	Average (veh/sec)	Standard deviation (veh/sec)
A	33	2.231	0.216
B	68	2.203	0.148

Method of the Test

Input and output flows of stop lines A and B were measured with a 10-pen recorder on which passage times of each vehicle were recorded. The relative offsets of the link of every 5 min were also recorded with the line printer as one of the computer outputs.

In order to test the optimality of the realized relative offsets, the optimum relative offsets of the link were obtained for sampled cycles through the procedures as follows:

(a) Several cycles were sampled.
(b) The input flows were plotted on accumulated number of vehicles vs time charts such as Fig. 1 for each of the stop lines and for each of the sampled cycles.
(c) Based on the known length of the red periods and the saturation flow rates, the assumed output curves were plotted on each of the charts for several values of the relative offset.
(d) The areas between the accumulated output and input curves were measured by planimeters. The areas are, as mentioned before, delay in a cycle.
(e) The aggregate delay of the link was calculated for several values of the relative offset by summing the delay values at the two stop lines.
(f) The aggregate delay values were plotted on delay vs relative offset charts for each of the sampled cycles. A smooth fitting curve was drawn on each of the charts and then the optimum relative offset of each of the sampled cycles was obtained as the value which gave the minimum aggregate delay.

Results of Offset Control Based on the Current Practice

An observation was carried out to demonstrate the actual behavior of on-line feedback control of offset based on the current practice where the pitch of offset shift is 2% of cycle length each time, offset is shifted once every 5 min and the threshold of differential coefficients is 2 vehicle-sec/per sec.

Figure 6 shows the results of the test. The round marks show the realized offsets, the X marks show the above obtained optimum offsets and triangular marks show the optimum offsets for only stop line B. The relative offset of 90% at the start of the test was given manually and was considered to be a fairly good offset because traffic which passes stop line B is inbound traffic and is heavier than the other at that time of day. The optimum relative offset for balanced traffic seems to be somewhere around 80% which was reached about an hour later.

It may be seen in Fig. 6 that the round marks are closer to the triangular marks rather than to the X marks. This means that the realized offsets were closer to the optimum offsets of stop line B than to that of the whole link or, in other words, the delay at stop line A is weighed lighter than that at stop line B. This is because the constants to normalize the detector counts by d_2 and the saturation flow rate at stop line A are given somewhat smaller than the real values on purpose in order that the relative offset of the link is not affected or disturbed too much by skipping of the third phase of Intersection 1. This phase, as mentioned earlier, is a right turn phase at Intersection 1 and is skipped when there is no right turn traffic. The fourth phase of the intersection, the phase for stop line A, starts 6 to 10 sec earlier when the third phase is skipped.

As an overall result of this test, it may be concluded that the method worked satisfactorily.

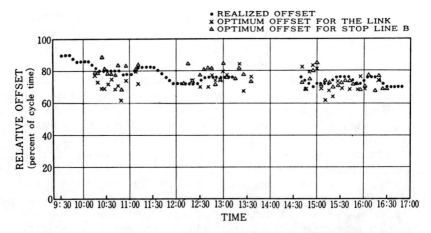

Fig. 6. Process of offset control with parameters in the current practice.

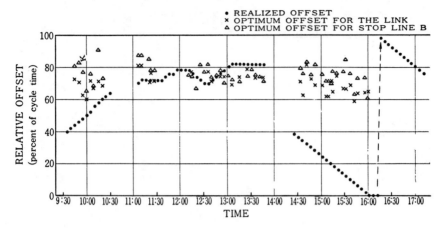

Fig. 7. Optimization process of the relative offset.

Optimization Process Starting from Adverse Offset

A second experiment was made to show how the method can optimize the relative offset of the link when the initial offset is not close to the optimum.

Figure 7 shows the process of approaching to the optimum offset starting from obviously adverse offsets. On-line feedback control based on the current practice was started from the initial offset of 40% while the optimum offset is assumed to be around 80%.

In the morning experiment, the relative offset increased consistently and became 70% about an hour later. In the afternoon experiment, however, starting from the initial offset of 40% the relative offset began to decrease and reached 80% about two hours and a half later. The exact reason why the relative offset went the other way around taking the distant way is not clear at present. Many factors such as waiting queue length, traffic volume, shape of input flow pattern, random fluctuation in traffic demand, detection errors and various combinations of them could possibly affect the direction of the optimization process. The method of offset control cannot find the shortest way to the optimum offset and therefore is not adequate for optimization from such extremely adverse initial offset as tried in the experiment.

It is encouraging, however, that the relative offset could finally reach the optimum region any way in both cases. Besides, it should be noticed that optimization from the initial conditions too far from the optimum point is not often necessary in the actual control process.

Offset Control with Parameters Different from the Current Practice

Another experiment was made with the control parameters different from those in the present practice in order to see how different the control character-

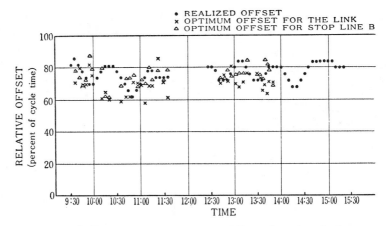

Fig. 8. Process of offset control with parameters different from those in the current practice.

istics are. The pitch of offset shift was increased to 4% and the threshold of the differential coefficient was also increased to 3 vehicle-sec/sec while they are 2% and 2 vehicle-sec/sec, respectively, in the current practice.

Figure 8 shows the results of the experiment. One can see that there is some cyclic fluctuation of offset in Fig. 8 as a sign of hunting presumably caused by too high a gain compared with the process shown in Fig. 6. Although there were of course many cycles where the offsets deviated from the optimum values, it cannot be immediately concluded that the wide variation of offsets in Fig. 8 is fully meaningless. It is possible that the control parameters used in this experiment are better than those in the current practice. It is necessary, however, to carry out more detailed and extensive studies on the characteristics of control behavior as well as traffic fluctuations in order to find a better combination of control parameters. A larger value of offset shifting pitch is desirable for quicker optimization of offsets and the optimization process shown in Fig. 7 would have been finished in about a half of the time if the shifting pitch of 4% were used.

Concluding Remarks

The method of on-line feedback control of offsets described in this paper has practical application for medium-heavy traffic conditions. It is important for satisfactory application of this method to use good constants for detector count coefficients and saturation flow rates. The pitch of offset shift, the frequency of offset shifting, and the threshold of differential coefficients used in the current practice are good for the practical use although there is still room for finding better combinations of such control parameters.

The method may also be applicable for automatic off-line long-range feedback optimization of offsets for conventional offset pattern selection control if on-line data on differential coefficients can be stored in the computer over several days.

Acknowledgment

It is a pleasure to acknowledge that many people paid attention to this theory and supported its adoption in the Tokyo Area Control System. The author especially wishes to thank Mr. Matsunaga of the Institute of Police Science who worked with the author from the pilot field experiments through to the practical application.

References

1. B. Wehner, General Report of Theme V: Area Control of Traffic, Eighth International Study Week in Traffic Engineering, Barcelona, 1966.
2. Institute of Civil Engineers, Area Control of Road Traffic, Proceedings of the Symposium, 1967.
3. M. Koshi, A possible method of program formation control of offset for area traffic control, *Seisankenkyu* 20, No. 3 (1968).
4. M. Koshi, A method on area control of traffic, *Seisankenkyu* 21, No. 10 (1969).
5. T. Nakahara, N. Yumoto, and A. Tanaka, Multi-Criterion Area Traffic Control System with Feedback Features, 1st International Symposium on Traffic Control, Versailles, 1970.
6. M. Koshi, The System of Area Traffic Control in Central Tokyo, Toyoto Review, 1970.

Optimal Synchronization of Traffic Signal Networks by Dynamic Programming

Nathan Gartner

Technion–Israel Institute of Technology, Haifa, Israel

Abstract

This paper presents a synchronization method for determination of optimal offsets in road traffic networks controlled by fixed-time signals. The method is based on Dynamic Programming and provides an optimal solution, independent of network layout, in a finite number of computation steps.

The mathematical model of the synchronization problem is introduced first. Definitions of the system's independent variables are given and the equations characterizing the constraints imposed on the offset across-variables are formulated. The dependent variables of the system are the cost functions associated with each link of the traffic network. The optimization target is a minimized economic objective function comprising the individual link functions and possibly including delay times as well as stops.

The problem is nonlinear (owing to the character of the link cost functions) and contains integer variables in the circuit constraint equations. The algorithm for its solution is based on partial minimizations in a multistage serial decision process. The process is concluded by determining a fundamental set of optimal offsets as well as the optimal value of the objective function. In order to overcome the "curse of dimensionality" inherent in multivariable problems, a Decision-Tree algorithm is employed for obtaining a minimization plan which is optimal in terms of the computational effort invested. The aim, in this case, is to minimize the required computer storage capacity and number of operations.

Introduction

The signal-program parameters affecting the traffic flow are: cycle time, green-time splits and offsets. One of the main objectives of traffic control is to determine these parameters in an optimal manner for the users of the traffic system, namely, to minimize the delay time as well as the number of stops. A coordinated traffic signal network requires a common cycle time for all signals, or one which is a submultiple of some master cycle. Since both throughput capacity and delay time increase with cycle length, the common cycle time should be the minimum required to satisfy the intersection with the largest traffic volume. The splits at each intersection are determined by the proportions of conflicting traffic streams. Thus, linking of the signals is obtained by appropriate selection of offsets throughout the network.

The conventional approach to the problem of traffic network control consists in adopting a master plan for coordination of the signal lights, analogous to the master plan of progression for an artery. Critical intersections and predominant routes in the network are identified and standard techniques are applied to them.

References p. 295

The remainder of the network is synchronized so as to approximate progressions as much as possible.

Modern approaches to the synchronization problem are based on selection of an appropriate objective function of the offsets throughout the network and attempting to minimize it, as first suggested by Gazis.[1] Some of these approaches are restricted to subnetworks of certain simple forms such as loopless arteries,[2] tree structures[3] or series-parallel networks,[4] whereas others are nonoptimal from a computational point of view.[5,6]

This paper presents an exact definition of the network synchronization problem and introduces an algorithm for its optimal solution. The optimization target is a minimized economic objective function comprising delay times as well as stops. The method is general, without restrictions on the form of the network, and the above-mentioned approaches can be derived from it as particular suboptimal cases.

Mathematical Model of Synchronization Problem

This section introduces the mathematical model of the synchronization problem of a signal-controlled traffic network. It defines the variables of the system, the constraining relations existing among them and the criterion of optimization.

System's Independent Variables

Consider a single directed traffic link connecting a pair of adjacent signalized nodes. Let S_i and S_j denote the signals at nodes i and j, respectively. Vehicle platoons released during the green time of S_i travel towards node j. Referring to Fig. 1, let

r_j = red time of S_j on link under study;

g_j = green time of S_j on link under study;

τ_{ij} = travel time of platoon head from S_i to S_j on connecting link;

γ_j = arrival time; time from beginning of green at S_j to terminal point of adjacent trajectory of platoon head, so that $-r_j < \gamma_j \leqslant g_j$;

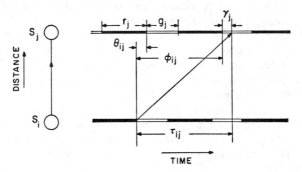

Fig. 1. Traffic signal parameters.

ϕ_{ij} = time from beginning of green at S_i to beginning of respective green at S_j so that the following identity is maintained:

$$\tau_{ij} - \gamma_j = \phi_{ij}, \tag{1}$$

θ_{ij} = relative phase (offset) of S_i and S_j, measured as the time from the beginning of green at S_i to the beginning of next green at S_j. We adopt the convention $0 \leqslant \theta_{ij} < 1$.

All quantities are expressed in cycles. A quantity having the dimensions of time can always be converted to cycles by dividing by the period. In this way we obtain $r_j + g_j = 1$ for all phases of the system's signals.

Link Cost Function

The purpose of a model is not to describe reality but to reduce the key features of that reality to more manageable forms for the purpose of decision making and control. Accordingly, we assume here a fluid traffic model with saturated flows, i.e., traffic volume at each node equals serving capability.

The system's independent variables are the cost functions associated with each link of the traffic network. These functions may comprise delay times as well as stoppages and depend only on the arrival times of the platoons at each signal. Newell has shown that this assumption approaches reality for high flows.[7]

Referring to Fig. 2a we obtain the following expressions for the average delay time per vehicle (per cycle) on link (i,j):

$$\delta_{ij}(\gamma_j) = \begin{cases} \dfrac{r_j}{g_j}\gamma_j & \text{for } 0 \leqslant \gamma_j < g_j \\ -\gamma_j & \text{for } -r_j < \gamma_j \leqslant 0 \end{cases}. \tag{2}$$

Disregarding the restriction on γ_j, the function is periodic as indicated by the interrupted prolongations. According to (1) we have,

$$\gamma_j = \tau_{ij} - \phi_{ij}. \tag{3}$$

Substitution of (3) in (2) leads to (Fig. 2b),

$$\delta_{ij}(\phi_{ij}) = \begin{cases} \dfrac{r_j}{g_j}(\tau_{ij} - \phi_{ij}) & \text{for } \tau_{ij} \geqslant \phi_{ij} > \tau_{ij} - g_j \\ \phi_{ij} - \tau_{ij} & \text{for } \tau_{ij} + r_j > \phi_{ij} \geqslant \tau_{ij} \end{cases} \tag{4}$$

Following Little,[8] let us define for arbitrary real x:

$$\text{int } [x] = \{\text{largest integer} \leqslant x\}, \tag{5}$$

$$\text{man } [x] = x - \text{int } [x]. \tag{6}$$

From Fig. 1 we conclude

$$\theta_{ij} = \text{man } [\phi_{ij}]. \tag{7}$$

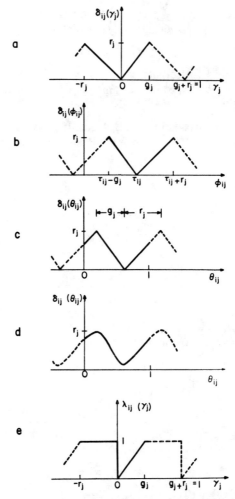

Fig. 2. Average cost functions (per cycle) for link (i, j): (a) average delay as function of arrival time; (b) average delay as function of offset time ϕ_{ij}; (c) average delay as function of offset time θ_{ij}; (d) average delay function with platoon dispersion; (e) average number of stops as function of arrival time.

The function $\delta_{ij}(\theta_{ij})$, obtained by substituting (7) in (4), is confined to the range of a single cycle (cf. Fig. 2c).

A link delay function for actual undersaturated traffic conditions, regarding platoon dispersal between signals, takes a form like that shown in Fig. 2d. Such functions can be obtained either by simulation[9] or by calculation.[10] Delays of secondary flows, caused by turning-in traffic, have to be calculated in a similar way and added appropriately to those of the main flows.

The second component of the cost function is the average number of stops per cycle $\lambda_{ij}(\gamma_j)$ (Fig. 2e). To obtain the total cost, both components must be combined according to the relative cost incurred by each of them. The following discussion is confined to delay functions.

Constraint Equations

The offset variables across the links of a signal-controlled traffic network have to comply with a set of circuit constraint equations.[11] These constraints stem from a physical requirement of synchronization to be obeyed by a coordinated signal network, namely, that *the algebraic sum of the offset variables along the links of each closed loop (circuit) of the signal network equals an integer number of cycles.* The formal statement of this requirement constitutes the circuit constraint equations.

A set of independent equations is described in matrix form as follows:

$$C_f \Theta = n_f, \tag{8}$$

where C_f is the fundamental circuit matrix of the network; Θ is a column vector of offsets defined across the links of the network; and n_f is a column vector of integer numbers associated with each of the fundamental circuit constraints.

Criterion of Optimization

In order to reduce costs we have to find an optimal set of offsets $\{\theta_{ij}^*\}$ which causes minimum delay time to traffic in the network. For each link (i,j) the total delay time is

$$d_{ij}(\theta_{ij}) = \sigma_{ij}\delta_{ij}(\theta_{ij}), \tag{9}$$

σ_{ij} denoting the traffic volume on that link. Hence, the objective function in a network of v nodes (vertices) will be,

$$\min D(\Theta) = \min_{\theta_{ij}} \sum_{i=1}^{v} \sum_{j=1}^{v} d_{ij}(\theta_{ij}) \qquad \text{for } i,j = 1, \ldots, v. \tag{10}$$

$d_{ij}(\theta_{ij}) \equiv 0$ if link (i,j) does not exist in the network.

Optimization Process

The network synchronization problem is summarized here as a nonlinear program:

    Given $d_{ij}(\theta_{ij})$ for $i,j = 1, \ldots, v$;
    find a set $\{\theta_{ij}^*\}$, *to*

$$\min \sum_{i=1}^{v} \sum_{j=1}^{v} d_{ij}(\theta_{ij}),$$

subject to

$$C_f \Theta = n_f,$$

where n_f is a vector of integer numbers and $0 \leqslant \theta_{ij} \leqslant 1$.

The program is nonlinear, owing to the character of the link cost functions, and contains integer variables. It would be unrealistic, even for simple networks, to search exhaustively all the system's independent variables. The algorithm presented in this section is based on partial minimizations in a multistage serial decision process. The approach is one of Dynamic Programming[12] based on Bellman's *Principle of Optimality*. In accordance with this principle, the signal network is disconnected in a sequence of progressive stages, each of them involving an exhaustive search on a small number of independent variables so that no optimal option is overlooked. In this way, operations are kept essentially additive rather than multiplicative. The process is concluded, after a finite number of computation steps, by determining an optimal set of offsets $\{\theta_{ij}^*\}$ as well as the optimal value of the objective function D^{**}:

$$D^{**} = D(\Theta^*) = \min_{\Theta} \{D(\Theta)\}. \tag{11}$$

In order to overcome the "curse of dimensionality" inherent in multivariable problems, both the objective function and the minimization procedure itself should be optimized.

Definitions

Graph-theoretic concepts are employed to describe the optimization process.[13,14] We denote by G the graph of the network, by $V(G)$ its vertex set and by $E(G)$ its edge set. Any subgraph H of G and its complement

$$\bar{H} = G - H \tag{12}$$

define a division of the vertices into three disjoint categories (Fig. 3a):

1. $V_I(H)$–*Interior vertices* of H incident to no edges of $\bar{H}$ (such as v_1, v_2, v_3).

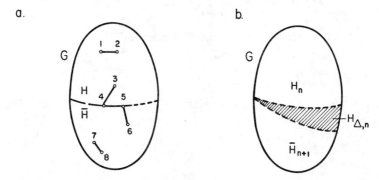

Fig. 3. (a) Disjoint categories of vertices and edges in G; (b) subgraphs at stage n of optimization process.

2. $V_A(H)$–*Connecting vertices* incident both to edges in H and $\overline{H}$ (such as v_4, v_5). This set constitutes an *articulation set*, its deletion from G disconnects all paths which exist between $V_I(H)$ to $V_E(H)$ leaving an unconnected subgraph.

3. $V_E(H)$–*Exterior vertices* of H, that is, interior vertices of $\overline{H}$ (such as v_6, v_7, v_8).

Analogous concepts are introduced for the edges $E(G)$:

1. $E_I(H)$–*Interior edges* [such as $(v_1, v_2), (v_3, v_4), (v_4, v_5)$],

$$E_I(H) = \{(v_i, v_j) \mid v_i, v_j \in V(H)\},$$

where

$$V(H) = V_I(H) \cup V_A(H).$$

2. $E_E(H)$–*Exterior edges* [such as $(v_5, v_6), (v_7, v_8)$],

$$E_E(H) = \{(v_i, v_j) \mid v_i, v_j \in V(\overline{H})\},$$

where

$$V(\overline{H}) = V_E(H) \cup V_A(H).$$

Let us arrange the vertices of G in the following order:

$$V_I(H) = \{v_1, v_2, \ldots, v_s\},$$
$$V_A(H) = \{v_{s+1}, v_{s+2}, \ldots, v_u\},$$
$$V_E(H) = \{v_{u+1}, v_{u+2}, \ldots, v_v\},$$

where $|V_I(H)| = s$, $|V_A(H)| = t$, $|V(H)| = u$, $|V(G)| = v$, and $s + t = u$.

Let η be the vector of offsets at the interior vertices $V_I(H)$,

$$\eta = (\eta_1, \eta_2, \ldots, \eta_s).$$

Let α be the vector of offsets at the connecting vertices $V_A(H)$,

$$\alpha = (\alpha_1, \alpha_2, \ldots, \alpha_t).$$

The total delay time associated with the interior vertices $E_I(H)$ is a function of the vector (η, α),

$$D(\eta, \alpha) = \sum_{i=1}^{u} \sum_{j=1}^{u} d_{ij}(\theta_{ij}). \tag{13}$$

(The offset across variable θ_{ij} is determined by the offset vertex variables θ_i and θ_j according to the relation $\theta_{ij} = \text{man} [\theta_i - \theta_j]$.)

For a given value of α, $\eta^*(\alpha)$ denotes the vector of offsets at the interior vertices $V_I(H)$ which minimizes the delay function $D(\eta, \alpha)$, i.e.,

$$D^*(\alpha) \leqslant D(\eta, \alpha),$$

where

$$D^*(\alpha) = D[\eta^*(\alpha), \alpha] = \min_{\eta} \{D(\eta, \alpha)\}. \tag{14}$$

The Algorithm

The multistage optimization process is described in this section in principle.

Initial Stage. At this stage we select a subgraph $H_1 \subset G$ containing at least one interior vertex. The total delay time contributed by the interior edges $E_I(H_1)$ is

$$D_1(\eta_1, \alpha_1) = \sum_{i=1}^{u_1} \sum_{j=1}^{u_1} d_{ij}(\theta_{ij}). \tag{15}$$

Partial minimization of this function, with respect to the offsets η_1 at the interior vertices $V_I(H_1)$ results in a function $D_1^*(\alpha_1)$ as follows:

$$\min_{\eta_1} \{D_1(\eta_1, \alpha_1)\} = D_1[\eta_1^*(\alpha_1), \alpha_1] = D_1^*(\alpha_1). \tag{16}$$

This function serves as an input to the first stage of the optimization sequence (cf. Fig. 4).

Typical Stage. At any stage n $(n = 1, \ldots, \mathcal{N})$ of the optimization process a subgraph $H_n \subset G$ is given and a function $D_n^*(\alpha_n)$ is known for all values of α_n. The purpose of this stage is to extend H_n to H_{n+1} (so that $H_n \subset\subset H_{n+1} \subset G$) and obtain the function $D_{n+1}^*(\alpha_{n+1})$ for all values of α_{n+1}.

Let us denote the subgraph added at stage n by $H_{\Delta,n}$ (Fig. 3b),

$$H_{\Delta,n} = H_{n+1} - H_n. \tag{17}$$

The vertex set $V(H_{\Delta,n})$ is composed of two classes:

1. $V_Y(H_{\Delta,n})$–interior vertices of H_{n+1}:

$$V_Y(H_{\Delta,n}) = V_X(H_{n+1}) - V_A(H_n), \tag{18}$$

where

$$V_X(H_{n+1}) = V_I(H_{n+1}) - V_I(H_n) \tag{19}$$

are the interior vertices added at this stage. The offsets at these vertices are denoted by vector χ_n,

$$\chi_n = (\chi_1, \chi_2, \ldots, \chi_{p_n}).$$

2. $V_B(H_{\Delta,n})$–connecting vertices of H_{n+1}:

$$V_B(H_{\Delta,n}) = V_A(H_{n+1}) - V_A(H_n). \tag{20}$$

The offsets associated with these vertices are denoted by vector β_n,

$$\beta_n = (\beta_1, \beta_2, \ldots, \beta_{q_n}).$$

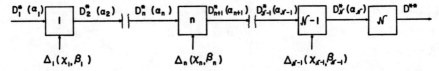

Fig. 4. Outline of the optimization process as a multistage serial decision system.

The input–output recurrent relation characterizing the optimization process at a typical stage n (Fig. 4) is derived below.

By definition we have

$$D_n^*(\alpha_n) = D_n\left[\eta^*(\alpha_n), \alpha_n\right] = \min_{\eta_n} \{D_n(\eta_n, \alpha_n)\}, \tag{21}$$

$$D_{n+1}^*(\alpha_{n+1}) = D_{n+1}\left[\eta^*(\alpha_{n+1}), \alpha_{n+1}\right] = \min_{\eta_{n+1}} \{D_{n+1}(\eta_{n+1}, \alpha_{n+1})\} \tag{22}$$

and

$$\eta_{n+1} = (\eta_n, \chi_n). \tag{23}$$

The delay function for H_{n+1} can be expressed by the delay function for H_n:

$$D_{n+1}(\eta_{n+1}, \alpha_{n+1}) = D_n(\eta_n, \alpha_n) + \Delta_n(\chi_n, \beta_n), \tag{24}$$

where

$$\Delta_n(\chi_n, \beta_n) = \sum_{i=s_n+1}^{u_{n+1}} \sum_{j=u_n+1}^{u_{n+1}} d_{ij}(\theta_{ij}). \tag{25}$$

Observing the circuit constraint equations given in (8) we can derive an unequivocal relation:

$$\alpha_n = f(\alpha_{n+1}, \chi_n). \tag{26}$$

Substituting (26) in (24) gives:

$$D_{n+1}(\eta_{n+1}, \alpha_{n+1}) = D_n\left[\eta_n, f(\alpha_{n+1}, \chi_n)\right] + \Delta_n(\chi_n, \beta_n). \tag{27}$$

Consequently we obtain

$$
\begin{aligned}
D_{n+1}^*(\alpha_{n+1}) &= \min_{\eta_{n+1}} \{D_{n+1}(\eta_{n+1}, \alpha_{n+1})\} \\
&= \min_{\eta_{n+1}} \{D_n\left[\eta_n, f(\alpha_{n+1}, \chi_n)\right] + \Delta_n(\chi_n, \beta_n)\} \\
&= \min_{\eta_n, \chi_n} \{D_n\left[\eta_n, f(\alpha_{n+1}, \chi_n)\right] + \Delta_n(\chi_n, \beta_n)\} \\
&= \min_{\chi_n} \left\{ \min_{\eta_n} \{D_n\left[\eta_n, f(\alpha_{n+1}, \chi_n)\right]\} + \Delta_n(\chi_n, \beta_n) \right\} \\
&= \min_{\chi_n} \{D_n^*\left[f(\alpha_{n+1}, \chi_n)\right] + \Delta_n(\chi_n, \beta_n)\}
\end{aligned}
$$

and finally,

$$D_{n+1}^*(\alpha_{n+1}) = \min_{\chi_n} \{D_n^*(\alpha_n) + \Delta_n(\chi_n, \beta_n)\}. \tag{28}$$

Terminal Optimization. The initial subgraph H_1 is extended gradually until the final stage $\mathcal{N}$ is reached, at which $H_{\mathcal{N}} \equiv G$. The process is illustrated as a multistage serial decision system in the functional diagram of Fig. 4. The input to stage $\mathcal{N}$, $D_{\mathcal{N}}^*(\alpha_{\mathcal{N}})$, is a partially minimized delay function comprising total delay on all network edges. The terminal optimization stage yields the minimal value D^{**} as follows:

$$\min_{\alpha_{\mathcal{N}}} \{D_{\mathcal{N}}^*(\alpha_{\mathcal{N}})\} = D_{\mathcal{N}}^*(\alpha_{\mathcal{N}}^*) = D(\eta^{**}) = D^{**}. \tag{29}$$

The set η^{**} determined at this stage is the sought optimal set of offset values at all vertices of the network.

Determination of an Optimization Plan

An optimization plan is uniquely defined by the sequence of connecting vertex sets $\{V_A(H_n)\}$. As noted above, this sequence has to be determined with a view to minimizing computation effort.

The computational procedure involves division of the offset value domain $(0, 1)$ into N equal intervals. The information input to any stage n of the optimization process comprises $(s_n + 1)N^{t_n-1}$ values of partially minimized delay times and their corresponding offsets at interior vertices. s_n denotes the number of interior vertices of H_n and t_n the number of connecting vertices. To obtain the output, $D_{n+1}^*(\alpha_{n+1})$, we have to perform $N^{t_{n+1}+p_n-1}$ search and computation operations. p_n denotes the number of interior vertices annexed at this stage.

To reduce storage demand we have to minimize

$$\hat{t} = \max_n \{t_n\}, \tag{30}$$

whereas to reduce running time we have to minimize both $\{t_n\}$ and $\{p_n\}$. A procedure for determining a plan which satisfies best these requirements is described below.

System Simplification

In order to obtain an efficient optimization plan we apply, at first, simplifying rules to degrade the complexity of the system.

Subgraphs of the network which are connected at a single articulation vertex can be treated separately. The reason is that, without reducing generality, we can assign zero value to the offset variable of one node.

In addition, series-parallel links are combined according to the following rules:

Rule 1: Parallel Combination. This is a trivial step, without optimization, in which an equivalent delay function is calculated for multiple links in parallel. The parallel links are represented, thereafter, by a single nonoriented equivalent link.

Referring to Fig. 5a, we obtain for the equivalent link:

$$d_3(\theta_{ij}) = d_1(\theta_{ij}) + d_2(1 - \theta_{ij}). \tag{31}$$

The relation between the pair of offsets is:

$$\theta_{ij} + \theta_{ji} = 1. \tag{32}$$

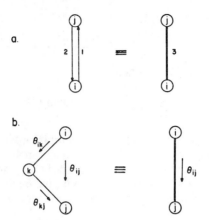

Fig. 5. Network simplification rules: (a) combination of parallel links; (b) combination of series links.

Rule 2: Series Combination. This is an elementary minimization step, in which an equivalent delay function is calculated for two links in series. This function enables us to replace the individual links by a single equivalent link connecting both terminal nodes thereby eliminating the common interior node.

We consider the two-link system (i, k) and (k, j) as a subgraph comprising a single interior node k and two connecting nodes i and j (Fig. 5b). The equivalent delay function is obtained by the following partial minimization:

$$d_3(\theta_{ij}) = \min_{\theta_{ik}} \{d_1(\theta_{ik}) + d_2(\theta_{kj})\}. \tag{33}$$

The offset constraining relation in this case is

$$\theta_{ik} + \theta_{kj} - \theta_{ij} = \text{integer}. \tag{34}$$

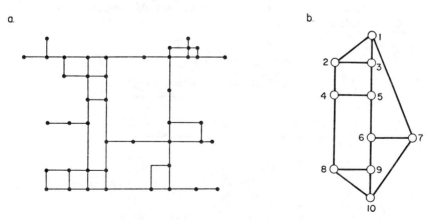

Fig. 6. (a) Example of signalized traffic network; (b) combined network of example in Fig. 6a.

Consequently,

$$\theta_{kj} = \operatorname{man} [\theta_{ij} - \theta_{ik}], \tag{35}$$

and by substituting in (33) we obtain

$$d_3(\theta_{ij}) = \min_{\theta_{ik}} \{d_1(\theta_{ik}) + d_2(\operatorname{man} [\theta_{ij} - \theta_{ik}])\}. \tag{36}$$

Repeated application of these rules yields optimizing offsets for *series-parallel networks*.[15] Non-series-parallel networks are only simplified by them, to yield a *combined network*. This network is characterized by a graph in which each vertex is of a degree not less than three. Figures 6a and 6b illustrate an actual traffic signal network and its corresponding combined form, respectively.

Decision Tree

The minimization stages of a combined network, i.e., the sequence of subgraphs $\{H_n\}$ for $n = 1, \ldots, \mathcal{N}$ are determined precisely by a Decision-Tree algorithm. An optimal solution is sought by partitioning the space of all feasible solutions into smaller and smaller subsets. A lower bound is determined for the "cost" of the solutions, in terms of computation effort, within each subset. Repeated branching and searching is applied, when necessary, until a feasible solution is found, the "cost" of which does not exceed the bound for any subset.

A decision-tree corresponding to the network of Fig. 6b is shown in Fig. 7. The upper node represents the single element of $V_A(H_0) = V_I(H_1)$ (zero stage). The node affiliated to it denotes the elements of the connecting set $V_A(H_1)$ (first stage). Subsequent nodes of the decision-tree represent the connecting sets obtained at subsequent stages. Each link originates from a box containing a vertex, the replacement of which produces the next connecting set. The numbers attached to the link represent the vertices of $V_X(H_n)$. An optimal sequence of stages, for the present example, is indicated by the bold-line path and is summarized in Table I:

TABLE I

Minimization Stages for Network of Fig. 6b.

Stage number	$V_A(H_n)$	$V_X(H_n)$
0	1	ϕ
1	2, 3, 7	1
2	2, 5, 7	3
3	4, 5, 7	2
4	4, 6, 7	5
5	8, 6, 7	4
6	8, 9, 7	6, 10

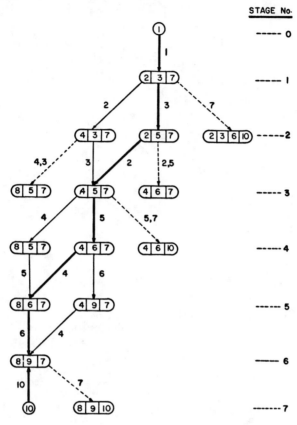

STAGE No.

Fig. 7. Decision-tree for combined network of Fig. 6b.

Additional optimal solutions may be obtained by following the uninterrupted light lines. The entire minimization process is drafted in Fig. 8a and accounted for in detail below.

It is convenient to consider the link delays to be a function of an integer, say k, where $0 \leqslant k \leqslant N - 1$ and adopt the notation $(X)_{\mathrm{mod}\,N} \equiv (X)_N$. Referring to Fig. 8a we obtain:

$$D_I(q, r) = \min_a \{ d_{12}[(a - q)_N] + d_{13}(a) + d_{17}[(a + r)_N] \} + d_{23}(q).$$

$$D_{II}(p, n) = \min_b \{ d_{35}(b) + D_I[(p - b)_N, (b + n)_N] \}.$$

$$D_{III}(m, n) = \min_c \{ d_{24}(c) + D_{II}[(c + m)_N, n] \} + d_{45}(m).$$

$$D_{IV}(l, k) = \min_d \{ d_{56}(d) + D_{III}[(l - d)_N, (d + k)_N] \} + d_{67}(k).$$

$$D_V(j, k) = \min_e \{ d_{48}(e) + D_{IV}[(e + j)_N, k] \}.$$

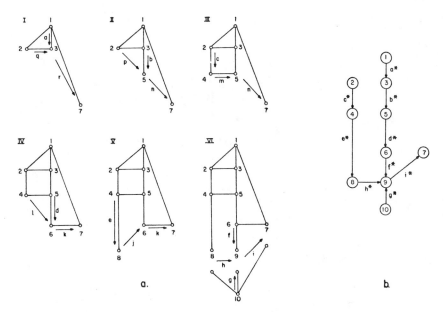

Fig. 8. (a) Sequence of sub-networks at all stages of optimization procedure for combined network of Fig. 6b; (b) tree configuration of optimized offsets.

$$D_{VI}(h, i) = \min_f \{d_{69}(f) + D_V [(h - f)_N, (f + i)_N]\} + d_{89}(h)$$
$$+ \min_g \{d_{109}(g) + d_{108} [(g - h)_N] + d_{107} [(g + i)_N]\}.$$

The minimal total delay time is obtained at stage VI:

$$D^{**} = \min_{h, i} \{D_{VI}(h, i)\}.$$

The fundamental set of offsets determined at this stage forms a tree configuration of the network (Fig. 8b). This set establishes an *optimal synchronization* of the signal network.

Conclusions

A Dynamic Programming method is introduced for determining an optimal set of offsets in synchronized signal-controlled traffic networks. The optimization plan is determined by a Decision-Tree algorithm, based on graph-theoretic concepts. The size of network that can be treated, on a given computing facility, depends on a quantity $\hat{t}$. This quantity denotes the maximal set of articulation vertices encountered during the minimization course of the combined network.

The procedures suggested by Inose et al.[3] and by Newell[2] can be described as special suboptimal cases of the general method presented in this paper for subnetworks with $\hat{T} = 1$. Similarly, the Combination Method[4] is obtained for subnetworks with $\hat{T} = 2$. Application of the proposed optimization method for actual signal networks permits, in some cases, up to 20% savings in delay times and stops compared with conventional coordination methods.

References

1. D. C. Gazis, Traffic control, time-space diagrams, and networks, *Traffic Control– Theory and Instrumentation* (T. R. Horton, Ed.), Plenum Press, New York, 1965.

2. G. F. Newell, Traffic Signal Synchronization for High Flows on a Two-Way Street, *Proc. Fourth Intern. Symp. on the Theory of Traffic Flow,* Karlsruhe, 1968.

3. H. Inose, H. Fujisaki and T. Hamada, Theory of road-traffic control based on macroscopic traffic model, *Electron. Lett.* **3** (8), 385–386 (1967).

4. J. A. Hillier, Appendix to Glasgow's experiment in area traffic control, *Traffic Eng. & Control* **7** (9), 569–571 (1966).

5. R. E. Allsop, Selection of offsets to minimize delay to traffic in a network controlled by fixed-time signals, *Transp. Sci.* **2**, 1–13 (1968).

6. I. Okutani, Synchronization of Traffic Signals in a Network for Loss Minimizing Offsets, *Proc. Fifth Intern. Symp. on the Theory of Traffic Flow and Transportation,* Berkeley, 1971.

7. G. F. Newell, Synchronization of traffic lights for high flow, *Q. Appl. Math.* **21** (4), 315–324 (1964).

8. J. D. C. Little, The synchronization of traffic signals by mixed-integer linear programming, *Operations Res.* **14** (4), 568–594 (1966).

9. J. A. Hillier and R. Rothery, The synchronization of traffic signals for minimum delay, *Transp. Sci.* **1**, 81–94 (1967).

10. R. E. Allsop, An Analysis of Delays to Vehicle Platoons at Traffic Signals, *Proc. Fourth Intern. Symp. on the Theory of Traffic Flow*, Karlsruhe 1968.

11. N. Gartner, Constraining relations among offsets in synchronized signal networks, *Transp. Sci.* **6**, 88–93 (1972).

12. R. E. Bellman, *Dynamic Programming*, Princeton University Press, New Jersey, 1957.

13. C. Berge, *The Theory of Graphs and its Applications,* Methuen, London, 1962.

14. O. Ore, *Theory of Graphs,* American Mathematical Society, Providence, Rhode Island, 1962.

15. R. J. Duffin, Topology of series-parallel networks, *J. Math. Anal. & Appl.* **10**, 303–318 (1965).

Synchronization of Traffic Signals in a Network for Loss Minimizing Offsets

Iwao Okutani

Department of Transportation Engineering, Kyoto University, Kyoto, Japan

Abstract

This paper develops some fundamental procedures for obtaining traffic signal offsets which minimize the total loss to traffic in a network. These procedures have some common assumptions: (1) all traffic signals in a network have a common period, (2) the red-green split is given at each signal, and (3) the loss to traffic generated in a link which has intersections at both ends of it depends only on the traffic flow on it and the offsets at those intersections, and, accordingly, is independent of offsets at any other intersections. That is, the loss to the flow q_{ab} between two adjacent intersections a and b is represented by a function $g(k_a, k_b, q_{ab})$ where k_a and k_b are the offsets at the intersections a and b, respectively.

It is verified through traffic simulation that the last assumption holds true if the traffic volume is more than about 600 vehicles per hour per lane and the form of the loss function $g(k_a, k_b, q_{ab})$ is represented as a cosine curve. Then the total loss to traffic in a network can be described as a function of the offset at each signal and the flow on each link. The objective is to find signal offsets that minimize it.

The mathematical procedures used in the paper are dynamic programming and the discrete maximum principle. In both cases, a network is divided into some sections that correspond to stages in a multistage decision process and the objective function or the total loss in the network is minimized in passing through those sections. In the former procedure the objective function is minimized by making a recurrence relation between the nth section and the $(n + 1)$th section in accordance with Bellman's principle of optimality and in the latter procedure the optimal solution is obtained through iterative calculation, using performance equations and a Hamiltonian function corresponding to the nth section.

As a numerical example, the optimal offset pattern is determined for a small network through the two procedures mentioned above. In addition, the minimum loss obtained by this optimization is compared with the total loss generated in the network when only traffic signals on main streets are optimally controlled, the network is synchronized under a one-way system, or the signal at each intersection is controlled individually.

Introduction

When the traffic volume exceeds some level over an entire network of intersections having traffic signals, the independent operation of each signal leads to a decrease in the capacity of the network. To avoid such inefficiency there is a control procedure which coordinates all signals through some method so as to reduce the loss to traffic in the network. The procedure is a traffic signal control system called area traffic control which has become a main problem of traffic control in many cities in the world, being supported by the rapid spread of computers.

References 311

When we say "an area traffic control procedure," we mean the method for rational determination of cycle time, split and offset at each signal. But the former two parameters can be determined rather independently at each intersection and so the determination of offset will be the main problem in the substantial sense of area traffic control whose purpose is to control the traffic flow by some combination of parameters at all signals. Therefore, in the present paper we focus attention on determining the optimum offset pattern for traffic signals in the network.

Some methods of synchronization of traffic signals already exist, such as the combination method,[1,2] the transyt method,[3] the method developed by Chang,[4] and the method for maximizing the width of through band developed by Little.[5]

In the present paper the author formulates the problem through the discrete maximum principle[6] and dynamic programming[7] and shows some numerical examples.

Assumptions

We adopt the loss to traffic for estimating the effectiveness of a control procedure and discuss the optimum area traffic control procedure or offset pattern which minimizes the total loss to traffic in a network. Here the loss to traffic is interpreted as anything that can be quantified, for example, the number of stops of vehicles, the travel time, the delay, or the queue length generated at an intersection. The important thing is that the criterion is an appropriate measure for estimating the effectiveness of the movement of traffic in a network. If the pattern of the traffic flow in the network varies with time, it may be desirable also to let this appropriate measure vary with time. Whence for convenience we use the word "loss" for designating the measure of the control effectiveness.

The main assumptions in the paper are as follows: (1) the red-green split is given at each signal in the network and all signals have a common cycle time which may be a cycle time determined by the most congested intersection. The amber light is disregarded. (2) The loss generated on a link having intersections at both ends depends only on the flow on the link and the offsets at those intersections, and, accordingly, is independent of offsets at any other intersections. That is, the loss to the flow q_{ab} between two adjacent intersections a and b is represented by a function $g(k_a, k_b, q_{ab})$ where k_a and k_b are the offsets at the intersections a and b, respectively.

Assumption 2 seems to be in conflict with the principles of area traffic control which aim to control the traffic, taking into account the relations among the intersections in the network, but it is considered to be realistic when the traffic becomes so heavy that area traffic control is required. It is not true, however, when the traffic is light.

This assumption is verified through traffic simulation as follows. We simulate traffic flow at a single one-way arterial street involving three signalized intersections a, b, and c. These intersections are 300 m apart and 50 m wide. The traffic flow q is generated at a point 300 m from intersection c and a car goes straightforward with no turning to either right or left at an intersection. Maximum speed is 50 km per hour. The cycle time and the red-green split are 100 sec and 50%, respectively.

The output of the simulation is the loss to traffic generated on the link between the intersections a and b as the offset of intersection c is varied while the relative offset between the intersections a and b is fixed at zero, and as the offset of intersection a is varied while the relative offset between the intersections b and c is fixed at zero.

The simulation is performed for two different car movement models; one is a fluid model and the other is a car-following model.

A. Fluid Model

The arrival rate of cars in a unit time is assumed to follow the Poisson distribution. The link is divided into many sections of 30 m length, and the movement of cars among these sections is simulated following May and Keller's density-speed equation, i.e.

$$v_i = v_{max} \cdot \left\{ 1 - \left(\frac{k_i}{220} \right)^{1.8} \right\}^{1.5}, \tag{1}$$

where

v_i = mean speed in the ith section (km/h);
v_{max} = maximum speed (=50 km/h);
k_i = traffic density of the ith section (veh./km).

We define the loss per car generated in the ith section, s_i, by

$$s_i = 1 - \frac{v_i}{v_{max}}. \tag{2}$$

Then the total loss in the ith section S_i can be represented as

$$S_i = \left(1 - \frac{v_i}{v_{max}} \right) k_i \cdot l_i, \tag{3}$$

where l_i = the length of the ith section (= 30 m).
Substitution of Eq. (1) into Eq. (3) yields

$$S_i = \left[1 - \left\{ 1 - \left(\frac{k_i}{220} \right)^{1.8} \right\}^{1.5} \right] k_i \cdot l_i. \tag{4}$$

By summing the S_i over a link we obtain the loss to traffic on the link.

The unit time of the simulation is 4 sec.

The results of the simulation are shown in Figs. 1 and 2 where S_{ab} represents the loss to traffic per cycle time generated on the link between the intersections a and b, and θ_{ab} and θ_{bc} are the relative offsets between the intersections corresponding to the suffixes,

$$\theta_{ab} = k_a - k_b, \qquad \theta_{bc} = k_b - k_c.$$

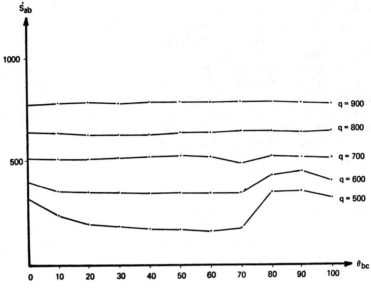

Fig. 1. $S_{ab} - \theta_{ab}$ curve.

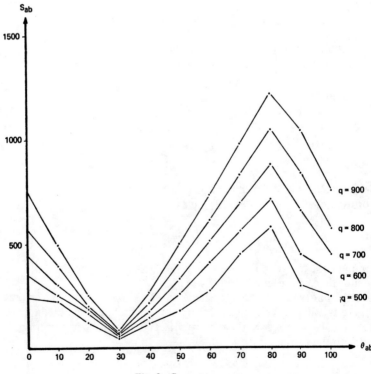

Fig. 2. $S_{ab} - \theta_{ab}$ curve.

From Fig. 1 we can see that S_{ab} is not affected very much by the relative offset θ_{bc}; it may be regarded as independent of θ_{bc} when the traffic volume q is more than about 600 vehicles per hour. From Fig. 2 it can be seen that θ_{ab} strongly influences S_{ab}. Hence, assumption 2 may be considered to hold for heavy traffic.

B. Car-following Model

We use the equation

$$x_j(t - T) - x_{j+1}(t - T) = T v_{j+1}(t) + b,$$

according to the car following theory[8], where $x_j(t)$ and $x_{j+1}(t)$ are the jth and $(j + 1)$th cars' coordinates at time t, respectively, $v_{j+1}(t)$ means the $(j + 1)$th car's speed at the same time, T, is the reaction time (chosen as 1 sec), and b is a constant assumed to be 6 m. The accelerating and decelerating rates are assumed to be 1.6 and 1.8 m/sec^2, respectively.

The arrival rate of cars is set uniform at the traffic generation point. The unit time of this simulation is 1 sec.

The results of the simulation give the same conclusions as A.

Procedure for Minimizing the Loss to Traffic

Procedure through the discrete maximum principle[9]

Here we analyse a grid network which has $(N + 1)$ intersections in the horizontal direction and M intersections in the vertical direction, and divide it into $(N + 1)$ sections 0, 1, 2, ..., N as shown in Fig. 3.

Let

$^h q_{i1}^n$: the traffic flow which flows from the intersection in the $(n - 1)$th section to the intersection in the nth section on the ith horizontal link.

$^h q_{i2}^n$: the opposite flow to $^h q_{i1}^n$.

$^v q_{i1}^n$: the flow which flows from the ith intersection to the $(i + 1)$th intersection on the ith vertical link in the nth section.

$^v q_{i2}^n$: the opposite folow to $^v q_{i1}^n$.

k_i^n: the offset at the ith intersection in the nth section, the time from a standard time to the beginning of the green time in the vertical direction or to the red time in the horizontal direction.

C: the common cycle time

$^h g_{i1}^n(k_i^{n-1}, k_i^n, {}^h q_{i1}^n)$: the loss suffered by $^h q_{i1}^n$ on the ith horizontal link in the nth section when the offset at the ith intersection in the $(n - 1)$th section is k_i^{n-1} and the offset at the ith intersection in the nth section is k_i^n.

$^h g_{i2}^n(k_i^{n-1}, k_i^n, {}^h q_{i2}^n)$: similarly the loss suffered by $^h q_{i2}^n$.

$^v g_{i1}^n(k_i^n, k_{i+1}^n, {}^v q_{i1}^n)$: the loss suffered by $^v q_{i1}^n$ on the ith vertical link when the offset at the ith intersection in the nth section is k_i^n and the offset at the $(i + 1)$th intersection in the same section is k_{i+1}^n.

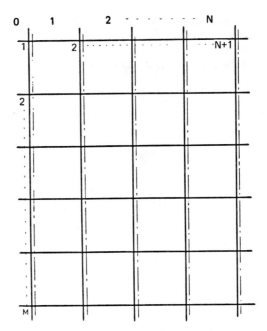

Fig. 3. Grid network.

$^{v}g_{i2}^{n}(k_{i}^{n}, k_{i+1}^{n}, {}^{v}q_{i2}^{n})$: similarly the loss suffered by $^{v}q_{i2}^{n}$. $(n = 0, 1, 2, ..., N;$
$i = 1, 2, ..., M)$.

Here

$$^{h}g_{i1}^{0}(k_{i}^{-1}, k_{i}^{0}, q_{i1}^{0}) = {}^{h}g_{i2}^{0}(k_{i}^{-1}, k_{i}^{0}, {}^{h}q_{i2}^{0}) = 0 \quad (i = 1, 2, ..., M),$$

$$^{v}g_{M1}^{n}(k_{M}^{n}, k_{M+1}^{n}, {}^{v}q_{M1}^{n}) = {}^{v}g_{M2}^{n}(k_{M}^{n}, k_{M+1}^{n}, {}^{v}q_{M2}^{n}) = 0 \quad (n = 0, 1, 2, ..., N).$$

The $^{h}g_{i\gamma}^{0}(\ )$ and $^{v}g_{M\gamma}^{n}(\ )(\gamma = 1, 2)$ are set to be zero because the loss generated on the exterior links of the network is not related to the control effectiveness of the network. They may be some constant values other than zeros.

The determination of the optimum control policy at each intersection that minimizes the total loss generated in the network comes to a mathematical problem of seeking k_{i}^{n} $(n = 0, 1, 2, ..., N; i = 1, 2, ..., M)$ so as to minimize the objective function represented by

$$F = \sum_{n=0}^{N} \left[\sum_{i=1}^{M} \{ {}^{h}g_{i1}^{n}(k_{i}^{n-1}, k_{i}^{n}, {}^{h}q_{i1}^{n}) + {}^{h}g_{i2}^{n}(k_{i}^{n-1}, k_{i}^{n}, {}^{h}q_{i2}^{n}) \} \right.$$

$$\left. + \sum_{i=1}^{M} \{ {}^{v}g_{i1}^{n}(k_{i}^{n}, k_{i+1}^{n}, {}^{v}q_{i1}^{n}) + {}^{v}g_{i2}^{n}(k_{i}^{n}, k_{i+1}^{n}, {}^{v}q_{i2}^{n}) \} \right]. \quad (5)$$

There are many mathematical techniques which can be used to minimize F

represented by Eq. (5). Here we use the discrete maximum principle developed by L. T. Fan and C. S. Wang which originated from Pontryagin's maximum principle. This discrete maximum principle by Fan and Wang is the generalized method of the discrete maximum principle which was published by S. Katz[10] in 1962.

Let the sections of the network correspond to the stages in a multistage decision process. The total loss generated in the network is minimized in passing through the $(N+1)$ stages which consist of state variables $g(k, k', q)$'s and decision variables k's.

However, it is impossible to apply the discrete maximum principle to the problem shown here as it is; it is necessary to change the problem to a standard type of problem of the discrete maximum principle by adding or modifying the state variables and the decision variables. For this the following variable is introduced as a state variable in the nth section.

$$x_1^n = x_1^{n-1} + \sum_{i=1}^{M} \{{}^hg_{i1}^n (k_i^{n-1}, k_i^n, {}^hq_{i1}^n) + {}^hg_{i2}^n (k_i^{n-1}, k_i^n, {}^hq_{i2}^n)\}$$

$$+ \sum_{i=1}^{M} \{{}^vg_{i1}^n (k_i^n, k_{i+1}^n, {}^vq_{i1}^n) + {}^vg_{i2}^n (k_i^n, k_{i+1}^n, {}^vq_{i2}^n)\} \quad (n = 1, 2, ..., N) \quad (6)$$

where

$$x_1^0 = \sum_{i=1}^{M} \{{}^vg_{i1}^0 (k_i^0, k_{i+1}^0, {}^vq_{i1}^0) + {}^vg_{i2}^0 (k_i^n, k_{i+1}^n, {}^vg_{i2}^0)\} . \quad (7)$$

This state variable x_1^n represents the accumulative loss from the 0th section to the nth section, thus the total loss generated over the whole of the network can be described by x_1^N.

Let

$$y_i^n = k_i^n \quad (8)$$

$$\theta_i^n = k_i^n - k_i^{n-1} \quad (-C \leqslant \theta_i^n \leqslant C), \quad (9)$$

then the state variable x_1^n given by Eq. (6) is written as

$$x_1^n = x_1^{n-1} + \sum_{i=1}^{M} \{{}^hg_{i1}^n (y_i^{n-1}, \theta_i^n + y_i^{n-1}, {}^hq_{i1}^n) + {}^hg_{i2}^n (y_i^{n-1}, \theta_i^n + y_i^{n-1}, {}^hq_{i2}^n)\}$$

$$+ \sum_{i=1}^{M} \{{}^vg_{i1}^n (\theta_i^n + y_i^{n-1}, \theta_{i+1}^n + y_{i+1}^{n-1}, {}^vq_{i1}^n)$$

$$+ {}^vg_{i2}^n (\theta_i^n + y_i^{n-1}, \theta_{i+1}^n + y_{i+1}^{n-1}, {}^vq_{i2}^n)\} . \quad (10)$$

In the above equation we regard y_i^n as a state variable and θ_i^n as a decision variable. From Eqs. (8) and (9) the performance equation associated with y_i^n

becomes

$$y_i^n = \theta_i^n + y_i^{n-1} \qquad (n = 1, 2, ..., N; i = 1, 2, ..., M). \tag{11}$$

Here y_i^0 is the offset k_i^0 $(i = 1, 2, ..., M)$ at the ith intersection in the 0th section. Thus the problem dealt with here is converted to the problem where decision variables θ_i^n's in each section are determined so as to minimize x_1^N subject to the performance Eqs. (10) and (11). This is the general problem which can be solved by the discrete maximum principle. The concrete method for solving the problem is as follows.

First we introduce the Hamiltonian function corresponding to the nth section

$$H^n = z_1^n x_1^n + \sum_{i=1}^{M} z_{i+1}^n y_i^n \qquad (n = 1, 2, ..., N), \tag{12}$$

where z_i^n is a covariant variable satisfying the equations

$$z_1^{n-1} = \frac{\partial H^n}{\partial x_1^{n-1}} \qquad (n = 1, 2, ..., N), \tag{13}$$

$$z_{i+1}^{n-1} = \frac{\partial H^n}{\partial v_i^{n-1}} \qquad (n = 1, 2, ..., N; i = 1, 2, ..., M). \tag{14}$$

Here the partial differential Eqs. (13) and (14) are interpreted as difference equations since the problem in the present paper is of discrete type.

Substitution of Eqs. (10) and (11) in the Hamiltonian function represented by Eq. (12) gives

$$
\begin{aligned}
H^n = z_1^n \Bigg[& x_1^{n-1} + \sum_{i=1}^{M} \{ {}^h g_{i1}^n (y_i^{n-1}, \theta_i^n + y_i^{n-1}, {}^h q_{i1}^n) \\
& + {}^h g_{i2}^n (y_i^{n-1}, \theta_i^n + y_i^{n-1}, {}^h q_{i2}^n) \} \\
& + \sum_{i=1}^{M} \{ {}^v g_{i1}^n (\theta_i^n + y_i^{n-1}, \theta_{i+1}^n + y_{i+1}^{n-1}, {}^v q_{i1}^n) \\
& + {}^v g_{i2}^n (\theta_i^n + y_i^{n-1}, \theta_{i+1}^n + y_{i+1}^{n-1}, {}^v q_{i2}^n) \} \Bigg] + \sum_{i=1}^{M} z_{i+1}^n (\theta_i^n + y_i^{n-1}).
\end{aligned} \tag{15}
$$

The following relations among the covariant variables are obtained from Eqs. (13) and (14) if we consider Eq. (15).

$$z_1^{n-1} = \frac{\partial H^n}{\partial x_1^{n-1}} = z_1^n. \tag{16}$$

On the other hand, the objective function F is described by

$$F = 1 \cdot x_1^N + 0 \cdot y_1^N + 0 \cdot y_2^N + \cdots + 0 \cdot y_M^N.$$

Hence the final condition of z becomes

$$z_1^N = 1 \quad \text{and} \quad z_{i+1}^N = 0 \quad (i = 1, 2, ..., M).$$ (17)

From Eqs. (16) and (17) the covariant variables z_1^n is uniquely determined as

$$z_1^n = 1 \quad (n = 1, 2, ..., N).$$

Therefore the Hamiltonian function eventually becomes

$$
\begin{aligned}
H^n = x_1^{n-1} &+ \sum_{i=1}^{M} \{{}^h g_{i1}^n \, (y_i^{n-1}, \theta_i^n + y_i^{n-1}, {}^h q_{i1}^n) \\
&+ {}^h g_{i2}^n \, (y_i^{n-1}, \theta_i^n + y_i^{n-1}, {}^h q_{i2}^n)\} \\
&+ \sum_{i=1}^{M} \{{}^v g_{i1}^n \, (\theta_i^n + y_i^{n-1}, \theta_{i+1}^n + y_{i+1}^{n-1}, {}^v q_{i1}^n) \\
&+ {}^v g_{i2}^n \, (\theta_i^n + y_i^{n-1}, \theta_{i+1}^n + y_{i+1}^{n-1}, {}^v q_{i2}^n)\} \\
&+ \sum_{i=1}^{M} z_{i+1}^n \, (\theta_i^n + y_i^{n-1}) \quad (n = 1, 2, ..., N).
\end{aligned}
$$ (18)

The optimum decision variables θ_i^n's in each section are obtained by finding values which minimize H^n represented by Eq. (18) in their domain. Here the domain of θ_i^n can be written as $-y_i^{n-1} \leqslant \theta_i^n \leqslant C - y_i^{n-1}$ since $y_i^n = \theta_i^n + y_i^{n-1}$ from Eq. (11) and $0 \leqslant y_i^n \leqslant C$.

We can solve the problem of minimizing H^n through applying dynamic programming, but here we omit the calculation process.

Now we have seen that the optimum offset at each intersection minimizing the total loss over the whole of the network can be determined through the discrete maximum principle. But in the above we show only the main points of the technique for determining the optimum offsets pattern. The concrete calculation steps are shown below.

Calculation Step

If we regard the problem dealt with here as a mathematical problem through the discrete maximum principle, it cannot be solved by using the usual method since the boundary conditions of the state variables, that is, the initial condition $x_1^0, y_1^0, y_2^0, ..., y_M^0$ and the final condition $x_1^N, y_1^N, y_2^N, ..., y_M^N$ are both free. Therefore here we consider the initial points to be fixed and determine the optimum solutions successively corresponding to any different sets of $x_1^0, y_1^0, y_2^0, ..., y_M^0$.

Step 1. Assume a set $y_1^0, y_2^0, ..., y_M^0$ and calculate x_1^0 by equation (7).

Step 2. Assume a set $x_1^N, y_1^N, y_2^N, ..., y_M^N$.

Step 3. Then $x_1^{N-1}, y_1^{N-1}, y_2^{N-1}, ..., y_M^{N-1}$ are calculated from equations (10) and (11). Calculate z_{i+1}^{N-1} $(i = 1, 2, ..., M)$ by equation (14) using those state variables in the $(N-1)$th section and determine the optimum decision variables $\overset{*}{\theta}_1^{N-1}, \overset{*}{\theta}_2^{N-1}, ..., \overset{*}{\theta}_M^{N-1}$.

Step 4. Iterate the same procedure in Step 3 until $x_1^1, y_1^1, y_2^1, ..., y_M^1$ and $\overset{*}{\theta}_1^1, \overset{*}{\theta}_2^1, ..., \overset{*}{\theta}_M^1$ are obtained.

Step 5. Calculate $x_1^0, y_1^0, y_2^0, ..., y_M^0$ from equations (10) and (11) using x_1^1, $y_i^1, \overset{*}{\theta}_i^1$ obtained at Step 4.

Step 6. If the state variables in the 0th section calculated at Step 5 are equal to the assumed values at Step 1, go to Step 7. Otherwise return to Step 2.

Step 7. Memorize the obtained values x_1^N and y_i^n $(n = 0, 1, 2, ..., N; i = 1, 2, ... M)$ and return to step 2 assuming new initial condition $y_1^0, y_2^0, ..., y_M^0$ and calculating x_1^0 by them.

Step 8. After calculating for all different initial sets $y_1^0, y_2^0, ..., y_M^0$ and x_1^0, compare x_1^N corresponding to each initial set and find out the minimum value and corresponding y_i^n $(n = 0, 1, 2, ..., N; i = 1, 2, ..., M)$. Then those state variables give the optimum offset pattern.

Procedure through dynamic programming

The same problem is solved through dynamic programming as follows:

Let $f_n(k_1^{n-1}, k_2^{n-1}, ..., k_M^{n-1})$ be the total loss to the traffic generated in the $(N-n+1)$ sections from the nth section to the Nth section when the offsets in those sections are optimized under the condition that the offsets at the intersections in the $(n-1)$th section are $(k_1^{n-1}, k_2^{n-1}, ..., k_M^{n-1})$. Now we seek to optimize the offsets over the whole of the network; that is, in accordance with the definition of f_n, we determine the offsets k_i^n $(i = 1, 2, ..., M; n = 0, 1, 2, ..., N)$ which correspond to f_0. For this we introduce the multistage decision process of dynamic programming.

First, the optimization of the Nth section is performed by the following equation.

$$f_N(k_1^{N-1}, k_2^{N-1}, ..., k_M^{N-1}) = \min_{k_1^N, k_2^N, ..., k_M^N} \left[\sum_{i=1}^{M} \{ {}^h g_{i1}^N (k_i^{N-1}, k_i^N, {}^h q_{i1}^N) \right.$$
$$+ {}^h g_{i2}^N (k_i^{N-1}, k_i^N, {}^h q_{i2}^N) \} + \sum_{i=1}^{M-1} \{ {}^v g_{i1}^N (k_i^N, k_{i+1}^N, {}^v q_{i2}^N)$$
$$\left. + {}^v g_{i2}^N (k_i^N, k_{i+1}^N, {}^v q_{i2}^N) \} \right]. \tag{19}$$

We can determine all the conditionally optimum offsets k_i^N $(i = 1, 2, ..., M)$ by utilizing the above equation. These offsets are denoted by $\hat{k}_i^N$ or $\hat{k}_i^N (k_1^{N-1}, k_2^{N-1}, ..., k_M^{N-1})$ in full.

Next, we optimize the $(N-1)$th section by using $f_N(k_1^{N-1}, k_2^{N-1}, ..., k_M^{N-1})$. That is, if we use the recurrence relation which follows the principle of optimality, f_{N-1} is expressed as

$$
f_{N-1}(k_1^{N-2}, k_2^{N-2}, ..., k_M^{N-2}) = \min_{k_1^{N-1}, k_2^{N-1}, ..., k_M^{N-1}} \left[\sum_{i=1}^{M} \right.
$$

$$
\{ {}^h g_{i1}^{N-1} (k_i^{N-2}, k_i^{N-1}, {}^h q_{i1}^{N-1}) + {}^h g_{i2}^{N-1} (k_i^{N-2}, k_i^{N-1}, {}^h q_{i2}^{N-1}) \}
$$

$$
+ \sum_{i=1}^{M-1} \{ {}^v g_{i1}^{N-1} (k_i^{N-1}, k_{i+1}^{N-1}, {}^v q_{i1}^{N-1})
$$

$$
+ {}^v g_{i2}^{N-1} (k_i^{N-1}, k_{i+1}^{N-1}, {}^v q_{i2}^{N-1}) \}
$$

$$
\left. + f_N(k_1^{N-1}, k_2^{N-1}, ..., k_M^{N-1}) \right] . \tag{20}
$$

From this equation all the conditionally optimum offsets $\dot{k}_i^{N-1}$'s ($i = 1, 2, ..., M$) in the ($N - 1$)th section are determined. Thus the optimization of the ($N - 1$)th section comes to an end.

Generally the optimization of the nth section is performed by the following equation.

$$
f_n(k_1^{n-1}, k_2^{n-1}, ..., k_M^{n-1}) = \min_{k_1^n, k_2^n, ..., k_M^n} \left[\sum_{i=1}^{M} \right.
$$

$$
\{ {}^h g_{i1}^n (k_i^{n-1}, k_i^n, {}^h q_{i1}^n) + {}^h g_{i2}^n (k_i^{n-1}, k_i^n, {}^h q_{i2}^n) \}
$$

$$
+ \sum_{i=1}^{M-1} \{ {}^v g_{i1}^n (k_i^n, k_{i+1}^n, {}^v q_{i1}^n)
$$

$$
+ {}^v g_{i2}^n (k_i^n, k_{i+1}^n, {}^v q_{i2}^n) \}
$$

$$
\left. + f_{n+1}(k_1^n, k_2^n, ..., k_M^n) \right] (n = 1, 2, ..., N - 1). \tag{21}
$$

From the above recurrence relation, the conditionally optimum offsets $\dot{k}_1^n, \dot{k}_2^n, ...,$ $\dot{k}_M^n$ at the intersections in the nth section are obtained for the given set of offsets $(k_1^{n-1}, k_2^{n-1}, ..., k_M^{n-1})$ in the ($n - 1$)th section.

If we proceed with the optimization process using Eq. (21), eventually we en-counter the following relation between the 0th section and the first section.

$$
f_0 = \min_{k_1^0, k_2^0, ..., k_M^0} \left[\sum_{i=1}^{M-1} \{ {}^v g_{i1}^0 (k_i^0, k_{i+1}^0, {}^v q_{i1}^0) + {}^v g_{i2}^0 (k_i^0, k_{i+1}^0, {}^v q_{i2}^0) \} \right.
$$

$$
\left. + f_1 (k_1^0, k_2^0, ..., k_M^0) \right] . \tag{22}
$$

The optimum offsets $k_1^0, k_2^0, ..., k_M^0$ obtained through Eq. (22) at this last stage are no more conditionally optimum offsets because there is no need to optimize the 0th section conditionally. These are absolutely optimum offsets designated as $\overset{*}{k}{}_i^0$.

After determining the optimum offsets in the 0th section as above, we can determine the absolutely optimum offsets in the first section $\overset{*}{k}{}_1^1, \overset{*}{k}{}_2^1, ..., \overset{*}{k}{}_M^1$ by choosing the conditionally optimum policies $\overset{.}{k}{}_1^1, \overset{.}{k}{}_2^1, ..., \overset{.}{k}{}_M^1$ which correspond to to $f_1(\overset{*}{k}{}_1^0, \overset{*}{k}{}_2^0, ..., \overset{*}{k}{}_M^0)$. In the same way the optimum offsets in the second section are determined by choosing $\overset{.}{k}{}_1^2, \overset{.}{k}{}_2^2, ..., \overset{.}{k}{}_M^2$ corresponding to $f_2(\overset{*}{k}{}_1^1, \overset{*}{k}{}_2^1, ..., \overset{*}{k}{}_M^1)$.

Thus all the optimum offsets in the nth section $\overset{*}{k}{}_1^n, \overset{*}{k}{}_2^n, ..., \overset{*}{k}{}_M^n$ ($n = 0, 1, 2, ..., N$) are obtained by investigating the conditionally optimum policies obtained before from the 0th section to the Nth section.

Hence a loss-minimizing set of offsets over the whole of the network is determined through dynamic programming.

Example

A grid network with nine signals at intersections is taken as a numerical example (Fig. 4). Every street is assumed to have 4 lanes. The numbers written around the network with arrows represent the traffic volume per lane per hour under two-way flow and the numbers beside the links represent the length of these links in meters. The numbers in circles and the numbers at the intersec-

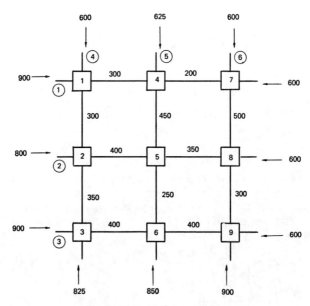

Fig. 4. Network for example.

tions label the streets and intersections, respectively. All intersections are 50 m wide. The cycle time and the split are 100 sec and 50% at all the intersections. The traffic flow is assumed to go straightforward without turning at intersections. The form of the loss function $g(\xi, \eta, q)$ can be determined by using the results of the traffic simulation previously mentioned.

From Fig. 2 we intuitively apply a cosine curve as the loss function, i.e.,

$$g(\xi, \eta, q) = a(q) - b(q) \cos \left[\frac{2\pi}{C} \{(\xi - \eta) - \overset{*}{\theta}\} \right], \qquad (23)$$

where $a(q)$ and $b(q)$ are quantities determined by q; and ξ and η are the offsets. $\overset{*}{\theta}$ is the optimal relative offset.

The $a(q)$ and $b(q)$ are represented by linear functions of q as in Fig. 5. Thus the form of the loss function becomes

$$g(\xi, \eta, q) = (0.893\, q - 150.5) - (0.800\, q - 149.5)$$

$$\cdot \cos \left[\frac{2\pi}{C} \{(\xi - \eta) - \overset{*}{\theta}\} \right]. \qquad (24)$$

For $\overset{*}{\theta}$, we take the travel time between adjacent intersections.

By using this loss function we calculate the total loss generated in the network shown in Fig. 4 for the following four cases: (1) the signal at each intersection is controlled individually, (2) only the offsets at the signals on some main arteries in the network are optimally set, (3) all the offsets over the network under two-way system are determined so as to minimize the total loss in the network through the discrete maximum principle and dynamic programming, and (4) the

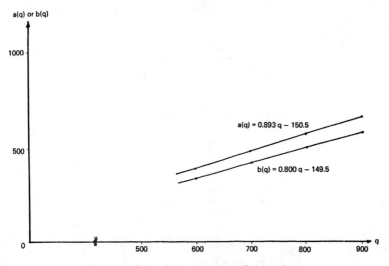

Fig. 5. Relation between $a\,(q)$ or $b\,(q)$ and q.

offsets over the network under one-way flow are synchronized optimally through the discrete maximum principle.

Case 1. The calculation was performed for 100 random offset patterns. The calculated total losses range from a minimum value of 19438 to a maximum value of 28292 with a mean value of 24181.

Case 2. Selected arteries where signals are synchronized are streets ①, ②, ③, and ⑥. The calculated minimum loss is 16561 which is 31.5% below the mean loss of Case 1.

Case 3. The loss to traffic in the network minimized through the discrete maximum principle is 13799 and the minimum loss through dynamic programming is 13737 which is 0.45% below the value through the discrete maximum principle, 17.0% below the total loss corresponding to main arteries synchronization and 43.2% below the mean total loss corresponding to the random offset pattern.

The calculation time required for optimization is 427 sec for the discrete maximum principle and 366 sec for dynamic programming, using FACOM 230-60.

Case 4. The direction of one-way is as follows. Streets ① and ③ are eastbound and street ② is westbound. Streets ④ and ⑥ are northbound and street ⑤ is southbound.

The eastbound traffic flow on the street ① under one-way flow is the sum of the eastbound traffic flow on the street ① and a half of the eastbound flow on the street ② in Fig. 4, and the eastbound flow on the street ③ under one-way flow is the sum of the eastbound flow on the street ③ and a half of the eastbound flow on the street ② in Fig. 4. On the other hand the westbound flow on the street ② under one-way flow is the sum of the westbound flow on the street ① and the street ③ in Fig. 4. It holds also for the southbound and northbound flow.

TABLE I

Obtained Offsets Pattern (sec)

| Number of intersection | Case 2 | Offset | | Case 4 |
| | | Case 3 | | |
		M.P.[a]	D.P.[b]	
1	0	0	0	0
2	60	60	60	90
3	10	30	10	60
4	40	50	30	30
5	80	0	90	60
6	60	0	50	90
7	40	90	70	60
8	10	50	40	30
9	70	40	0	10

[a]M.P.−Maximum Principle.
[b]D.P.−Dynamic Programming.

The result of the optimization through the discrete maximum principle is that the minimum loss in the network under one-way flow is 6827 which is 50.3% below the minimum loss under two-way flow.

The obtained optimal offset patterns are shown in Table I.

Conclusion

In the present paper we have optimized the offset pattern for a grid network but we can determine the optimum offset pattern for networks with diversified form through the same methods if only we let $g(\xi, \eta, q)$ be constant on some links which must be removed from the grid network to construct a given network.

For applying the procedures shown here to an actual network it may be necessary to determine the form of the loss function, taking into account the effects of the intersections' capacity, bus and taxi stops, pedestrians, etc.

Acknowledgment

The author acknowledges the encouragements and the suggestions given him by Prof. Eiji Kometani and Prof. Tsuna Sasaki of Kyoto University.

References

1. J. A. Hiller, Appendix to Glasgow's experiment in area traffic control, *Traffic Eng. Control* 7 (9), 565–571 (1966).

2. R. E. Allsop, Selection of offsets to minimize delay to traffic in a network controlled by fixed-time signals, *Transportation Sci.* 2 (1) (February 1968).

3. J. A. Hillier and Joyce Holroyd, The Glasgow experiment in area traffic control, *Traffic Eng.* (October 1968).

4. A. Chang, Synchronization of traffic signals in grid networks, *IBM J.* (July 1967).

5. J. D. Little, The synchronization of traffic signals by mixed-integer linear programming, *Operations Res.* 14 568–594 (1966).

6. L. T. Fan and C. S. Wang, *The Discrete Maximum Principle,* John Wiley & Sons, Inc., New York, 1964.

7. R. E. Bellman and S. E. Dreyfus, *Applied Dynamic Programming,* Princeton University Press, Princeton, N.J., 1962.

8. T. Sasaki, *Traffic Flow Theory,* Gijyutsu Shoin, Tokyo, 1965.

9. I. Okutani, Determination of loss-minimizing offsets pattern through the maximum principle, *Proc. JSCE.* No. 174 (1970-2).

10. S. Katz, A discrete version of Pontryagin's maximum principle, *J. Electron. Control* 13 (2) (1962-8).

11. I. Okutani, Fundamental aspect for area traffic control, *Traffic Eng.* 3 (4) (Tokyo, 1968-7).

Network Flow Model of the Australia-Europe Container Service

K. J. Noble and R. B. Potts
Mathematics Department, University of Adelaide, Adelaide, South Australia

Abstract

The Australia-Europe container system is formulated as a network flow model. The minimization of the cost of the movement of empty containers and the total container inventory is expressed as a linear programming problem. This is essentially a two-commodity flow problem because of the necessity to distinguish two types of containers—general, used for dry cargo, and insulated, used primarily for refrigerated cargo. The problem is solved using an efficient heuristic method based on the special structure of the network. A computer program for this method and a program for simulating the flow of containers through the network is used to validate the model and answer many practical questions concerning the operation of the container system. Realistic data have been supplied by the shipping consortia operating the system.

Introduction

The rapid recent development of theories of vehicular traffic flow has resulted, in part, from a realization that modern road systems have greatly simplified traffic movement, rendering it much more amenable to theoretical analysis. This system simplification has also taken place in shipping transportation. In the ten years since the first traffic flow symposium was held, irregular haphazard fragmented conventional shipping has, in part, been replaced by regular scheduled integrated container shipping.

This paper formulates the Australia-Europe container system as a network flow model, with emphasis on the efficient control of the flow of containers. The approach taken is essentially an OR one in which we are concerned not with the mathematical analysis for its own sake but for its practical application to real-life situations. For this reason, we have based our analysis largely on historical data for the Australia-Europe container service, although some use has been made of predictive data. The network flow approach employed here would be applicable not only to other container trades[1] but also to similar systems such as the containerized mail service analysed by Horn.[2]

Australia-Europe Container Shipping System

The service inaugurated in March, 1969 by Associated Container Transportation (ACT) and Overseas Containers Ltd. (OCL) was designed to provide a

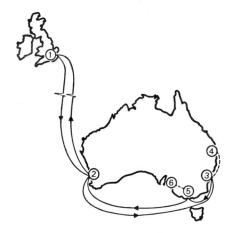

Fig. 1. Australia-Europe container system. Nodes 1, 2, 3, 4, 5, 6 represent Tilbury,
Fremantle, Sydney, Brisbane, Melbourne, and Adelaide, respectively. The solid lines
represent the container ship movements. The dashed line from node 3 to node 4
represents a feedership service between Sydney and Brisbane, and the dashed line from
node 5 to node 6 represents a rail service between Melbourne and Adelaide.

regular schedule of cellular container ships between the United Kingdom and
Australia with one port of call, Tilbury in the UK and three ports of call,
Sydney, Melbourne, and Fremantle in Australia (Fig. 1). With a speed of about
22 knots, the container ships complete a round voyage in about 70 days, ad-
hering closely to the following schedule:

	Port	Node	Day
Leaving	Tilbury	1	1
Arriving	Fremantle	2	23
Arriving	Sydney	3	28
Arriving	Melbourne	5	33
Arriving	Fremantle	2	39
Arriving	Tilbury	1	62
Leaving	Tilbury	1	71

It is a feature of container ship operations that the time a ship spends in port is
minimal—less than half a day in Fremantle at each Southbound and Northbound
call, about 3 days in each of Sydney and Melbourne, and about 9 days at Tilbury.
As new ships have been phased into the service, the frequency of sailings has
been increased and in 1970 was about one a week with an improvement to one
every 5 or 6 days expected when all fourteen ships planned for the service are
operational.

    The containers used in the service are mainly standard 20-ft containers of two
types—general containers used for dry cargo, and insulated containers designed
especially for refrigerated cargo (or reefer cargo as it is commonly called) but

used also for dry cargo. Each cellular ship was designed to have a capacity of about 1300 containers, including about 350 insulated containers, although subsequent modifications have enabled these capacities to be increased.

In Australia, terminal facilities for unloading and loading containers are provided at Fremantle, Sydney, and Melbourne, where depots for packing and unpacking of containers are also provided. In addition, Brisbane has a terminal-depot served by a coastal feedership to Sydney, and Adelaide a depot with a direct rail link to the Melbourne terminal.

Empty containers can be stockpiled at terminals or depots, and because there are local and international imbalances in full container movements, interstockpile empty container movements are required. Typical of these movements of empty containers (empties) are: Sydney to UK by containership; Sydney to Brisbane by feedership; and Melbourne to Adelaide by rail.

The service provided by the container ships—regular, periodic sailings with few ports of call—is a marked contrast to the conventional shipping service which uses many ports of call in Australia and the UK and which does not adhere to regular schedules. It is the basic simplicity of the container service which makes it readily amenable to mathematical and computer analysis.

Network Flow Model

Network

The movement of full and empty containers may be represented as flow in a network of nodes and links; and the efficient control of this movement may be expressed as the problem of minimizing the cost of flow in this network.

The nodes of the network generally represent stockpiles for containers, and the links of the network represent transport modes between the stockpiles. When a container is packed with export cargo, an empty container is consumed and a full container is produced. Conversely, when an imported full container is unpacked, a full container is consumed and an empty container is produced. This production and consumption of full and empty containers is assumed to occur only at nodes of the network, and the containers (full or empty) may be transported only along the network links.

The detail with which the network is chosen depends on the uses to be made of the network model. For many applications, we have found it adequate to regard each port of call as a container stockpile, represented by a single node. Thus in Fig. 2, node 5 represents the Melbourne stockpile while the link from node 3 to node 1 represents the possibility of a container movement from the Sydney stockpile to the Tilbury stockpile.

For other purposes, a more detailed network, as indicated in Fig. 3, is necessary. Each port of call is now represented by two nodes, one representing the container ship in port and the other the container stockpile. This detail helps to distinguish between containers which are loaded and unloaded at a port and those which are simply in transit.

Since we are concerned with the dynamic flow of containers throughout the

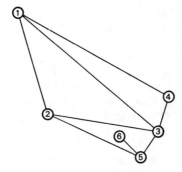

Fig. 2. Australia-Europe container network. Nodes 1, 2, 3, 4, 5, 6 represent container stockpiles at Tilbury, Fremantle, Sydney, Brisbane, Melbourne, and Adelaide, respectively. Links between nodes represent possible container movements (for clarity only a few links have been drawn).

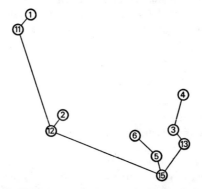

Fig. 3. Detailed Australia-Europe container network. Nodes 1, 2, 3, 4, 5, 6 represent container stockpiles at Tilbury, Fremantle, Sydney, Brisbane, Melbourne, and Adelaide, respectively. Nodes 11, 12, 13, 15 represent the containership in port at Tilbury, Fremantle, Sydney, and Melbourne, respectively.

network, it is necessary to attach to each link a travel time. Because of the inherent regular periodic structure of the container service it is convenient to measure the travel time in units corresponding to the interval between successive ships. Thus a travel time of 3 units for a weekly container service means a travel time of 3 weeks. For convenience, we assume in our description that the service is weekly.

When used in reference to full container movements, the term "travel time" is interpreted as follows. The process involved in sending cargo from Tilbury to Fremantle (say) is quite inflexible. An empty container is taken from the Tilbury stockpile, is packed with the cargo, and is loaded on the container ship at Tilbury. When the ship arrives at Fremantle, the imported full container is unpacked and the resultant empty is placed on the Fremantle stockpile. The time occupied by the complete process is called the "travel time" for the full container movement from Tilbury to Fremantle, and might be 5 weeks compared with (say) 3 weeks for the Tilbury-Fremantle empty movement.

The movement of containers over time is represented as flow in a dynamic network. For example, Fig. 4 is the 10-week dynamic version of Fig. 2, with links illustrating possible container movements. Since the travel time for full containers from Tilbury to Fremantle is 5 weeks, the complete dynamic version of the network contains a link from node 1 in week 1 to node 2 in week 6, from node 1 in week 2 to node 2 in week 7, and so on; these links are shown as solid lines in Fig. 4. Similarly the travel time for empty containers from Sydney to Brisbane is 1 week, and the complete dynamic network contains a link from node 3 in week 1 to node 4 in week 2, from node 3 in week 2 to node 4 in week 3, and so on; and these links are shown as dashed lines in Fig. 4. To allow for containers being held at Melbourne from week 8 to week 9, there is a link (dotted line) from node 5 in week 8 to node 5 in week 9; and the dotted line from node 6 in week 4 to node 6 in week 5 allows for containers being held at Adelaide from week 4 to week 5. Thus it is possible to represent as flows along links of the dynamic network not only full and empty container movements, but also the holding of containers at stockpiles.

In formulating the dynamic network, some judgement is needed in interpreting container movements as occurring from one week to another week, but the sensitivity of the results to the assumptions made can easily be tested. A more accurate day-to-day network would be useful for problems of detailed container control, but would be far too complex for the applications we have in mind.

Cargo movements are taken as a given input to our network model, and represented in the dynamic network as the flows of given numbers of full containers. Because of cargo imbalances, it is necessary to hold empty containers in stockpiles and move them between stockpiles, and these possibilities are

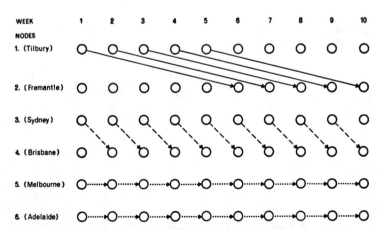

Fig. 4. 10-period dynamic version of Fig. 2. Solid lines represent full container movements, dashed lines represent empty container movements, and dotted lines represent stockholding at stockpiles. Only a few links are shown.

represented in the dynamic network as flows which have to be determined. In determining these optimum flows of empty containers, our criterion for optimization is the minimization of the total cost of container inventory and empty movements, and each link of the dynamic network is assigned a unit cost. The container inventory cost reflects the weekly cost of owning a container, and so for example the unit inventory cost assigned to a link with travel time 5 weeks is 5 times the weekly inventory cost per container. Having accounted for the inventory cost on all links, we account for empty movement costs by assigning each empty movement link the appropriate unit empty movement cost. Thus in Fig. 4 the overall unit cost assigned to the Sydney-Brisbane empty movement link is the sum of the weekly inventory cost per container and the unit empty movement cost from Sydney to Brisbane.

In some applications we shall need to distinguish between general and insulated containers, and regard them as two commodities in our network flow model. The flow of these two commodities will be interdependent because dry cargo may be packed into insulated containers.

Mathematical Notation

In the following list of the mathematical notation to be used, subscripts i, j take values $1, 2, 3, \ldots$ and refer to nodes or stockpiles i, j. The variable t, taking values $1, 2, 3, \ldots$ refers to the week number. The superscript k signifies the two container types, $k = 1$ for general and $k = 2$ for insulated containers.

$a_{ij}(t), b_{ij}(t)$ = number of container loads of dry, reefer cargo to be dispatched in week t from stockpile i to stockpile j. (1)

$\bar{a}_{ij}(t)$ = number of container loads of dry cargo (to be dispatched in week t from i to j) which could be packed in insulated containers. (2)

$e_{ij}^k(t), f_{ij}^k(t)$ = number of empty, full containers of type k to be dispatched in week t from i to j. (3)

$g_i^k(t)$ = total number of containers of type k added to the system at stockpile i in week t (negative values of $g_i^k(t)$ signify containers taken from the system). (4)

$u_{ij}(t)$ = maximum number of empties (general + insulated) which can be dispatched in week t from i to j. (5)

$n^k(t)$ = total number of containers of type k in the system in week t. (6)

s_{ij}, t_{ij} = number of weeks for an empty, full container movement from i to j. (7)

c_{ij} = unit cost for moving an empty container from i to j. (8)

α^k = unit weekly inventory cost for containers of type k. (9)

C = total cost of container inventory and interstockpile empty movements. (10)

For convenience we interpret $e_{ii}^k(t)$ as the number of containers of type k held at stockpile i during week t (after arrivals from and departures to other stockpiles) and take $s_{ii} = 1$. The total number of containers in week t is then given by those held at stockpiles and those moving empty plus those moving full in week t, so that

$$n^k(t) = \sum_{i,j} \sum_{\tau} e_{ij}^k(t - \tau) + \sum_{i,j} \sum_{\tau}' f_{ij}^k(t - \tau). \tag{11}$$

Here the summation is taken over links of the dynamic network (such as in Fig. 4) with τ varying over the range $\tau = 0$ to $s_{ij} - 1$ in the summation Σ_τ, and τ varying over the range $\tau = 0$ to $t_{ij} - 1$ in the summation Σ_τ'. The summation in (11) is readily interpreted on the dynamic network (such as Fig. 4) as the sum over all links which originate in week t or earlier and terminate in week $(t + 1)$ or later.

Linear Programming Formulation

The problem of minimizing the total cost of container flows in the dynamic network can be formulated as the following linear program:

$$\text{Minimize } C = \sum_t \sum_k \left(\sum_{i,j} c_{ij} \, e_{ij}^k(t) + \alpha^k n^k(t) \right) \tag{12}$$

$$\text{subject to } e_{ij}^k(t) \geqslant 0, \tag{13}$$

$$\sum_k e_{ij}^k(t) \leqslant u_{ij}(t), \tag{14}$$

$$\sum_k f_{ij}^k(t) = a_{ij}(t) + b_{ij}(t), \tag{15}$$

$$b_{ij}(t) \leqslant f_{ij}^2(t) \leqslant b_{ij}(t) + \bar{a}_{ij}(t), \tag{16}$$

$$\sum_j \{e_{ij}^k(t) + f_{ij}^k(t)\} - \sum_j \{e_{ji}^k(t - s_{ji}) + f_{ji}^k(t - t_{ji})\} = g_i^k(t). \tag{17}$$

In obtaining the objective function (12), in which $n^k(t)$ is given by (11), the usual and adequate assumption of linear costs is made. The variables to be determined are $e_{ij}^k(t)$ and $f_{ij}^k(t)$, all other quantities being given.

Capacitated links in the network represented by the constraint equation (14), can be interpreted in different ways. For example, $u_{ii}(t)$ could represent the capacity of the stockpile at i and would be independent of t unless changes in the stockpile were made. Between one port of call i and the next port of call j, $u_{ij}(t)$ could represent the number of empties which can be carried on the container ship on the leg i to j. The space for empties on each leg could be calculated from the given full container flows and the ship capacities. Although reefer cargo can be carried only in certain cells of the ship, there is no restriction on the placement of empties. In many applications of our model, it has been possible to ignore link capacities.

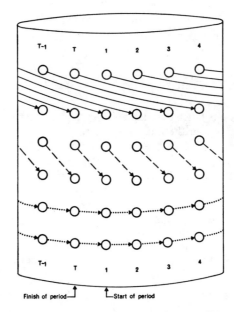

Fig. 5. T-period dynamic network with cyclic boundary condition. Network can be thought of as drawn on the surface of a cylinder, so that week T is adjacent to week 1.

The above LP formulation does not specify initial and terminating conditions. For many applications, we have found it convenient and adequate to impose a cyclic boundary condition, as used by Horn.[2] A natural cycle period of one year suggests itself, but for generality we let

$$T = \text{number of weeks in the cycle period.} \qquad (18)$$

The variable t then assumes the values $1, 2, \ldots T$ and, for example, $u_{ij}(-2)$ is taken equal to $u_{ij}(T - 2)$. The cyclic boundary condition means, in effect, that the T-period dynamic network is wrapped around a cylinder so that weeks T and 1 are adjacent, as indicated in Fig. 5.

If the cyclic boundary condition is used, a necessary condition on the given quantities $g_i^k(t)$ is that

$$\sum_t \sum_i g_i^k(t) = 0, \qquad (19)$$

so that the total number of containers added to the system is equal to the total number taken out.

Heuristic Solution

No efficient solution techniques are known for solving general two-commodity flow problems on large networks, but for the Australia-Europe container network flow model formulated above, an excellent heuristic procedure is available which reduces the two-commodity problem to two one-commodity problems. Each of these can be solved using an algorithm such as the efficient out-of-kilter algorithm.[3]

The cargo imbalances in the Australia-Europe system provide the motivation for the heuristic solution procedure. Australia exports much more reefer cargo than she imports, and imports more dry cargo than she exports. Northbound insulated containers are packed with reefer cargo only, and if dry cargo is not packed in Southbound insulated containers, most of these travel empty. Then because of the dry cargo imbalance, many general containers must travel empty Northbound. But if the insulated containers are packed with dry cargo South-bound this reduces the Northbound movement of empty general containers and the general container inventory. A considerable saving in inventory and empty movement costs is achieved.

There are two phases in the heuristic solution procedure. In Phase I, the dry cargo and general containers are excluded and attention focussed on the movement of the reefer cargo in the insulated containers. In preparation for Phase II, travel times for moving empty insulated containers are taken the same as for full containers, i.e., we take $s_{ij} = t_{ij}$. In Phase II, the movement of the general cargo is considered and the optimum movements of the empty general containers analyzed allowing for the empty insulated containers available from Phase I. For each phase of the procedure, the LP is a one-commodity flow problem which is readily solved and the final solution at the end of Phase II is certainly a feasible solution of the original two-commodity problem. Because of the structure of the container network being considered and the particular cargo imbalance patterns, it turns out to be a very good solution.

The two phases of the heuristic procedure can be described mathematically as follows:

Phase I. Let

$$a_{ij}(t) = e_{ij}^1(t) = f_{ij}^1(t) = g_i^1(t) = 0 \tag{20}$$

and

$$s_{ij} = t_{ij}. \tag{21}$$

Note that the capacity constraint (14) becomes

$$e_{ij}^2(t) \leqslant u_{ij}(t) \tag{22}$$

and (15) and (16) simplify to

$$f_{ij}^2(t) = b_{ij}(t). \tag{23}$$

By solving the resulting one-commodity flow problem determine $e_{ij}^2(t)$ and hence

$$\bar{f}_{ij}^2(t) = \min \{e_{ij}^2(t), \bar{a}_{ij}(t)\} \tag{24}$$

and

$$\bar{e}_{ij}^2(t) = e_{ij}^2(t) - \bar{f}_{ij}^2(t) \tag{25}$$

which represent, respectively, the number of empty insulated containers available for packing with dry cargo and the number of insulated containers which will still remain empty.

Phase II. Let

$$e_{ij}^2(t) = \bar{e}_{ij}^2(t), \tag{26}$$

$$f_{ij}^2(t) = b_{ij}(t) + \bar{f}_{ij}^2(t), \tag{27}$$

and solve the resulting one-commodity problem to determine $e_{ij}^1(t), f_{ij}^1(t)$; this is the required heuristic solution.

A slight improvement in the Phase I procedure is possible. After determining $f_{ij}^2(t)$ from (27), we may repeat Phase I to determine $e_{ij}^2(t)$ subject to (20), (22), and (27) and with the s_{ij} set at the empty movement travel times. This may reduce the inventory of insulated containers.

The fact that the heuristic procedure provides a good solution may be attributed to the nature of the cargo imbalances in the Australia-Europe system. In other container systems, different heuristics may be required.

Computer Programs

Computer programs for solving the one-commodity flow problem have been written for a CDC 6400 computer using the out-of-kilter algorithm. Allowance has been made for adding nodes to the network and incorporating additional links.

In addition, a deterministic SIMSCRIPT program has been written to simulate the container system. A finer time scale is employed than is possible in the LP approach, but empty container movements must be specified as input. The SIMSCRIPT program output indicates the number of containers on hand at stockpiles, when and where shortages occur, and the mean travel times for container movements. The simulation program complements the LP program, and the two used together have proved successful in a variety of applications of the model.

Container Movement and Cost Data

Container Flows

An analysis of historical data over the period February-August 1970 and covering 20 voyages yielded the average full container flows and major empty flows listed in Tables I and II. Because these figures relate to a comparatively short period during the growth of the container service, they should not be taken as an indication of later operations. In particular, the empty movements were atypical because in the period analyzed many containers were being positioned in Australia for other trades. More typically, large numbers of these containers would be returned empty to Tilbury.

Table I illustrates the cargo imbalances for both dry and reefer cargo. The imbalance at Melbourne, for example, is a net import of 90 container loads of

TABLE I

Average Cargo Figures (Container Loads) per
Voyage over Period February-August 1970

	Dry cargo	Reefer cargo
Import		
Tilbury-Sydney	500	2
Tilbury-Melbourne	480	3
Tilbury-Adelaide	65	1
Tilbury-Brisbane	85	1
Tilbury-Fremantle	80	1
Total	1210	8
Export		
Sydney-Tilbury	180	20
Melbourne-Tilbury	390	110
Adelaide-Tilbury	70	20
Brisbane-Tilbury	90	20
Fremantle-Tilbury	70	50
Total	800	220

TABLE II

Major Flows of Empty General Containers
(Average Number per Voyage) over Period
February-August 1970

Link	Number
Sydney-Tilbury	100
Melbourne-Tilbury	40
Sydney-Brisbane	5
Melbourne-Adelaide	5

dry cargo and a net export of 107 container loads of reefer cargo. Table II shows the major flows of empty general containers. There are no major flows of empty insulated containers because they are imported to Australia packed with dry cargo.

Ship Capacity for Empty Containers

The ship capacity for empty containers can be calculated in the following way. We assume an "average" ship with a capacity of 1300 containers and carrying the full container loads listed in Table I (except that for simplicity we ignore the import of any reefer cargo). We calculate the number of full containers carried on each leg of the voyage and hence deduce the capacities for empties. The average ship carries 1210 full containers on the Tilbury-Fremantle leg and, therefore, $1210 - 80 = 1130$ full containers on the Fremantle-Sydney leg (Fremantle exports are loaded when the ship calls *Northbound* at Fremantle). The situation at Sydney is a little more complicated. The number of full containers discharged is 500 for Sydney and 85 for Brisbane and the number loaded is 180 + 20 = 200 from Sydney and 90 + 20 = 110 from Brisbane. The number of full containers on the Sydney-Melbourne leg is therefore 1130 – 585 + 310 = 855. A similar calculation gives 900 full containers on the Melbourne-Fremantle leg and 1020 full containers on the Fremantle-Tilbury leg. These results are summarized in Table III together with the ship capacities. It will be noted

TABLE III

Ship Capacity for Empty Containers (Using Data in Table I)

Ship leg	No. of fulls on board	Ship capacity for empties
Tilbury-Fremantle	1210	90
Fremantle-Sydney	1130	170
Sydney-Melbourne	855	445
Melbourne-Fremantle	900	400
Fremantle-Tilbury	1020	280

that there is little space for empties on the Tilbury-Fremantle leg, for a load factor exceeding 90% is achieved on Southbound voyages. In the reverse direction, the load factor is nearly 80%.

For a proper representation of ship capacities and empty movement costs, the detail in the network illustrated in Fig. 3 is necessary.

Cost Data

Inventory costs and empty movement costs are needed as inputs to the model, and the following estimates (in Australian dollars) were used:

>   *Inventory costs* (per week per container)
>   for general containers $4
>   for insulated containers $8
>   *Empty movement costs* (per container)
>   between any two ports via container ship $40
>   between Melbourne and Adelaide via rail $30
>   between Sydney and Brisbane via feeder ship $80.

There are alternative transport modes (e.g., rail, feeder ship) between ports in Australia but these are much more costly than the container ship and are used only in emergencies.

For a proper representation of these costs, the detailed network (Fig. 3) is needed. In fact the $40 unit cost for any port to port transport of an empty container is comprised of a loading charge of $20 at the origin port and an unloading charge of $20 at the destination port. Thus in Fig. 3, links such as (2, 12), (3, 13) representing loading and unloading are ascribed costs of $20 while links such as (13, 15) representing sea legs are given zero costs.

Model Applications

Validation

The network flow model of the Australia-Europe container system has been validated by conducting numerous runs with the LP and simulation computer programs and verifying that the results were acceptable as a description of how the system operates. For example, the LP program was run with estimates of containerized cargo for 1970 and optimal empty movements were as expected (see Table IV). Sydney, a large producer of empties, supplies Brisbane, Melbourne and Fremantle with the empties they consume, and the remainder are returned to Tilbury. Empties from Melbourne satisfy the Adelaide demand.

TABLE IV

Summary Results for Run of LP Program
Using 1970 Cargo Estimates

	General containers	Insulated containers
Container inventory	14,250	4,150
Cost of empty movements	$600,000	$10,000
Average weekly empty movements		
Sydney–Tilbury	160	negligible
Sydney–Melbourne	65	"
Sydney–Fremantle	40	"
Melbourne–Adelaide	35	"
Sydney–Brisbane	10	"

TABLE V

Mean Travel Times for Full Container Movements
Obtained in Simulation Runs

Link	Normal ship schedule	Alternate ships omit Fremantle Southbound	Difference
Tilbury–Fremantle	37.2	40.5	+3.3
Tilbury–Adelaide	51.2	50.8	−0.4
Tilbury–Melbourne	48.8	48.4	−0.4
Tilbury–Sydney	45.2	44.5	−0.7
Tilbury–Brisbane	49.6	49.6	0.0
Fremantle–Tilbury	37.5	37.3	−0.2
Adelaide–Tilbury	46.7	46.6	−0.1
Melbourne–Tilbury	43.8	43.9	+0.1
Sydney–Tilbury	47.9	47.8	−0.1
Brisbane–Tilbury	53.1	53.0	−0.1

A somewhat different test of the model was provided by using the simulation program to investigate the effect of altering the routing of the container ships by omitting the Southbound call at Fremantle on alternate voyages. The simulation program, in contrast to the LP program which uses a week as a basic unit, is sensitive to a slight alteration in travel times and Table V compares the results for the normal and revised schedules. As expected, the mean travel times from Tilbury to Adelaide, Melbourne and Sydney decrease by about half a day at the expense of an increase of about $3\frac{1}{2}$ days in the travel time to Fremantle. Travel times to Brisbane and times on the Northbound sailings are unchanged.

Weekly Cargo Fluctuations

The network flow model has been applied to the problem of assessing the effect of weekly fluctuations in cargo movements. It is not evident *a priori* whether such fluctuations would necessitate larger container stocks and more empty movements. For an average monthly flow of 1600 full containers from Tilbury to Sydney, we considered three possible weekly flows:

	Full container flows in week				
Case	1	2	3	4	Total
A	400	400	400	400	1600
B	480	320	480	320	1600
C	320	480	480	320	1600

For the three different weekly movements, the container inventory and overall empty movements were almost identical. The weekly fluctuations (in contrast to seasonal fluctuations) are too rapid to influence the operation of the system because at any one time about 7 or 8 consecutive weekly batches of containers are in transit.

Packing Dry Cargo in Insulated Containers

The sensitivity of the container inventory and empty container flows to the availability of dry cargo suitable for packing in insulated containers is quite dramatic. Our model has been used to illustrate this by doing computer runs using various values for the assumed percentage of dry cargo suitable for packing in insulated containers. The results (again based on 1970 cargo predictions) are tabulated for the extreme values 0 and 100% in Table VI, and indicate an overall reduction in cost from $6.2 to $5.1 million. As indicated in Fig. 6, this reduction is achieved when the percentage of suitable dry cargo is 25% for then almost all available empty containers will be used.

Sensitivity to Cost and Travel Time Data

Table VI illustrates what proves to be a very significant feature of the cost structure of the container flows in the Australia-Europe system. The inventory

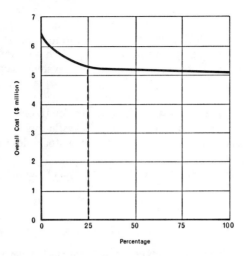

Fig. 6. Overall cost of container inventory and empty movements as a function of the percentage of dry cargo assumed suitable for packing in general containers. Once the 25% level is reached virtually no cost improvement is obtained.

costs are very much greater, in total, than the cost of moving empties. This helps simplify the analysis of different situations and suggests, for example, that the container flow patterns do not depend sensitively on the assumed unit costs, a fact which can be readily verified.

The total costs are also insensitive to the empty movement travel times because the empties comprise only a small fraction of the total container inventory. A change in travel times for full container movements does alter the total costs. An increase in average travel time from eight to nine weeks increases the container inventory by about an eighth.

TABLE VI

Cost Comparison Where 0%, 100% of Dry Cargo Is
Suitable for Packing in Insulated Containers

	0%		100%	
	General	Insulated	General	Insulated
Container inventory	16,950	4,100	14,250	4,150
Container inventory cost (50 weeks)	$3,390,000	$1,640,000	$2,850,000	$1,660,000
Cost of empty movements (50 weeks)	$740,000	$430,000	$600,000	$10,000

Other Applications

The network flow model has also been applied to many other aspects of the Australia-Europe container system and to answer such questions as:

(i) When to inject additional containers into the system to cope with a growth in trade;

(ii) What savings would accrue if the container consortia pooled containers;

(iii) Whether containers which would otherwise be travelling empty around the Australian coast should be used to carry domestic cargo;

(iv) What stocks of containers should be held at stockpiles to meet unexpected demands;

(v) What would be the effects on container flows in Australia if inland stockpiles were established.

Acknowledgment

This paper reports some of the results of a project sponsored by the Australian Commonwealth Department of Trade and Industry, and the authors acknowledge the continued interest and help of the Secretary, Sir Alan Westerman, and other officers of his Department. A major portion of the data we have used was supplied by Associated Container Transportation (Australia) Ltd., and Overseas Containers (Australia) Ltd., and these firms were generous with their assistance in many ways. The SIMSCRIPT program was written by Mr. R. Aust. The first author acknowledges the support of a CSIRO Postgraduate Studentship.

References

1. F. L. Weldon, Cargo containerization in the west coast-Hawaiian trade, *Operations Res.* 6, 649–670 (1958).
2. W. A. Horn, Determining optimal container inventory and routing, *Transportation Sci.* 5, 225–231 (1971).
3. L. R. Ford and D. R. Fulkerson, *Flows in Networks*, p. 162, Princeton University Press, Princeton, N.J., 1962.

A Theoretical Study of Bus and Car Travel in Central London

F. V. Webster and R. H. Oldfield
Road Research Laboratory, Department of the Environment, Crowthorne, Berkshire, England

Abstract

A theoretical study has been carried out to see what the effect would be of varying the size of bus in the central London bus fleet and what effect the degree of private car restraint would have on the optimum bus size and on overall travel costs in the area.

For each size of bus considered the overall travel times for cars, buses, goods vehicles and taxis were calculated and costed. To these were added the operating costs of all vehicles on the network. This procedure was carried out for each size of bus in turn and the whole process repeated for each level of private car restraint considered, ranging from present conditions up to full restraint where no cars are left on the network.

Considering, initially, time and operating costs only, an optimum bus size for present conditions in central London was found to be a 55-seater double-deck bus and for conditions of complete restraint of private car traffic, a 40-seater bus. In these calculations (where time and operating costs only are considered) the greatest community savings occur under full restraint. However, if estimates of subjective costs which drivers bear on transferring to bus travel because of loss of privacy, convenience and comfort, etc., are introduced into the calculations, a level of restraint at rather less than half of all private cars is found to give the greatest community savings, and at this level of restraint the optimum bus size is again a 40-seater.

The calculations showed that the mean direct journey speed (origin to destination as the crow flies) of all bus and car travellers would be expected to rise from 6 km/h at present to $7\frac{1}{2}$ km/h when about half of the present cars are restrained and to 8 km/h under complete restraint.

Complete restraint of private traffic would be expected to reduce total time and operating costs of the central London street network by about £20 m per year (18%). If the subjective costs mentioned above are included, full restraint would be expected to increase total community costs to a level somewhat higher than the present level, and an optimum level of restraint as mentioned above is found at which the overall community saving would be expected to be about £10 million per year.

Aim of Study and Criteria Used

This theoretical study is a continuation of one reported in Refs. 1 and 2.

Its first aim is to find the optimum size of bus for use in the central area of London (an area of 32.4 km^2) under various conditions of private car restraint. One criterion for optimum is that the sum of the time costs of all travellers and the operating costs of all vehicles on the average main road in the network is minimized. This criterion takes no account of some people's preference for using their own car even when time and operating costs are greater by so doing; thus, another criterion is to include in the optimization process the subjective costs of

References pp. 345-346

drivers' preference for using their own cars (called mode-specific costs). Both methods are considered in this study.

The second aim of the study is to investigate the community saving resulting from a transference to bus travel of some or all of present car users in central London.

The Advantages of Different Size Buses

Large buses have the advantages that fewer are needed for a given load. Thus, both fares and road congestion due to buses are less and speeds are higher. Small buses, however, have the advantage that the route coverage and frequency of service can be improved because so many more of them are required to carry the load, so that walking and waiting times are reduced; and since the number of passengers getting off and on a given bus at each stop is less, stopped time is reduced.

With each criterion used here, there is an optimum situation in which the best balance is obtained between the relevant variables, viz., number of passengers per bus, speed of cars and buses on the network, walking and waiting time for buses and operating costs of all vehicles.

Method Used When Mode-Specific Costs Are Excluded

The study was divided into several parts, each one corresponding to a particular level of restraint of private car traffic, which varied from no restraint (i.e., present conditions*) up to full restraint, i.e., no private cars at all. For each level of restraint, each of the following sizes of bus was studied in turn, it being assumed that for any one system all buses were identical: 14, 20, 30, 40, 50, 60, 70, 80, 100 seaters.

The buses were assumed to be occupied to 65% of seating capacity during peak periods, but for off-peak periods the computer program sought the level of occupancy which minimized total time and operating costs (see Section 4 of Ref. 5).

The distance between bus stops and the density of routes (relative to present density) were also optimized for each situation studied.

For each bus size in turn, calculations were made of the mean direct journey speed† for both bus and car travellers, separately for peak and off-peak periods. These were obtained by, firstly, estimating the speeds on the network from a speed-flow curve shown in Fig. 1 which is based on observed speeds on main roads in central London.[3-5] The network speeds were then modified to include parking time for car travel, and bus-stop time and walking and waiting times for

---

*No restraint, in this paper, means no more restraint than exists at present.
†The direct journey speed is the straight-line distance from origin to destination divided by the total time taken, including walking and waiting for bus travellers and parking time for car travellers.

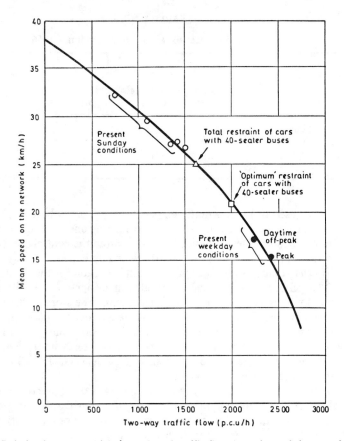

Fig. 1. Relation between vehicles' speeds and traffic flow on main roads in central London.

bus travel. These latter times were obtained from actual studies carried out in central London.[6]

Having calculated the total journey times for all travellers they were then costed and added to the operating costs of all vehicles on the network to find the overall travel cost per kilometer of network per hour. This was then repeated for each bus size in turn until all the selected bus systems had been costed.

The whole process was then repeated for several different levels of private car restraint.

Basic Data and Assumptions

(1) *Flows*. Vehicular and passenger flows in central London in the year 1966 (taken as "present" conditions) are given in Table I.

It was assumed that the total number of bus and car travellers in the area at

TABLE I

Vehicular and Passenger Flows in Central London[3] – 1966

		Average two-way flow of vehicles per hour	Passenger occupancy	Passengers per hour	Percentage of passengers using car or bus
Peak	buses	127	38	4826	74
	cars	1196	1.45	1734	26
Off-peak	buses	101	18	1818	60
	cars	828	1.45	1201	40

present (taken as 1966) is constant, i.e., people may transfer between bus and car but no new journeys are generated and no existing ones are suppressed.

(2) *Journey length.* Mean length of journey within the central area (origin to destination as the crow flies) by car is 4 km and by bus is 2.7 km.[7,8]

(3) *Bus type.* Since the study is providing information for the future it is assumed that all buses are one-man operated with automatic fare-collection facilities and with separate doors for entry and exit. The stopped time at bus stops is given as $5 + 2.3 b$ seconds, where b is the number of boarding passengers. Alighting is assumed to take place at the same time as boarding, never lengthening the stopped time. Small buses with fewer than 40 seats are single-deckers, and large buses with more than 60 seats are double-deckers. The 40 to 60 seater range can be either single or double-deckers.

(4) *Bus pcu values.* Each bus type considered has a pcu value (computed in terms of its effect on saturation flow at traffic signals) which is a function of its plan area; for convenience, it is expressed in terms of its nominal seating capacity. The pcu value is different for single and double-deckers and is affected by the degree of restraint of cars. It also takes into account the effect on the traffic of stopping at bus stops. The model for estimating pcu values was calibrated from saturation flow measurements of London Transport buses at signal-controlled intersections. Computed pcu's of the various bus types are shown in Fig. 2.

(5) *Bus occupancy.* It was assumed that each bus, during peak periods, had the maximum practicable occupancy, taking into account the heavily loaded and lightly loaded directions of the routes. This was taken to be 65%, on average, of the seating capacity (the present average peak-period loading factor for central London).

During off-peak periods the occupancy was assumed to be that value which minimized the sum of the time costs and operating costs of all vehicles on the street network. This was obtained by running the computer program many times for each bus size–car restraint situation considered; on each occasion the off-peak bus occupancy was increased by 1 passenger per bus until the optimum was found, subject to the condition that this occupancy is always lower or equal

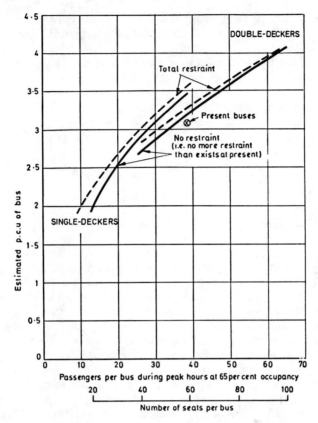

Fig. 2. Computed pcu values of buses of various types and sizes.

to the peak-time occupancy so that the chance of a bus passenger being left behind at a bus stop in the off-peak period is no higher than for peak times.

(6) *Overall costs.* The total costs on the network are summarized as follows: (i) bus passengers' waiting, walking and travel times (converted into monetary terms) plus bus fares; (ii) car travellers' parking time and travel time costs plus parking charges and running costs; (iii) operating costs of taxis and goods vehicles.

Mean costs on the network, measured in £/km/h were derived after assuming that one week's travel consisted of 20 h of peak conditions and 64 h of off-peak conditions.

(7) *Value of time.* The value of travellers' time while riding on a bus or car is taken at £0.175/h and the part of the journey spent walking or waiting for a bus as £0.44/h. These figures are based broadly on figures recommended by the Department of the Environment and are weighted according to the number of people travelling on business and in their own time.

(8) *Crew costs.* Bus crew wages rise with the number of crews required in accordance with a power law which increases the wage by 33% for a doubling of the number of crews required.

(9) *Bus fares.* Bus fares reflect the full operating costs, which are based on data supplied in the *Commercial Motor Journal.*[9]

(10) *Operating costs.* Operating costs in new pence per vehicle-km of cars, taxis and goods vehicles are as follows[10]:

Cars (running costs only): $0.8 + \dfrac{16}{v}$

Taxis: $0.6 + \dfrac{47}{v}$

Goods vehicles: $1.1 + \dfrac{71}{v}$

where v is the speed in km/hour.

Method Used When Mode-Specific Costs Are Included

The overall costs under conditions of restraint as described above do not take account of any in-built subjective resistance of car riders to transferring to bus travel; in other words, even if the time and operating costs by bus were lower than when travelling by car, many present car users would still prefer to use their cars. This in-built resistance may arise from a desire for privacy, extra comfort, door-to-door convenience or a number of other factors. It is now well accepted that time and operating costs can be given fairly meaningful average values but these other factors are extremely difficult to evaluate. They are described in this paper as "mode-specific costs"; they have sometimes been referred to as "subjective costs." To take account in this theoretical exercise of this subjective resistance it has been assumed that a system of charging for the use of congested roads is in operation and that car travellers change their mode of travel when the road pricing charges plus their other travel costs exceed the travel costs by bus plus their mode-specific costs. An elasticity of demand for travel by car (as opposed to travel by bus) of unity was assumed, i.e., if the overall travel costs by car increase by 1% (with no increase in bus costs), then 1% of car travellers transfer to buses. This assumption of elasticity of demand implies that the last car driver will not change over to bus travel until the road pricing charge is infinite. For practical reasons an arbitrary cutoff value has to be assumed. To test the sensitivity of the results to variations of this cutoff value, calculations have been made over a wide range of values; this sensitivity test is described in the next section. It will be seen later that without the inclusion of "mode-specific costs" a situation of complete restraint of private cars produces the maximum community savings in total costs, but with the inclusion of these costs a considerably lower amount of restraint produces the optimum situation.

It can be argued, also, that the environment suffers as a result of intensive use of the street network by private cars (noise, pollution, visual intrusion, danger, etc.) and that these costs should also be considered. It is by no means clear what values should be attached to these effects and this aspect of the problem is not considered further in this paper.

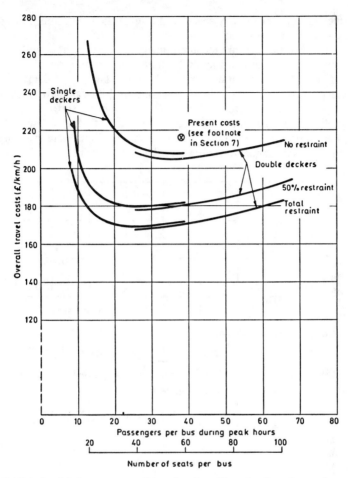

Fig. 3. Relationship between overall travel costs and bus size (ignoring mode–specific costs.)

Optimum Size of Bus for Various Levels of Restraint

The total time and operating costs of all vehicles using the network for zero, half, and full restraint conditions are shown in Fig. 3. No mode-specific costs (as defined earlier) are included in the calculations for these curves. For present conditions, the optimum bus can be seen to be a 55-seater double-decker. For conditions of half and full restraint the optimum bus is a 40-seater double-decker. If it were considered that double-deck construction were inappropriate for buses with fewer than 55 seats the optimum bus sizes for half and full re-straint conditions would be a 40 and a 35 seater single-decker, respectively. If the values of time assumed are increased by 50%, indicating more importance being attached to faster journey speeds rather than the level of bus fares and car running costs, the optimum size is reduced by about 20%.

When mode-specific costs are included, the curve for present conditions remains unchanged because in this study these costs are used only in relation to changes in mode from present conditions. With an elasticity of demand for private car travel (as opposed to bus travel) of unity, the calculations show that overall costs are minimized when about 40% of private car traffic is restrained and the optimum bus size under these conditions is a 40-seater. The optimum degree of car restraint is rather more (55%) when small buses are used and rather less (35%) when very large buses are used. These curves are shown in Fig. 4,

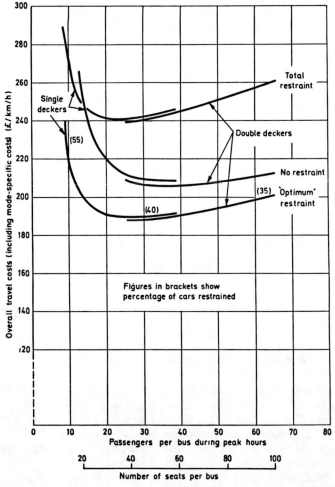

Fig. 4. Relationship between overall travel costs and bus size (mode–specific costs included.)

together with a curve for 100% restraint. When mode-specific costs are excluded, full restraint gives the least overall costs; with these costs included, full restraint gives the highest overall costs. The precise level of the full restraint curve depends on the cutoff value assumed for the road pricing charge which is necessary to make the last car user transfer to bus travel. In Fig. 4 the cutoff is assumed to be at a value of 10 times the bus travel costs. If the cutoff value, expressed as a ratio (r) of the sum of bus travel cost and mode-specific costs to bus travel cost is made equal to 100, the curve of full restraint in Fig. 4 moves up about £100 per km per hour, a very considerable change. Figure 5 shows the effect of varying the cutoff ratio, r, on the overall costs at various levels of restraint. It can be seen that at the optimum level of restraint of 40%, the cutoff ratio has a negligible effect but at higher degrees of restraint the effect becomes important.

All the curves in both Figs. 3 and 4 indicate that overall costs rise rapidly when

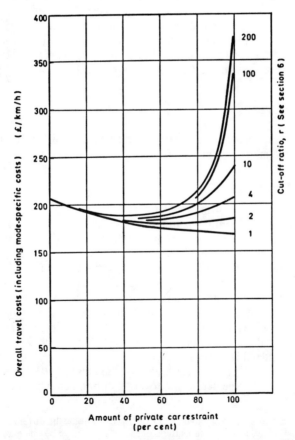

Fig. 5. Relationship between overall travel costs (including mode–specific costs) **and** amount of private car restraint using a 40-seater bus.

very small buses are used; at the other end of the scale very large buses also increase total travel costs, particularly under conditions of full restraint.

Effect of Transferring Present Car Riders to Bus Travel

The curves in Fig. 3 suggest that, if no account is taken of mode-specific costs, there would be large savings if all present car users could be persuaded to travel by bus instead of by car. The total costs on the network would fall from an estimated £206[‡]/km/hour at present to £169/km/hour, a decrease of 18%. For a 14-h day with 300 days in the year, and, say, 130 km of main road in central London, the overall savings would be about £20 million per annum. If mode-specific costs are included, the saving in total costs at the optimum level of restraint compared with the present no-restraint situation, would be £18 per km per hour. This is about half the saving under full restraint when ignoring mode-specific costs.

This saving of £18/km/hour corresponds to an overall annual saving of about £10 million. There are arguments both for and against the inclusion of mode-specific costs in the final assessment. Fortunately, as far as the determination of optimum bus size is concerned it does not affect the choice very greatly.

Overall Travel Speeds

Mean direct speeds for bus and car journeys when the different bus systems are in operation are shown in Figs. 6 and 7.

Peak-Period Speeds

Direct speeds for bus journeys under present peak-period conditions are about $5\frac{1}{2}$ km/h. With "optimum restraint" and a 40-seater double-decker bus system in operation (the optimum bus size), journey speed would rise to about 7 km/h; a system of smaller buses, under these conditions, could give slightly higher speeds, but fares would have to be increased (or the service run at a loss) as this would not be the optimum bus system; with 100-seater buses, speeds would be about 15% lower at just under 6 km/h. With complete restraint, bus speeds could rise to about 8 km/h with the optimum 40-seater bus system and to about $8\frac{1}{2}$ km/h with a 20-seater minibus system, but, again, at increased cost. The 100-seater under these conditions (if loaded to 65% occupancy in peak periods) would give considerably lower speeds.

Figure 7 shows the speeds of car travellers; the present peak-hour direct speed of 8 km/h would be expected to rise to $11\frac{1}{2}$ km/h under "optimum restraint"

[‡]Actual present costs are estimated as £215 per km per hour; the difference is mainly because in the theoretical calculations the buses were assumed to be one-man, two-door types, the route density and number of stops per km were optimised and the walking and waiting times were based on random journeys rather than actual journeys.

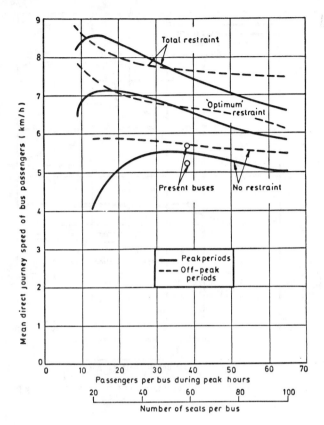

Fig. 6. Direct journey speed (origin to destination) of bus passengers during peak hours and off-peak hours.

and to over 20 km/h under complete restraint. If, however, a 20-seater minibus system were used this would fall to about 17 km/h.

Table II gives some figures for peak-period speeds on the street network and also for overall origin-to-destination speeds. It can be seen that generally the speeds on the network are about twice the mean direct speeds, particularly for bus passengers. The highest speeds are enjoyed by those who still drive their cars during severe restraint of private traffic. Those present car travellers who transfer to bus travel have much lower speeds than those who remain as car travellers but it is interesting to see that the overall speed of those who transfer to bus under full-restraint conditions ($8\frac{1}{2}$ km/h) is higher than the present speed by car (8 km/h). Present bus passengers benefit the most, increasing their overall speed by 40% when the restraint is increased from zero to 100%. The reason why present car users transferring to bus travel have higher journey speeds than present bus passengers is because their journeys tend to be longer, hence waiting time is a smaller proportion of the total time.

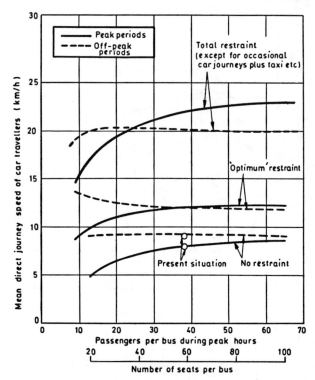

Fig. 7. Direct journey speed (origin to destination) of car travellers during peak hours and off-peak hours.

It is somewhat discouraging that direct journey speeds by bus are not very high in absolute terms even under complete restraint, and it appears virtually impossible to raise direct speeds by bus for journeys of average length to over 9 or 10 km/h on the existing main-road network of central London using conventional operating techniques and with no special priority on the street system.

Off-Peak Period Speeds

Mean direct journey speeds in the off-peak periods are shown by the broken lines in Figs. 6 and 7. With 'optimum-restraint' and using the bus that yields the minimum total cost (40-seater double-decker), the direct bus journey speed in the off-peak is practically the same as in the peak, viz., 7 km/h (see Fig. 6), while with complete restraint and with the same size of bus the direct bus journey speed is marginally lower during the off-peak compared with peak times. As in the peak, bus journey speeds could be made higher if smaller buses were used but at increased cost. There is not such a pronounced falling-off in speed when larger buses are used as was observed for peak-period conditions; this arose be-

TABLE II

**Speeds of Travel During Peak Periods for Different Levels of
Private Car Restraint**

	"No-restraint (with 55-seaters)	"Optimum restraint" (with 40-seaters)	"Full restraint" (with 40-seaters)
On the street network (km/h)			
Vehicular speed	15	19½	25½
Scheduled bus speed (including bus stops)	11½	14½	17½
Direct speeds (km/h) (origin to destination)			
Present car user, still using car[a]	8	11½	(21)[b]
Present car user, transferring to bus	(6)	7½	8½
Present bus passenger	5½	7	7½
Overall average traveller	6	7½	8

[a]Parking time of 10, 5, and 0 min assumed for zero, optimum, and full restraint conditions, respectively.
[b]This speed refers to those few cars which remain on the network, e.g., police cars, taxis, etc.

cause the computer program optimised bus occupancy in the off-peak period, whereas in the peak period a constant load factor of 65% was assumed. This assumption was made because there was some evidence to suggest that buses should be loaded as fully as possible during peak periods. In computations currently being carried out peak-period occupancy is also being optimised.

Figure 7 shows that during the off-peak period, direct journey speeds of car travellers are almost independent of the size of the bus used for a given amount of car restraint, the speeds being 9 km/h and 20 km/h at no restraint and total restraint, respectively. It is interesting to note from Fig. 7 that when very large buses are used under complete car restraint, the occasional car journey in the off-peak period is some 3 km/h slower than in the peak. This is due to the lower pcu flow on the road during the peak hours brought about by having fewer buses to carry the passenger load. This suggests, as mentioned above, that the peak-period occupancy of 65% is not necessarily the optimum one, particularly for very large buses.

Table III gives some figures for off-peak speeds on the street network as well as origin-to-destination speeds. Present speeds on the network of both buses and cars are about 20% higher than in the peak period, and direct speeds are roughly 10% higher than peak-period speeds. Under optimum restraint the difference has very largely disappeared and it has completely disappeared under full restraint.

TABLE III

Speeds of Travel During Off-Peak Periods for Different Levels of Private Car Restraint

	"No-restraint (with 55-seaters)	"Optimum restraint" (with 40-seaters)	"Full restraint" (with 40-seaters)
On the street network (km/h)			
Vehicular speed	18	$21\frac{1}{2}$	25
Scheduled bus speed (including bus stops)	14	16	18
Direct speeds (km/h) (origin to destination)			
Present car user, still using car[a]	9	$12\frac{1}{2}$	$(20\frac{1}{2})$[b]
Present car user, transferring to bus	$(6\frac{1}{2})$	$7\frac{1}{2}$	$8\frac{1}{2}$
Present bus passenger	6	7	$7\frac{1}{2}$
Overall average traveller	7	$7\frac{1}{2}$	8

[a]Parking time of 10, 5, and 0 min assumed for zero, optimum, and full restraint conditions, respectively.
[b]This speed refers to those few cars which remain on the network, e.g., police cars, taxis, etc.

Costs to Bus Passengers

Since it is assumed that buses are not subsidised, the fares equate with the costs of running the bus service. Bus fares have been calculated in pence/km and are shown for zero and full restraint in Fig. 8. As the buses become smaller the fare rises but not steeply until the size has fallen to below 30-seaters. There seems little to be gained by having very large buses from this point of view. For a given bus size, the greater the degree of car restraint, the lower the fare. For zero and full restraint the bus fares, with optimum size buses in each case, would be 1.85 pence (0.77 new pence) and 1.77 pence (0.74 new pence) per km, respectively.

Frequency of Bus Stops

It was mentioned in Section 3 that when the total cost of the network was calculated using a particular size of bus and a given amount of car restraint, the bus stop spacing and relative bus route densities were optimised so as to obtain the minimum cost.

The relationship between bus size and bus stop spacing is shown in Fig. 9 for conditions of no car restraint and also for complete restraint. It can be seen that if small buses are used under present conditions, the optimum bus stop

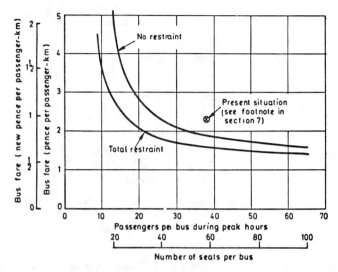

Fig. 8. Relationship between bus fare and bus size.

frequency falls to one stop per km. The reason is that the high bus flows resulting from using small buses cause less interference with other traffic when their stopping places are kept to a minimum; this is particularly valuable under present conditions in central London where there is very little reserve capacity. With complete car restraint there is enough reserve capacity on the road to enable minibuses to stop with nearly the same frequency as the larger buses without interfering too drastically with the journey speeds of other traffic. Figure 9 shows that with complete restraint the bus stop frequency is practically independent of bus size.

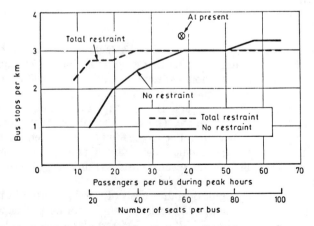

Fig. 9. Relationship between bus size and optimum bus stop frequency.

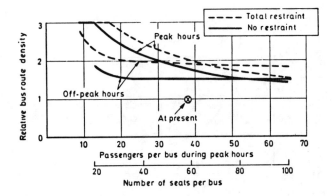

Fig. 10. Relationship between bus size and relative bus route density.

Density of Bus Routes

The relationship between bus size and the relative bus route density is shown for zero and complete restraint conditions in Fig. 10 separately for peak and off-peak periods. Using present size 55-seater double-deckers, Fig. 10 suggests that the density of bus routes should be increased if the waiting and walking times of an average bus passenger are to be minimized. This result should be used with caution since the data for these walking and waiting times were based on random journeys in central London[6]; they may not be representative of actual journeys. However, the general shapes of the curves are believed to be valid; these curves show that there should be more routes when buses are smaller and when restraint of private cars is more severe. They also indicate that more routes are required in peak periods than in off-peak periods, but this may not always be acceptable to both operator and passenger alike.

Conclusions

1. The optimum size of bus varies with the degree of restraint imposed on private traffic—a 55-seater double-decker is optimum for present conditions of no restraint, and a 40-seater double-decker or a 35 seater single-decker for complete restraint. This result ignores mode-specific costs, i.e., the costs of drivers' preference for using their own cars. If it is accepted that the subjective costs of inconvenience, lack of privacy, etc., (the mode-specific costs) suffered by a car user changing to bus travel should be included in the assessment, then there is an optimum level of car restraint which is given by minimizing time costs plus operating costs plus these mode-specific costs. At this level of restraint, the optimum bus size is a 40-seater single- or double-decker.

2. The estimated time plus operating costs of travel in central London per km of network per hour is about £206 under present conditions. This would be expected to fall to £179 (drop of 13%) for 50% restraint and to £169 (drop of

18%) for full restraint. On the network as a whole this would amount to a saving of £20 million per annum in time and operating costs (ignoring mode-specific costs) if full restraint were imposed.

3. If mode-specific costs are included, the £206 per km per hour would be expected to fall to £188 (drop of 9%) at the optimum level of restraint (this is when about 40% of cars are restrained). Full restraint would be expected to increase the overall costs to between £240 and £340 per km per hour depending on the relevant assumptions. The overall savings in community costs at the optimum level of restraint when mode-specific costs are included would amount to about £10 million per year. It can be argued that environmental costs (noise, danger, pollution, etc.), should also be included in the assessments. As yet this has not been done. Inclusion of such costs would undoubtedly increase the overall savings under restraint conditions.

4. Overall speeds (as the crow flies) of all travellers during peak periods would be expected to increase with increasing restraint from 6 km/h at present to $7\frac{1}{2}$ km/h under 40 to 50% restraint and to 8 km/h under full restraint. Present bus passengers' speeds would increase from $5\frac{1}{2}$ to $7\frac{1}{2}$ km/h, whereas present car travellers who changed to bus travel would experience initially no change in speed at the higher levels of car restraint. Of course, those who remain with their cars would enjoy speeds of up to about 20 km/h at almost full restraint.

5. Bus fares would be expected to be slightly less with more private car restraint, though they could rise if a system of buses, smaller than optimum for the particular level of restraint, is used, but these smaller buses would have some benefit in increasing bus journey speeds.

6. The optimum spacing of bus stops in central London was found to depend on bus size, being about 3 per km over most of the range but falling to 1 per km for small buses under present no-restraint conditions.

7. The optimum density of routes was also found to vary with bus size and the degree of car restraint. More routes should be provided as smaller buses are used and as restraint of private cars becomes more severe. More routes are required in peak periods compared with off-peak periods.

Acknowledgment

This paper is contributed by permission of the Director of Road Research. Crown Copyright 1971. Published by permission of the Controller of Her Brittanic Majesty's Stationery Office.

References

1. F. V. Webster, A theoretical estimate of the effect of London car commuters transferring to bus travel, Ministry of Transport, RRL Report LR 165, Crowthorne, 1968.

2. D. J. Lyons, Bus travel in town centres, *Traffic Eng. Control* **11** (1), 20-23, (1969).

3. Greater London Council, Central London Traffic Survey April-May 1966, Research Report No HT/RD-4, October 1966.

4. R. J. Smeed and J. G. Wardrop, An exploratory comparison of the advantages of car and buses for travel in urban areas, *Inst. Trans. J.* **30** (9) (March 1964).
5. J. M. Thomson, Speeds and flows of traffic in central London—Sunday traffic survey, *Traffic Eng. Control* 8 (11), 672-6 (1967).
6. E. M. Holroyd and D. A. Scraggs, Journey times by car and bus in central London, *Traffic Eng. Control* 6 (3), 169-173 (1964).
7. R. F. F. Dawson and J. G. Wardrop, Passenger-mileage by road in Greater London, Department of Scientific and Industrial Research, *Road Research Technical Paper* No 59. London, 1962 (HM Stationery Office).
8. London Transport Board, Annual Report and Accounts for the year ended 31 December 1966, London (Stationery Office).
9. Commercial Motor, *Tables of Operating Costs for Goods and Passenger Vehicles 1966*, Temple Press Books, London.
10. R. F. F. Dawson, Vehicle operating costs and cost of road accidents in 1967, Ministry of Transport, RRL Technical Note TN 360, Crowthorne, 1967.

Public Transportation Line Positions and Headways for Minimum Cost*

Bernard F. Byrne[†] and Vukan R. Vuchic

Towne School of Civil and Mechanical Engineering, University of Pennsylvania, Philadelphia, Pennsylvania

The Problem

The basic elements which must be determined in planning public transportation service in an area are positions of lines, headways and number of vehicles (fleet size). The general trade-off in this planning is between the cost of service to the operator (system cost) and the cost of travel to users (derived in this model from travel time). This study presents a method for deriving the line positions, headways and fleet size which minimize the total system and user costs.

The study analyzes a rectangular urban area from which passengers travel to and from the CBD, which is defined as one side of the area.

Background

The optimization of various physical and operational aspects of public transportation systems has been the subject of several recent studies. Black[2] and Creighton et al.[3] analyzed the optimum division between two types of service in a single corridor. Vuchic[4] analyzed optimal station locations for two different criteria. Creighton et al.,[5] in another paper, analyzed the densities of freeways and arterials to find that which minimized construction and user costs. Subsequently, Holroyd[6] optimized the spacing of transit lines on a square grid, assuming uniform headways and a uniform trip end density. Holroyd discovered that the optimum spacing and headways occur whenever the cost of "effective" walking time,[‡] the cost of operation and the cost of waiting time are equal. Later, McClenahan[7] also did some work on this problem, utilizing a somewhat different set of assumptions and relying on the computer for numerical solutions.

---

*Excerpt from a thesis[1] submitted in partial fulfillment of the requirements for the Ph.D. degree at the University of Pennsylvania, Philadelphia, Pennsylvania. The research was partially supported by the Urban Mass Transportation Administration, Grant No. URT-PA8(70), through the Transportation Studies Center at the University of Pennsylvania.
  †Presently at De Leuw Cather & Co., San Francisco, California.
  ‡Effective walking time was defined as the difference between actual travel time and travel time if the trip was made by transit from door to door.

References p. 360

The Model

The Assumptions

Suppose that there exists a rectangular area of length L and width D with a gridiron pattern of streets. A coordinate system is superimposed on the area so that the transit lines are parallel to the y axis. The CBD, to which all passengers commute, spans the width D parallel to the x axis. This area model is presented in Fig. 1: x_i represents the position of line i, while x'_i is the boundary between the passenger sheds of lines i and $i + 1$ (to be defined later).

The trip end density is defined by a function $P(x, y)$. This function represents the number of passengers per unit area per unit time t_0 at the point (x, y) desiring service to the CBD. Further, it is assumed that demand during t_0 is uniform, i.e., that the analysis is done for a homogeneous period, such as rush hour, base period, etc. This function is also referred to as the population density or the passenger density.

The costs of the whole transporting process in the model, which are to be minimized, are divided into system operating costs and passenger travel costs. The latter are computed as passenger travel time multiplied by the value of time

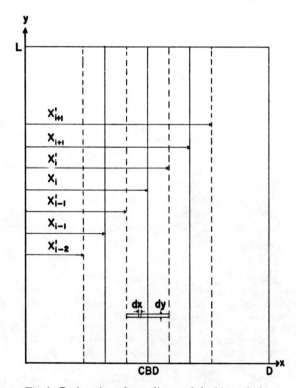

Fig. 1. Designation of area, lines and shed boundaries.

assumed for the respective travel component. The components are: access to
lines, waiting time, and riding time. Unit values of individual components, in
dollars per passenger hour, are designated as: access to the lines $-\gamma_a$, waiting $-\gamma_w$,
riding (travel on the system) $-\gamma_r$. The cost of the transit service operation is assumed to be γ_0 dollars per vehicle hour.

The transit operating speed V_0 is assumed to be known. Headways h_i are uniform on each line, but not necessarily equal for all lines.

The waiting time of passengers is assumed to be proportional to the headway
with a constant of proportionality $c, 0 < c < 1$. The time of access parallel to
lines is neglected, assuming a considerable density of stops. The travel time of
all passengers within the CBD is not considered since it does not affect this
service directly.

Passengers at the point (x, y) choose the line that will give them the least cost
trip to the CBD. Consequently, there exists a line, designated x'_i, between transit
lines i and $i + 1$, such that the costs of going by either line are equal. Based on
the preceding definitions the total cost of traveling from x_i to the CBD via line
i will be

$$C_i = \frac{\gamma_a}{V_a}(x'_i - x_i) + c\gamma_w h_i + \frac{\gamma_r y}{V_0}.$$

The cost of travel via line $i + 1$ will be similar, the access distance being
$(x_{i+1} - x'_i)$. By equating the two cost expressions, the boundary is defined as

$$x'_i = \frac{x_{i+1} + x_i}{2} + \frac{k_1}{2}(h_{i+1} - h_i), \tag{1}$$

where $k_1 = cV_a\gamma_w/\gamma_a$. The first and last boundaries are defined as $x'_0 = 0$ and
$x'_n = D$.

Objective Function

The objective is to find the line positions and headways which minimize the
total cost Z of transporting persons from the study area to the CBD, defined as

$$Z = C_0 + C_u,$$

where C_0 is the system cost and C_u is user cost.

The operating cost for each line during the period t_0 is the product of the
number of vehicles N', the unit cost of vehicle operation γ_0 and t_0. The number
of vehicles can be expressed as:

$$N' = \sum_{i=1}^{n} \frac{T}{h_i}, \tag{2}$$

where T is the cycle time (equal for all lines). The total system operating cost is
the sum of the products over all n lines

$$C_0 = \gamma_0 t_0 T \sum_{i=1}^{n} \frac{1}{h_i}. \tag{3}$$

The expressions for individual user costs will be derived by observing first the population in a small element of the served area: $P(x, y)\, dxdy$. The access cost for this population will be

$$dC_{a_i} = \frac{\gamma_a}{V_a}\, |x - x_i|\, P(x, y)\, dxdy.$$

The access cost for the whole line i will then be

$$C_{a_i} = \int_0^L \int_{x'_{i-1}}^{x'_i} \frac{\gamma_a}{V_a}\, |x - x_i|\, P(x, y)\, dxdy. \tag{4}$$

The waiting time cost C_w and riding time cost C_r for line i can be obtained in a similar way. All these costs are then summed over all n lines to yield the total cost of transportation in the area:

$$Z = \sum_{i=1}^{n} \left[\frac{\gamma_0 t_0 T}{h_i} + \int_0^L \int_{x'_{i-1}}^{x'_i} \left(\frac{\gamma_a}{V_a}\, |x - x_i| + c\gamma_w h_i + \frac{\gamma_r y}{V_0} \right) P(x, y)\, dxdy \right]. \tag{5}$$

The optimum solution will be that one which minimizes this total cost.

Solution

Line Positions and Headways

To find the optimal line positions, the objective function (5) is differentiated with respect to x_i utilizing Leibniz's rule and set equal to zero. The expression thus obtained, simplified by introducing (1), is:

$$\int_0^L \int_{x'_{i-1}}^{x_i} P(x, y)\, dxdy = \int_0^L \int_{x_i}^{x'_i} P(x, y)\, dxdy. \tag{6}$$

This equation shows that each line should be positioned such that the populations using it from each side are equal. If, for simplicity, the cumulative function $F(t)$ is introduced:

$$F(t) = \int_0^t \int_0^L P(x, y)\, dydx; \tag{7}$$

Eq. (6) can be rewritten as

$$F(x_i) - F(x'_{i-1}) = F(x'_i) - F(x_i). \tag{8}$$

To derive the optimal headways the objective function is differentiated with respect to h_i. Setting $\partial Z / \partial h_i$ equal to zero and solving for h_i yields

$$h_i = \{ \gamma_0 t_0 T / c\gamma_w\ [F(x'_i) - F(x'_{i-1})] \}^{1/2} \tag{9}$$

This expression shows that, similar to Holroyd's[6] findings in a different situation, for the optimal headway the passenger waiting cost should be equal to the system operating cost.

Since x_i' is a function of x_i and h_i, Eqs. (6) and (9) must be used to derive the optimal line positions and headways. One starts the algorithm by assuming a value for x_1 and finding x_1' from Eq. (6). Knowing x_1', h_1 is found from Eq. (9). One can then assume a value for x_2, determine x_2' from Eq. (6) and h_2 from Eq. (9). The value for h_2 is substituted into (1) and a new value found for x_2. If this value agrees with the assumed value within some predetermined degree of accuracy, then the derived x_2, x_2', and h_2 are correct. If not, the new value is used as the assumed value for x_2 and the process repeated until the result converges. This process is repeated with x_3, etc., until x_n' is found. If $x_n' = D$ within some predetermined degree of accuracy, then the entire solution has been found. If not, x_1 is changed and the process is repeated until this end boundary condition is satisfied.

Fleet Size Constraint

In some cases this analysis may be required for an existing situation for which the number of vehicles is fixed, say N. Then the number of vehicles required for the service cannot be greater than N.

$$\sum_{i=1}^{n} \frac{T}{h_i} \leqslant N. \tag{10}$$

For the case in which the optimal fleet is greater than N, (10) will be an equality. It may then be incorporated in the objective function using the Lagrangian multiplier technique.

$$Z = C_0 + C_u + \lambda \left(\sum_{i=1}^{n} \frac{T}{h_i} - N \right). \tag{11}$$

Since the last member of this expression does not contain x_i, the condition (6) for line positions also holds for this case. The optimum headways are obtained by differentiating (11) with respect to h_i. Setting the obtained expression equal to zero and solving for h_i yields

$$h_i = \{ (\gamma_0 t_0 + \lambda) \, T/c\gamma_w \, [F(x_i') - F(x_{i-1}')] \}^{1/2}. \tag{12}$$

If (12) is substituted into (10) (as an equality), solved for λ, and λ is introduced back into (12), one obtains

$$h_i = T \sum_{j=1}^{n} [F(x_j') - F(x_{j-1}')]^{1/2} / N[F(x_i') - F(x_{i-1}')]^{1/2}. \tag{13}$$

This expression can be better interpreted through its inverse, i.e., frequency instead of headway. One can see from such an expression that the frequency of service on each line is proportional to the ratio of the square root of the served population to the sum of square routes of populations on all lines.

With Eq. (13) the optimum number of vehicles can be obtained by Eq. (2). Since the constraint (10) is an inequality, the unconstrained solution should be derived first; if the number of vehicles obtained is less than the available fleet, that solution is optimal. If it is greater, then new headways must be determined from Eq. (13) and substituted in Eq. (1) to find new boundaries and the new boundaries introduced into Eq. (6) for new line positions and back into Eq. (13) for new headways. The process is repeated until the solution converges.

Optimum Number of Lines

To obtain the optimum number of lines, the region will be considered to have a variable width x''. First, the total cost will be computed for 1 and 2 lines in the area and the behavior of the costs for the two cases will be examined. As an initial step, therefore, expressions for the costs with optimum line positions and headways will be defined. To derive the access cost, Eq. (8) is introduced into Eq. (4); after rearranging, the expression is

$$C_{a_i} = \frac{\gamma_a}{V_a} \int_0^L \left[\int_{x_i}^{x_i'} xP(x,y)\,dx - \int_{x_{i-1}'}^{x_i} xP(x,y)\,dx \right] dy. \qquad (14)$$

Since $F(t)$, defined above, is a monotone nondecreasing function,

$$t_2\left[F(t_2) - F(t_1)\right] \geqslant \int_0^L \int_{t_1}^{t_2} xP(x,y)\,dx\,dy \geqslant t_1\left[F(t_2) - F(t_1)\right]. \qquad (15)$$

The operating time cost C_{0i} is obtained by substituting Eq. (9) into Eq. (3); waiting time cost is likewise obtained from Eq. (9) and Eq. (5). If the sum of these two costs is designated as C_{bi} and the constant $k = 2(c\gamma_w\gamma_0 t_0 T)^{1/2}$ is introduced, one obtains

$$C_{bi} = k\left[F(x_i') - F(x_{i-1}')\right]^{1/2}. \qquad (16)$$

If Z_i is the total cost with i lines in the area of width x'', it will be shown that $Z_1 < Z_2$ as $x'' \to 0$ and $Z_2 < Z_1$ and $x'' \to \infty$. Let $\bar{x}_1$ be the optimal position of the line in the case with only one line; and x_1, x_2, and x_1', the positions and boundary, respectively, in the case there are two lines. The following relationships can be obtained from Eq. (8):

$$F(x'') = 2F(\bar{x}_1), \qquad (17)$$

$$F(x'') = 2F(x_2) - F(x_1'), \qquad (18)$$

and

$$F(x_1') = 2F(x_1). \qquad (19)$$

Considering the case in which $Z_1 < Z_2$ and assuming $x_1' > \bar{x}_1$, it may be shown that, because of the additivity of integrals,

$$Z_2 - Z_1 = \frac{2\gamma_a}{V_a} \int_0^L \left[-\int_{x_1'}^{x_2} xP(x, y)\,dx + \int_{x_1}^{\bar{x}_1} xP(x, y)\,dx \right] dy$$

$$+ k\{[F(x'') - F(x_1')]^{1/2} + [F(x_1')]^{1/2} - [F(x'')]^{1/2}\}.$$

From (15), using Eqs. (17)-(19) and rearranging, we obtain

$$Z_2 - Z_1 \geqslant \frac{k\{[F(x'') - F(x_1')]\,F(x_1')\}^{1/2}}{[F(x'') - F(x_1')]^{1/2} + [F(x_1')]^{1/2} + [F(x'')]^{1/2}}$$

$$- \frac{\gamma_a}{V_a}(x_2 - \bar{x}_1)\,[F(x'') - F(x')].$$

Since $x_1' > \bar{x}_1$ by assumption and F is monotone nondecreasing,

$$Z_2 - Z_1 \geqslant [F(x'') - F(x_1')]^{1/2} \left\{ \frac{k}{2 + \sqrt{2}} - \frac{\gamma_a}{V_a}(x_2 - \bar{x}_1)\,[F(x'') - F(x_1')]^{1/2} \right\}.$$

$$(20)$$

As $x'' \to 0$, the population $F(x'') - F(x_1')$ may be made arbitrarily small so that in (20) the second term within the braces may be made smaller than the first one. Therefore, in this case, $Z_1 < Z_2$. If, on the other hand, $\bar{x}_1 > x_1'$, we may obtain in a similar manner:

$$Z_2 - Z_1 \geqslant [F(x_1')]^{1/2} \left\{ \frac{k}{2 + \sqrt{2}} - \frac{\gamma_a}{V_a}(\bar{x}_1 - x_1')\,[F(x_1')]^{1/2} \right\}.$$

Using a similar argument as previously, namely, that as $x'' \to 0$, $F(x_1')$ may be made arbitrarily small, the second term in the braces may be made smaller than the first one. Therefore, $Z_1 < Z_2$ for small x''.

The proof that when x'' is large, $Z_2 < Z_1$ can be formulated in a similar way. First the expression for $Z_1 - Z_2$ is derived for $x_1' < \bar{x}_1$ and shown to be greater than zero. Then the process is repeated for $\bar{x}_1 < x_1'$.

Since $Z_1 < Z_2$ for small x'' and $Z_2 < Z_1$ for large x'', there must exist some value of x'', say x_1°, for which $Z_1 = Z_2$. For $x'' < x_1^\circ$, only one line is optimal. For $x'' > x_1^\circ$, two lines are optimal. Since, at larger values of x'', access time cost dominates waiting time cost and since a greater number of lines reduces the access time cost, it seems reasonable to suppose that there exists a point x_2° at which $Z_2 = Z_3$, and that for $x'' > x_2^\circ$ three lines are optimal, and so on. This is illustrated in Fig. 2, which shows the total transportation cost Z as a function of corridor width x'' for different numbers of lines.

For computing the optimum number of lines for an area with a given width D, two alternative approaches are available. From the foregoing proofs one method would be started with x'' small enough for one line to be optimal. Z_1 and Z_2

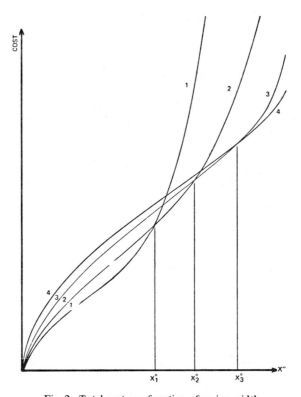

Fig. 2. Total cost as a function of region width.

would then be calculated while successively increasing x'' until $Z_1 > Z_2$. Starting there, Z_2 and Z_3 would be calculated while increasing x'' until $Z_2 > Z_3$. The
process would be repeated until $x'' = D$. The second method would be started
by computing the total cost for an estimated number of lines. The computation
for adjacent numbers of lines would be made searching for the number for which
Z is minimized.

The first method is illustrated in the subsequent example.

Special Case of Uniform Population Density

If the population density is assumed to be uniform, i.e., $P(x, y) = P$, then a
number of simplifications occur. Introducing this into Eq. (6) and carrying out
the integration yields

$$x_i' - x_i = x_i - x_{i-1}'. \tag{21}$$

Similarly the expression for headways (9) simplifies:

$$h_i = [\gamma_0 t_0 \, T/c\gamma_w PL(x_i' - x_{i-1}')]^{1/2}. \tag{22}$$

Since the population distribution is uniform, it is assumed that headways on all lines are equal. With this assumption, Eq. (1) is substituted into Eq. (21). Because $x_0' = 0$, a recursive solution may be developed. Since $x_n' = D$, it may be shown that

$$x_i = (2i - 1) D/2n. \tag{23}$$

If this is used in Eq. (1) as modified and substituted into (22), it follows that

$$h_i = \cdots = h_i = \cdots = h_n = (n\gamma_0 t_0 \, T/c\gamma_w PLD)^{1/2}. \tag{24}$$

Thus, the assumption of equal headways is borne out.

In order to find the optimum number of lines, the assumed population density and results, Eqs. (23) and (24), are substituted into Eq. (5) to produce the objective function,

$$Z = \gamma_a PLD^2/4V_a n + 2(nPLDc\gamma_w \, \gamma_0 t_0 \, T)^{1/2} + \gamma_r PL^2 D/2V_0.$$

Letting

$$A = \gamma_a PLD/4V_a, \quad B = 2(PLDc\gamma_w \, \gamma_0 t_0 \, T)^{1/2}, \quad \text{and} \quad E = \gamma_r PLD/2V_0,$$

n_0 is wanted such that

$$A/(n_0 - \tfrac{1}{2}) + B(n_0 - \tfrac{1}{2})^{1/2} > A/n_0 + B(n_0)^{1/2} < A/(n_0 + \tfrac{1}{2}) + B(n_0 + \tfrac{1}{2})^{1/2}.$$

Manipulating this expression, we obtain

$$n_0 = [\{1 + D(\gamma_a^2 PL/2V_a^2 \, c\gamma_w \gamma_0 t_0 T)^{1/3}\}/2], \tag{25}$$

where $[u]$ designates the largest integer $\leqslant u$.

In the case that the fleet size is restrained, Eq. (21) remains unchanged. However, Eq. (22) is superceded by Eq. (13). Since headways are equal, the same positions as before remain optimal and

$$h_1 = \cdots = h_i = \cdots = h_n = nT/N. \tag{26}$$

In order to find the optimum number of lines for this case, (23) and (26) are substituted into (1) and (5) to produce the objective function,

$$Z = \gamma_a PLD^2/4V_a n + c\gamma_w PLDnT/N + \gamma_r PL^2 D/2V_0 + N\gamma_0 t_0.$$

Letting

$$B' = c\gamma_w PLDT/N \quad \text{and} \quad E' = \gamma_r PL^2 D/2V_0 + N\gamma_0 t_0,$$

n_0 is wanted such that

$$A/(n_0 - \tfrac{1}{2}) + B'(n_0 - \tfrac{1}{2}) > A/n_0 + B'n_0 < A/(n_0 + \tfrac{1}{2}) + B'(n_0 + \tfrac{1}{2}).$$

Transformations similar to the preceding case lead to

$$n_0 = [\{1 + (\gamma_a DN/V_a c\gamma_w T)^{1/2}\}/2], \tag{27}$$

where again $[u]$ means the largest integer $\leqslant u$.

Example

In order to illustrate the solutions derived in this paper, a number of examples have been performed. Due to space limitations, only an example based on data from an area in South Philadelphia will be presented here. For clarity a number of illustrations are used.

The values of parameters used are:

$$V_a = 3 \text{ mi/h} \qquad \gamma_a = \$7.00/\text{per.-h}$$
$$V_0 = 10 \text{ mi/h} \qquad \gamma_w = \$7.00/\text{per.-h}$$
$$c = 1/2 \qquad \gamma_r = \$2.50/\text{per.-h}$$
$$T_L = 15 \text{ min} \qquad \gamma_0 = \$11.00/\text{veh.-h}$$

The values of time selected are based on those derived by Lisco[8] and Local Government Operational Research Unit.[9] Since these are, perhaps, the parameters whose values are most open to question, certain parts of the examples have been solved for several different values of time.

The area selected for the example is in South Philadelphia, bounded on the north by South Street, on the east by Delaware Avenue, on the south by Pattison Avenue and on the west by 26th Street. It is an approximately rectangular area with a width of 2.22 and a length of 2.73 miles. The x axis of the area runs from the west to the east, y axis from north to south.

Data on the number of transit trips going from this area to the CBD were obtained from the Delaware Valley Regional Planning Commission. The available data, from a 1960 survey, were in terms of daily trips for each district and peak hour trips for the whole area. A K factor derived from these data for the whole area was applied to each district to obtain the peak hour trips ($t_0 = 1$ h). In or-

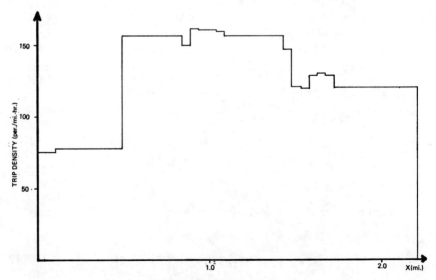

Fig. 3. South Philadelphia trip density function.

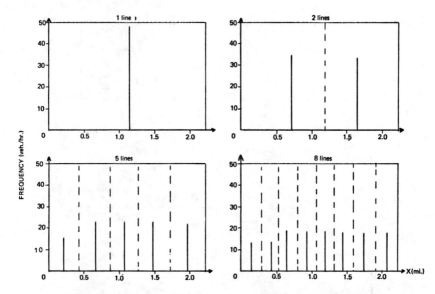

Fig. 4. Optimum line positions and frequencies (South Philadelphia).

der to obtain trip end densities, the number of trips going to the CBD were accumulated for each district and divided by the area of the district, measured from district maps. Accumulating the population density along y produces the population density function along x, shown in Fig. 3. The results of computations for this example are summarized here.

The optimal positions and frequencies for several given numbers of lines are shown in Fig. 4. In the diagrams showing these solutions, the solid lines represent the positions of the transit lines. The height of the lines is proportional to the frequency on the line. The dashed lines represent the boundaries; note that the frequencies are generally lowest on the west side and highest in the middle, with the east side having the intermediate values. This follows the population density function (Fig. 3), as would be expected.

The optimal line position and frequencies for a given fleet size are shown in Fig. 5 for 6 lines (in this case the optimal number of vehicles is 93). Comparing the two diagrams in Fig. 5 it is apparent that constraining the fleet size to only 50 vehicles causes the optimum positions to shift very slightly toward the areas with lower population density. This is caused by changes in frequency which affect the locations of the boundaries, which in turn affect the line positions. The frequencies show a greater effect of fleet constraint than the line positions, as Fig. 6 shows. It is interesting that the lower frequencies occur in the areas of lower population densities.

The optimum number of lines was found by varying the width of the served region x'', as demonstrated previously. To examine how the optimal number of lines changes with the width of the served region, a diagram of total cost (user and operator) as a function of x'' for different numbers of lines has been made.

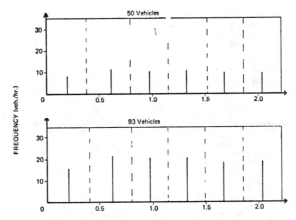

Fig. 5. Optimum line positions and frequencies for a given fleet size (South Philadelphia).

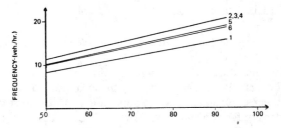

Fig. 6. Optimum frequencies as a function of fleet size (South Philadelphia).

This diagram, shown in Fig. 7, indicates that, as predicted by preceding theoretical derivations, each number of lines is optimal for a certain range of region width. The optimal number of lines increases with region width, while the sections of widths which are optimal for each successive number of lines varies due to varying population densities. The flatness of the intersections of the curves in the diagram indicates that the transition between optimum numbers of lines is

TABLE I

Optimum Number of Lines and Number of Vehicles for Different Values of Time South Philadelphia Example

Value of time (\$/per. hr)	7	6	5	4	3	2	1
Optimum number of lines	7	6	6	6	5	5	4
Optimum number of vehicles	100	86	78	70	55	45	28

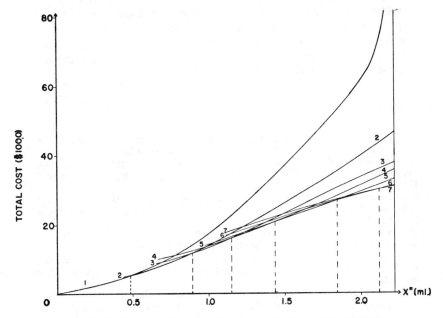

Fig. 7. Total cost as a function of region width (South Philadelphia).

gradual, i.e., that the total cost is not affected significantly if, e.g., 4 or 6 lines are used in the region where 5 are optimal.

The sensitivity of the optimal number of lines and the optimal number of vehicles to the value of access and waiting time is shown in Table I. When the value of time changes by a factor of seven, the optimal number of lines changes only by a factor of less than two. It is also interesting that for the same change in the value of time the optimum fleet size changes only by a factor of less than four. Thus, neither the optimum number of lines, nor the optimum fleet size are particularly sensitive to the value of time.

Conclusions

Using the model developed in this paper, the optimum line position is obtained when the populations using the line from each side are equal. The optimum headway occurs when the waiting time cost is equal to the operating cost. When the fleet size is restrained, the distribution of vehicles to lines is proportional to the square roots of the populations on each line, not linearly to the population function itself, as might be expected. In the special case of a uniform population density lines are uniformly placed and have equal headways. Also, the vehicle allocation to lines is uniform for both an optimal and a restrained fleet size.

The performed examples, one of which is presented in this paper, indicate that the model and solutions are applicable to real situations. Naturally, the derived relationships and optimal solutions are limited in their realistic validity by the assumptions of the model; in practice a number of other factors must be included in the analysis of existing and planned services. Nevertheless, the results and algorithms presented here may be used as a valuable component in planning public transportation routes and operations.

References

1. B. F. Byrne, *Public Transportation Line Positions and Headways for Minimum User and System Cost,* Ph.D. Dissertation, University of Pennsylvania, Philadelphia, Pa. 1971.
2. A. Black, A Method for Determining the Optimal Division of Express and Local Rail Transit Service, *Bulletin 342,* pp. 106–120, Highway Research Board, Washington, D.C., 1964.
3. R. L. Creighton, D. I. Gooding, G. C. Hemmens, and J. E. Fidler, Optimum Investment in Two Mode Transportation Systems, *Record 47,* Highway Research Board, Washington, D.C., 1964.
4. V. R. Vuchic, *Interstation Spacing for Line-Haul Passenger Transportation,* Ph.D. Dissertation, ITTE, University of California, Berkeley, Calif., 1966.
5. R. L. Creighton, I. Hoch, M. Schneider, and H. Joseph, Estimating Efficient Spacing for Arterials and Expressways, *Bulletin 253,* pp. 1–43, Highway Research Board, Washington, D.C., 1960.
6. E. M. Holroyd, Optimum bus service: A theoretical model for a large uniform urban area, in *Vehicular Traffic Science* (Edie, Ed.), pp. 308–328, American Elsevier Publishing Company, New York, 1967.
7. J. McClenahan, *Optimizing Transit Service,* Ph.D. Dissertation, University of Pennsylvania, Philadelphia, Pa., 1971.
8. T. E. Lisco, The Value of Commuters Travel Time—A Study in Urban Transportation, *Record 245,* p. 36, Highway Research Board, Washington, D.C., 1968.
9. L. G. O. R. U. (Local Government Operational Research Unit), Modal Split, *Report No. C32,* Reading, England, August 1968.

Study of a Collective Taxi System

P. H. Fargier and M. Cohen
Operational Research and Data Processing Department,
Institut de Recherche des Transports, Arcueil, France

Introduction

Collective Taxis

This is a means of surface public transport operated under various forms in some countries, in particular in Latin America and in the Middle East.
We state two of these forms:

(a) The first, using a fixed route, generalizes the bus principle: cars (with a capacity varying from that of a private car to that of a minibus) travel on a line (some deviations are sometimes authorized); but they do not stop at regular points, only on request, to load or unload—such a system exists, for instance, in Brazil.

(b) The second operates on entirely free itineraries. Cars travel freely, in a town for instance, and stop anywhere to pick up and discharge customers. Drivers have the right to refuse customers whose destination does not fit in with that of the passengers already on board.

The following study refers to the latter system.

Problems to be Solved

(a) *Tariff.* Two different kinds of rates may be considered:
A fixed tariff, whatever the origin and destination.

A so-called "proportional" tariff, which could take the following form in practice: An area served by collective taxis would be divided into zones, with a fixed tariff within each zone, but varying from one zone to another (dearer for more congested zones as the commercial speed is lower). Furthermore, an origin–destination matrix would give the cost of transfer from one zone to another.

These two systems may be combined: A basic charge plus a proportional tariff, for instance.

(b) *Quality of Service.* It should be evaluated from two points of view:
Average deviations to be borne by customers.

Customers waiting time, not only on an average, but also in probability. These two parameters are themselves dependent on the following parameters: Number of taxis, in operation, Capacity of these taxis, and Geographical distribution of taxis in relation to the local demand.

Purpose of the Present Paper

The present study is a first attempt to obtain elements which may help to solve the above problems. A theoretical approach is complemented by a simulation method.

The theoretical approach makes it possible to obtain formulas in the case of an infinite "homogeneous" network (the meaning given to the word "homogeneous" will be defined in a later section), and to envisage their extension in more general cases.

However, these formulas involve certain parameters which are inadequately known, related to the behavior of taxi drivers and, in addition, possibilities of extension are uncertain and, in any case, insufficient.

Simulations carried out so far, refer to one vehicle only. It is possible, on the basis of the results achieved, to define the probable policy of the taxi drivers, and to carry out a first check of the theoretical formulas, for any network and traffic demand.

A. Theoretical Study of "Homogeneous" Networks

An attempt is made in this section to solve the problem of collective taxis in the rather theoretical case of a so-called "homogeneous" network, defined by the assumptions given below. This case, which is relatively simple, seems to be of some interest, as far as applications are concerned, and could, furthermore, be used as a basis for a more thorough study. Some of the assumptions could, moreover, be easily modified.

Assumptions Which Define An Homogeneous Network—Notations

(a) A network is considered composed, either solely of one-way streets, or two-way streets, on which identical taxis travel at the same speed v (time lost in picking up and discharging customers is not taken into consideration, for even empty taxis will travel at the same speed); there is an average of ω taxis per unit of length; the maximum number of passengers is n.

(b) There is a probability $a\,dx\,dt$ that a customer will appear in a distance dx during a time dt; a is independent of the place, the time, and the rate of loading the taxi; arrivals of the various customers are independent of one another; customers arrive separately, each customer waits until a taxi comes and picks him up.

(c) When a taxi holding i passengers passes before a customer, the likelihood of acceptance is p_i; we have

$$1 = p_0 > p_1 \geqslant \cdots \geqslant p_{n-1} > p_n = 0.$$

The p_i's are assumed to be independent of the point considered and the length of the journey.

(d) Knowing that a taxi approaches the stretch dx with i passengers, the probability that a passenger will get out on this stretch will be assumed independent of the "history" of the taxi, and equal to $i(dx/r_i l)$. In this formula, l is the average journey length that the customers would travel if each one took the shortest route, and r_i a coefficient $\geqslant 1$ resulting from extensions of journeys due to deviations.

(e) The network is such, and destinations are distributed in such a way, that the average frequency of taxis is the same at every point. The moments taxis pass any given point form a Poisson distribution.

(f) Under stable conditions, the probability P_i that a taxi, passing a given point, has i passengers, is the same everywhere, and is independent of the "history" of this point (it is implied therefore that routes are sufficiently entangled that if no taxi has come to a given point for a long time, the first which arrives will not necessarily be carrying more passengers).

Note on These Assumptions. The values of the p_i and r_i, which are data in this case, are, of course, not well known a priori, and their determination itself requires a study. It will be seen (later) that simulation provides a preliminary solution to this problem.

Mathematical Solution

The data are the variables $v, \omega, n, a, p_i, l, r_i$. Stable conditions are assumed. It is fairly easy to see that, in this case:

The average taxi frequency at one point is ωv.

An average of a/ω passengers enter each taxi during each unit of time (every customer waiting is picked up sooner or later by a taxi; for each length unit there are ω taxis, and a new customers arrive during each unit of time).

If the unit of distance is called kilometer (km), each taxi performs an average of al/ω "useful" passenger-km per unit of time ("useful" meaning that deviations are not taken into account) and $al\bar{r}/\omega$ real passenger-km, $\bar{r}$ being a certain average of the r_i. (The value of $\bar{r}$ may be determined by reasoning as follows: on the average, during unit of time, a taxi travels vP_i units of distance with i passengers, during which $vP_i i/r_i l$ customers are discharged, the average number of customers discharged per unit of time being a/ω, we have

$$\frac{a}{\omega} = \sum_i \frac{vP_i i}{r_i l}$$

On the other hand, the number of real passenger/km travelled per unit of time is

$$\frac{al\bar{r}}{\omega} = v \sum_i P_i i,$$

then

$$\frac{a}{\omega} = \frac{\sum_i viP_i}{l\bar{r}}.$$

When equalizing these two values of a/ω, we have

$$\frac{1}{r} = \sum_i iP_i \frac{1}{r_i} \bigg/ \sum_i iP_i.$$

When a taxi is full to capacity, its performance per unit of time is νn real passenger-km. The average taxi loading rate is therefore $a l \bar{r}/\omega \nu n$. This average rate cannot exceed 1. As $\bar{r}$ itself must remain lower than r_n, there can be a steady state only if $a l \bar{r}_n/\omega \nu n < 1$.

The mathematical problem is the calculation of the P_i's; knowledge of these will, in effect, make it possible to answer questions, in particular regarding customer waiting time.

When a customer sees a taxi coming, the probability that he will be able to board is

$$\pi = \sum_{i=0}^{n} P_i p_i.$$

The average frequency being $\omega \nu$, the waiting time follows an exponential law of parameter $\pi \omega \nu$ (density $\pi \omega \nu e^{-\pi \omega \nu t}$, average $1/\pi \omega \nu$). The probability that two customers will be waiting simultaneously on a distance dx at a given time, is of the second degree in dx. Therefore, the probability of a customer waiting is

$$\int_{t=0}^{\infty} Pr \left\{ \begin{array}{l} \text{a customer is waiting since} \\ \text{a time } Te(t, t+dt) \text{ and has} \\ \text{not taken a taxi yet} \end{array} \right\} dxdt = dx \int_{0}^{\infty} a e^{-\pi \omega \nu t}\, dt = \frac{adx}{\pi \omega \nu}.$$

It is admitted (Assumption a) that these probabilities were independent of each other for the successive distances dx travelled by a taxi (this is a consequence of the route tangle).

If at moment t a taxi is carrying i passengers, at moment $t + dt$ it will hold:

$$(i - 1) \text{ with a probability } \frac{iv}{r_i l}\, dt \quad \text{if } i > 0;$$

$$(i + 1) \text{ with a probability } p_i \frac{a}{\pi \omega \nu}\, vdt = p_i \frac{a}{\pi \omega}\, dt \quad \text{if } i < n;$$

$$i \text{ with a probability } 1 - \frac{iv}{r_i l}\, dt - p_i \frac{a}{\pi \omega}\, dt \quad \text{if } 0 < i < n;$$

$$0 \text{ with a probability } 1 - p_0 \frac{a}{\pi \omega}\, dt \quad \text{if } i = 0;$$

$$n \text{ with a probability } 1 - \frac{nv}{r_n l}\, dt \quad \text{if } i = n.$$

As we are considering the case of the steady state, the P_i's have the same value in t and $t + dt$; from which it may be deduced:

$$P_0 = P_1 \frac{v}{r_1 l} dt + P_0 \left(1 - p_0 \frac{a}{\pi \omega} dt\right)$$

$$1 < i < n: \ P_i = P_{i+1} \frac{i+1}{r_{i+1} l} v dt + P_i \left(1 - \frac{iv}{r_i l} dt - p_i \frac{a}{\pi \omega} dt\right) + P_{i-1} \, p_{i-1} \frac{adt}{\pi \omega}$$

$$P_n = P_n \left(1 - \frac{nv}{r_n l} dt - p_n \frac{a}{\pi \omega} dt\right) + P_{n-1} \, p_{n-1} \frac{adt}{\pi \omega}.$$

Or, after simplifying and dividing by dt:

$$P_1 \frac{v}{r_1 l} - P_0 p_0 \frac{a}{\pi \omega} = 0$$

$$1 < i < n: \ P_{i+1} \frac{(i+1)v}{r_{i+1} l} - P_i \left(\frac{iv}{r_i l} + p_i \frac{a}{\pi \omega}\right) + P_{i-1} p_{i-1} \frac{a}{\pi \omega} = 0.$$

By adding the i first relations, member to member, we have

$$0 \leqslant i < n \ P_{i+1} \frac{i+1}{r_{i+1} l} v - P_i p_i \frac{a}{\pi \omega} = 0$$

or

$$\frac{P_{i+1}}{P_i} = \frac{alp_i \, r_{i+1}}{(i+1) v \omega \pi}.$$

When, for $0 \leqslant i \leqslant n$

$$P_i = P_0 \frac{1}{i!} \left(\frac{al}{v \omega \pi}\right)^i (p_0 p_1 \ldots p_{i-1})(r_1 r_2 \ldots r_i).$$

If we let

$$f_i = \frac{1}{i!} \left(\frac{al}{v \omega}\right)^i (p_0 p_1 \ldots p_{i-1})(r_1 r_2 \ldots r_i) \ \text{ for } i > 0 \ \text{ and } f_0 = 1,$$

then

$$P_i = P_0 \frac{f_i}{\pi^i},$$

$$1 = \sum_{i=0}^{n} P_i = P_0 \sum_{i=0}^{n} \frac{f_i}{\pi^i},$$

$$\pi = \sum_{i=0}^{n} P_i p_i = P_0 \sum_{i=0}^{n} \frac{f_i p_i}{\pi^i}.$$

Therefore,

$$\pi = \frac{\displaystyle\sum_{i=0}^{n} \frac{f_i p_i}{\pi^i}}{\displaystyle\sum_{i=0}^{n} \frac{f_i}{\pi^i}}, \qquad P_j = \frac{\displaystyle\frac{f_j}{\pi^j}}{\displaystyle\sum_{i=0}^{n} \frac{f_i}{\pi^i}} \tag{1}$$

π may be calculated with the first equation. The right side of the equation represents the average p_i weighted by f_i/i. Now, the p_i are smaller than or equal to 1, and decrease as i increases. When π increases, the relative importance of the highest p_i increases. For $\pi = 0$, only p_n, which is zero, occurs. Thus, the right side is an increasing function of π, the value of which is 0 for $\pi = 0$ and smaller than 1 for $\pi = 1$. Its first derivative for $\pi = 0$ is

$$\frac{f_{n-1}\, p_{n-1}}{f_n} = \frac{nv\pi}{al\, p_{n-1}\, r_n}\, p_{n-1} = \frac{nv\omega}{alr_n}.$$

This expression is > 1 (condition for a steady state given above). The equation in π will thus have at least one solution between 0 and 1. It must be possible to demonstrate directly that it has only one. Let us be satisfied, for the time being, with the following indirect reasoning: If there were several solutions in π, this would mean that there were several stable distributions; now, the evolution in the number of customers of a taxi forms a birth and death process, and the general properties of this type of process show that there can only be one stable distribution.

Possible Calculations with the Aid of these Formulas

The above formulas make it possible, with an iterative method, to calculate the system's characteristics, knowing the demand (a) and the supply (ω). But quite frequently, it will be rather the determination of supply necessary to satisfy the demand with a certain quality of service which will be sought; more specifically, one will wish to calculate ω, knowing (a) and giving the value of the average waiting time w admissible. Now, $w = 1/v\omega\pi$. The calculations are relatively simple.

Supposing that

$$k_i = \frac{l^i}{i}\, p_0 p_1 \ldots p_{i-1}\, r_1 r_2 \ldots r_i \quad \text{for } i > 0 \quad \text{and } k_0 = 1,$$

we have

$$\pi = \frac{\displaystyle\sum_{i=0}^{n} k_i p_i a^i w^i}{\displaystyle\sum_{i=0}^{n} k_i a^i w^i} \quad \text{and} \quad \omega = \frac{1}{vw\pi} = \frac{1 \displaystyle\sum_{i=0}^{n} k_i a^i w^i}{vw \displaystyle\sum_{i=0}^{n} k_i p_i a^i w^i}. \tag{2}$$

If a tariff which is independent of the distance is applied, and if it is desired that the taxi should receive τ francs per unit of distance travelled (even when empty), the tariff must be equal to $\tau(\omega v/a)$ (in effect, an average of $a/v\omega$ passengers are picked up per unit of length). If the tariff is proportional to the direct distance d, it must be equal to $\tau(\omega v/a)\,(d/l)$.

B. Possible Applications and Extensions of the "Homogeneous Network" Model

The assumptions of the model studies in section A are not very realistic. It may however be hoped that the formulas obtained (or at least the methods used) will serve the purposes of approximation, either as they stand, or after modification, in more general conditions which are nearer reality. This possibility will now be examined.

The arguments developed in this section make no claim to be absolutely rigorous, but rather suggest approximate formulas, which could be tested later by means of simulation.

Unlimited Homogeneous Urban Web, Constantly Balanced Exchanges Between Zones, Regular Network

This situation is hardly likely, but it is interesting to discover the shortcomings of the model, even in this case. Then the main assumptions are verified.

No preferential direction or arc.

It is in the taxi's interest to look for customers, therefore to maintain the same speed, even when empty.

However,

Assumption (f) supposes a fairly tight network.

p_i's and r_i's could be dependent variables of π (if demand is low, taxis will be more willing to accept deviations)

A taxi may change its route on sight of a potential customer; customers may have a tendency to wait at junctions; if a customer is already waiting, it is in the next one's interest to go wait up-stream; one may wonder whether stopping points should not be provided in practice; if there were a sufficient number of these, the model would remain applicable.

Some customers arrive in groups.

The driver trying to maximize his income, may have a tendency to refuse long journeys if the tariff is independent of length; the customers' waiting times will then be a function of the length of the journey.

The model gives the elements for choosing the optimum density of the network on which the taxis will travel (if the density increases, the average waiting time increases, but the ultimate journey decreases).

Possibilities of Extending the Model

In general, values $a, \omega, v,$ and l at each point of the network, as well as the destination distribution for every point of origin, may be considered as data. It will be assumed, moreover, that the p_i's and r_i's have already been evaluated. Under such conditions, if ω were known at every point, expressions linking the P_i and dP_i/dx values at every point (dP_i/dx being the differential of P_i in relation to the curvilinear abscissa of the point on the road) could be obtained for the steady state (that is to say by supposing $dP_i/dt = 0$), by the method used in section A. These relations should make it possible, at least theoretically, to calculate the P_i at every point.

ω at every point is not known at the outset, but only the total number of taxis in the network served. A major problem is, therefore, to determine how the taxis are distributed from one place to another. In order to achieve this, one may take as a basis the fact that, in steady state, $d\omega/dt$ is zero at every point, or, which is equivalent, that for any portion of the network, the number of incoming taxis is equal, on an average, to the number of outgoing taxis (per time unit). This leads to relations which must make it possible, at least theoretically, to determine ω at every point (to establish these relations, the destination distributions, known for every origin point will be used, and assumptions as to taxi driver behavior will be made: rules for acceptance of customers, choice of route, etc.).

In practice, the method presented briefly above seems very cumbersome. That is why only the simple specific case of what we called the isotropic zones was studied, and approximate indications for more general cases were given. For this purpose, parameters a and ω may conveniently be replaced by the following:

A = average number of customers arriving, per time and surface unit;

Ω = average number of taxis per surface unit;

D = network density, i.e., length of the network per surface unit.

In the case of homogeneous networks, A, Ω, D, are independent of the place and time. It is sufficient to suppose $a = A/D$ and $\omega = \Omega/D$. It may be noted that π and the P_i's are independent of D; on other hand, w is proportional to D.

Note. The preceding formulas cannot be applied directly when customers are waiting at special ranks provided for this purpose; in which case:

Δ = number of ranks per surface unit;

I = average distance between ranks.

From the point of view of the taxis, as well as from that of customers waiting, it may be admitted, as an approximation, that each rank is "spread" over a length I. Therefore, in the preceding formulas, it is sufficient to replace D by ΔI.

"Isotropic" Zone Delimited by a "Reflecting" Border

(a) Concepts of "Reflecting" Border and of "Isotropism." If we consider what happens in a selected area in an unlimited homogeneous urban web, it is observed that:

Taxis with a random number of customers, the probabilities of which are the P_i's, disappear and appear on the border of this area.

Customers are picked up and discharged, within the area, according to the laws described previously.

Customers appearing in the area have destinations in all directions with the same probability; this characteristic may be called *isotropy*.

Reciprocally, when there is an isotropic homogeneous area, the border of which absorbs taxis reaching it, and releases others in accordance with the above law, the steady state within that area will be the same as that found for an unlimited area. The name of *reflecting border* may be given to a border with such properties (because, statistically, it releases as many taxis as it absorbs, the numbers of passengers and directions being governed by the same laws for taxis absorbed and released).

(b) Case of Two Isotropic Homogeneous Areas Separated by a Common Border. The number of taxis absorbed (or released) per border length unit and per time unit only depends, in a first approximation, on Ωv, i.e., the density of taxis multiplied by their speed (roads are supposed not to be winding, as this would lead to a reduction in the speed to be taken into account here).

Therefore, and still approximately, if two adjacent areas have Ωv of the same value, taxis cross the border in both directions in equal numbers. If, moreover, the r_i's, the p_i's and the product Al have identical values in both areas, then the P_i's will also be identical, and the common border will act as a reflecting border for each of the two areas. We will then have, for both areas together, a steady rate resulting from the juxtaposition of the steady states of each one considered separately. This state is stable, because if an area has too many taxis, the number of crossings at the border will be unbalanced to its disadvantage.

The result obtained (as an approximation) is therefore that:

Taxis are distributed in such a way that Ωv has the same value in each of the two areas.

The P_i's are the same in both areas, and correspond to the steady state of each area considered separately.

This is on condition that

Each area is homogeneous and isotropic;

p_i's and r_i's are the same in both areas (it seems that this can be generally admitted);

Al has the same value in each one of the two areas.

The condition concerning Al is arbitrary a priori. It is, however, close to assumptions used in opportunity models: the denser an area, the more travellers have a chance to find a satisfactory destination. In effect, A is proportional to density, and l to the probability that a journey will end on a given short distance beyond the point already reached by the passenger. However, the assumption of a real compensation between variations of A and l is doubtful (opportunity models would lead rather to Al^2 constant), and could only be considered as an approximation.

(c) Isotropic Area Bounded by a Reflecting Border. The reasonings of the above section (b) are, of course, applicable to the case of an *isotropic* area bounded by a reflecting border, within which $A, D, v,$ and l vary continuously, the product Al, the p_i's and the r_i's remaining constant. The results are thus as follows:

Taxis are distributed in such a way that $v\Omega$ has the same value everywhere (value which may be calculated if the total number of taxis is known).

The P_i's and π have the same values everywhere; these values may be obtained through the homogeneous network formulas, from $v\Omega, Al$, the p_i's and the r_i's.

$w = \dfrac{D}{\pi\Omega v}$ varies from one point to another, in proportion to D.

Notes. D is only known approximately if the network is irregular.

It was indicated in the above section (b) that if roads were winding, the value of v taken into account must be reduced.

(d) Practical Value of Considering Such Areas. There are three chief assumptions; the likelihood of which must be discussed:

Al is constant; it was indicated in the above section (b) how this condition may be considered.

The area is bounded by a reflecting border. This assumption may seem fantastic; it could, however, be verified relatively when the operation of collective taxis is limited to a certain area of the town (which seems fairly reasonable). When arriving at the border of this area, the taxi will discharge its customers who will then use other modes of transport, and will pick up new groups of customers who have arrived by means of these other modes.

The area is isotropic. This condition seems to be the least realistic of the three, except in very central areas. In effect, most of the customers coming from a given district go towards the center, which modifies the taxi distribution.

In conclusion, assumptions presented here, and in particular that of isotropism, only seem applicable in specific situations (taxis serving a central area only). In the following paragraph, an attempt will be made to give pointers on more general cases.

Pointers Concerning More General Cases

Cases where both the following conditions are fulfilled are considered here:

(a) The various parameters (A, l, D, v) vary from one point to another sufficiently slowly for the homogeneous network formulas to be admitted at every point. It is, however, recalled that, in order to be able to apply these formulas, it will first be necessary to determine the value of Ω at each point, that is to say to know the distribution of taxis, of which only the total number is known.

(b) The exchange is balanced, that is to say that for two random areas, A and B, the average exchanges from A to B, and from B to A, are equal. This assumption is generally valid, if a full day is considered, in view of the return journeys. On the other hand, for periods of an hour, it will not always be valid (it will be more valid at slack periods than at peak hours).

Let us consider any imaginary line dividing the area served by taxis into two parts. The exchange being balanced, there will be the same number of passengers crossing this line in both directions. This equality is also found for taxis in a steady state (otherwise there would be an accumulation of taxis on one side of the line). We, therefore, obtain this first result; the average occupation rate of taxis crossing a line is the same in both directions.

In order to obtain more precise results, let us consider a small area; it holds, at a given time, a number of taxis, and a number of passengers inside these taxis. During a short period of time δt, some of these passengers will leave this area in various directions. They will be distributed among these directions according to proportions which are on average known in advance (they result from the origin-destination matrix; it may be admitted that the effects of deviation are self-compensating). We shall make the following assumption: *As a first approximation, taxis leaving the area will be distributed among the various directions, in the same manner as the passengers.*

We shall discuss later the validity of this assumption, of which we shall first examine the effects: at a given point, the average occupation rate is independent of the taxi directions and, in view of the balance of exchanges between areas, it is deduced that the average occupation rate is the same at every point. This condition is sufficient to determine the geographical distribution of taxis, thus the value taken by Ω. Moreover, it shows that the homogeneous network formulas may be applied directly to the area of activity of the taxis, taken in its entirety.

Discussion of the Assumption on the Distribution of Taxis Leaving an Area. This assumption is based, in the first place, on a number of vague ideas. Each taxi has all the more chances of meeting a customer going in a given direction as there are passengers in that direction; furthermore, the various traffic currents must cross each other sufficiently so that the occupation rate is the same in all directions.

However, a closer analysis should take into account the behavior of the taxi driver: criteria of acceptance or refusal of customers, choice of itineraries. Now, the taxi driver will try to maximize his profit. With a fixed tariff system, for instance, he will prefer to go towards areas where the journeys are short. On the other hand, a proportional tariff, increased by a fixed charge, to compensate time lost in picking up and discharging, would strengthen the validity of our assumption. If the occupation rate was higher in an area, it would be in the taxi drivers' interest to go there in preference, which would reduce the gap.

In case of unbalanced exchange, the above results could be extended in the following manner: The average occupation rate would still be the same at every point, but at each point, it would depend on the direction, and it is for each direction that formulas based on the homogeneous network case should be used.

In conclusion, it may be assumed that, under fairly general conditions, taxis are distributed in such a way that the average occupation rates are the same at every point. If the exchange is balanced, the homogeneous network formulas would be directly applicable to the total area of activity of the taxis. In case of unbalanced exchange, a more detailed study, which would distinguish the various directions, would be necessary.

C. Simulation Studies

The theoretical approach outlined in sections A and B, although interesting, has proved insufficient, for two kinds of reasons:

The assumptions adopted differ from real conditions, and cases in which the formulas obtained may be considered as valid, at least approximately, are insufficiently known.

Even when these formulas are valid, some of the parameters (p_i's and r_i's) related to the behavior of the taxi drivers and the customers, are unknown.

Simulations are undertaken in order to have means of studying any situation, of determining the taxi drivers' and customers' behavior (and therefore of knowing the p_i's and r_i's), and of discovering whether or not the theoretical formulas may be used in fairly general conditions.

The simulations described below represent a first stage only. They already provide interesting results, but they need to be pursued and developed.

Description of Simulations Carried Out

(a) **Choice of Model.** The problem is, in the first stage, to define the taxi drivers' strategy and, in a second phase, to operate the model in order to deduce the value of parameters, such as waiting time, drivers' earnings, average deviations imposed on customers, distribution of taxis in a town, etc.

Two different kinds of models were envisaged: A model with N taxis on a given network, confronted with a given demand observed constantly at every point; a model with only one taxi on a given network, confronted with the demand left by the other taxis, observed at the point where the taxi passes by.

The second solution was adopted for the following reasons:

It constitutes a first stage which gives a better view before launching out into the construction of a more elaborated model.

It is possible, with this model, to vary the strategy of a taxi driver, that of the competitors remaining unchanged, which corresponds to the question we are asking; in effect, to vary the policies of all taxis simultaneously in the same way, would not allow to give the optimum policy for each taxi considered separately.

Finally, this model is much easier to program, and was therefore likely to be achieved sooner.

On the other hand, it introduces the notion of "Residual Demand" which is difficult to evaluate. Also, it is not possible with this model to evaluate all the main parameters directly: this is, in particular, the case of waiting time, as customers are only taken into account when a taxis passes by the point considered. However, this model should only be considered as a first step, and none of the results are final.

(b) **Operation of the System.** The town is represented in the form of a network. The picking up and discharging points are situated exclusively at ranks which are the nodes of the network. This does not mean that these ranks exist effectively—each rank represents an area.

The simulation is by events. An event consists of the movement of a taxi from one rank to the next, on the route followed by the taxi. At each rank, any passengers who have arrived at their destination get out. A number N of customers waiting is generated according to a probability distribution which is a datum of the simulation.

When N has been determined, each of the N customers is examined in turn. His destination is determined at random (with the aid of an origin–destination matrix). According to this destination, and to the history of the passengers already on board of the taxi, customers are accepted or refused. When all the waiting customers have been examined, we move to the next rank.

The criteria of acceptance and refusal concern the customers (deviations must be below a certain threshold) on the one hand, and the taxi driver (profit expected must be above a certain threshold) on the other hand.

Simulations have been carried out with two different networks, for various levels of demand, with fixed and proportional tariffs. Thresholds governing the criteria of acceptance and refusal of customers were made to vary.

Results Achieved

(a) **Evaluation of Parameters.** As far as acceptance and refusal of customers is concerned, it was observed that if the sole criterion of the taxi driver's profit is taken into account, large deviations (which may exceed 1000%) are often imposed upon the customers. On the other hand, if it is stipulated that these deviations should remain within reasonable limits, the loss born by the taxi driver is not great. It may therefore be assumed that, in practice, a situation in which deviations are not excessive will be created, either because of the regulations, or spontaneously; and thresholds of acceptance and refusal of customers were selected accordingly.

One of the aims of the simulations was also to evaluate the p_i's and r_i's.

It is recalled that p_i is the probability that a customer may board a taxi in which there are already i passengers, and that r_i is defined by the fact that $i \, dx/r_i l$ is the probability that, when a taxi approaches the stretch dx with i passengers, one of them will get out on this stretch.

In the simulation, the value of p_i is the proportion of customers picked up when the taxi holds i passengers, and the value of r_i is obtained by stating that $i/r_i l$ is equal to the ratio of the number of times a passenger gets out, leaving in the taxi $(i - 1)$ passengers, to the total distance travelled with i passengers on board.

Fairly stable values were obtained, the averages of which are given in Table I.

TABLE I

i	0	1	2	3	4	5	6
p_i	1	0.73	0.57	0.51	0.48	0.45	0
r_i	1	1.1	1.1	1.2	1.3	1.4	1.7

(b) Comparison with the Homogeneous Network Formulas. The simulations were carried out under fairly artificial conditions, since it is not the real traffic demand that was supposed, but the "residual" demand; that is to say, the number of customers waiting at a rank at a random moment. This is a random number; let $\bar{N}$ be its mathematical expectation. We have $\bar{N} = \alpha w$, calling α the rate of arrival of customers at a rank per time and w the average waiting time of a customer.

Let v be the average number of ranks visited by a taxi, when travelling along a unit of length. It is normal to suppose an equivalent for parameter a of the homogeneous network: $a = v\alpha$. We have thus $aw = v\alpha w = v\bar{N}$. $\bar{N}$ is a datum. If v may be evaluated, formulas giving P_i and π in relation to aw may be applied. This is what we will do for one of the networks studied, and an average value of $\bar{N}$. We will then compare the results with those of the simulation.

Evaluation of v. If all distances between adjacent ranks were identical, v would be equal to the inverse of this common distance. In the network which is considered here, this distance is very variable, therefore an average must be taken.

If it is admitted that the sections all have the same chance of being used, whatever their length, the following must be supposed:

$$v = \frac{1}{\text{average length of sections}} = \frac{1.77}{1000}.$$

If it is admitted that the sections all have the same chance of being used, whatever their length, the following must be supposed:

$$v = \frac{1}{\text{average length of each section}} = \frac{2.1}{1000}.$$

A priori, the second assumption seems to be nearer the mark. A route has more chance of being used if, for a given length, it is provided with more ranks. Simulation results show that v depends slightly on the values of the various parameters, its average value being in the range of $2/1000$.

Calculation of π and the P_i's. The following is considered: $v = 2/1000$, $N = 1.125$, p_i's and r_i's values given above, and $l = 2.137$ (value observed in simulation).

We have, therefore, $aw = Nv = 2.50/1000$; π and P_i may be calculated by the following formulas:

$$\pi = \frac{\sum\limits_{i=0} k_i p_i (aw)^i}{\sum\limits_{i=0}^{6} k_i (aw)^i}, \qquad P_i = \frac{k_i (aw)^i}{\sum\limits_{i=0}^{6} k_i (aw)^i},$$

in which

$$k_i = \frac{l^i}{i!} p_0 p_1 \dots p_{i-1} r_1 r_2 \dots r_i \quad \text{for } i > 0 \quad \text{and } k_0 = 1.$$

TABLE II

	π	p_0	p_1	p_2	p_3	p_4	p_5	p_6
Calculation	0.51	0.02	0.12	0.22	0.25	0.19	0.12	0.07
Simulation	0.52	0.02	0.11	0.25	0.26	0.20	0.11	0.05

Table II compares the results of this calculation with the values obtained in simulation.

It is not of course possible to draw a final general conclusion from a single example, but as the concordance between the results of this calculation and those of the simulation suggests that the homogeneous network formulas are usable in fairly general cases.

Notes. When, as has been done so far, the residual demand is taken, the corresponding real demand cannot be stated in advance. On the other hand, the latter may be found, as *a result* of simulation, as a function of the number M of taxis operating in the network. In effect, it may be admitted that each taxi serves (in probability) the same number of customers on each line. Therefore, traffic demand from rank I to rank J, which would have led to the given residual demand, is equal to M times the number of customers taken effectively from I to J by the taxi. The average waiting time at a rank is then equal to $\bar{N}$ divided by M times the number of customers accepted by the taxi at this rank per time unit. The average waiting time for all the network is equal to

$$w = \frac{\bar{N} \times \text{number of ranks}}{M \times \text{number of customers accepted by a taxi per time unit}}.$$

Conclusion

The study presented in this report should not be considered as terminated. It has, however, in its present form, furnished results which should not be neglected.

In the first place, we have gathered pointers as to the likely behavior of taxi drivers and clients, and developed a method for describing this behavior. Then, we have a simulation tool which, although far from providing all the desired information, makes it possible, nevertheless, to explore various types of networks and demand profiles. Finally, we have obtained in a theoretical case formulas which give a first answer to various questions arising (for a given demand, calculation of average waiting time, knowing the number of taxis or, on the contrary, of the necessary number of taxis if maximum waiting times and tariffs are set). In view of the considerations under Section B, and the verification by simulation, it may be hoped that these formulas will remain approximately valid under fairly general conditions.

However, these results remain insufficient from many points of view:

Regarding behavior, tests are to be continued, and above all, the possibility of a driver preferring to go to areas where earning prospects are better, has not been considered.

The simulation program does not allow for consideration of all taxis confronted with a given demand.* Moreover, the fact of only considering the case with ranks—in a relatively small number—restricts the scope of the results. In effect, waiting time seems to be very sensitive to the number of stops.

The theoretical formulas obtained have been insufficiently tested, and could not in any case cover all situations. They are certainly no longer valid if traffic is appreciably unbalanced, or if demand is concentrated on certain points, or when the drivers give preference to certain areas because of better earning prospects.

---

*Since this report was written, a program to simulate the behavior of all the taxis confronted by a given demand has been carried out. As an example, it has been applied to the town of Grenoble (about 300,000 inhabitants).

Investigation of Sorting and Train Formation Schemes for a Railroad Hump Yard

M. W. Siddiqee
Stanford Research Institute, Menlo Park, California

Abstract

Four schemes are investigated for systematically assigning the tracks and sorting various cars and batches to form new outbound trains in a railroad hump yard. The suitability and applicability of each scheme are discussed, and guidelines are recommended for improving the process of sorting and the formations of trains in a railroad hump yard.

Introduction

In a railroad hump yard, each incoming train typically contains randomly distributed cars destined for various geographical zones. In the hump yard the incoming trains are pushed over a hump. As each car reaches the hump, it is manually disconnected from the preceding car and directed to a preassigned sorting track. All the cars destined for a particular zone must be grouped together to form a batch. Batches are later coupled together to form new outbound trains. A typical hump yard configuration is shown in Fig. 1.

In U. S. hump yards, typically, 8 to 10 outbound trains, each with 50 to 100 cars grouped in 1 to 10 batches, are required to be formed from 8 to 10 incoming trains, each containing randomly distributed cars. Frequently, the number of available sorting tracks is considerably less than the number of desired batches, and these tracks are limited in length. Under such circumstances the process of sorting, batch formation, and outbound train formation may become an involved and complicated operation, resulting in excessive rehumping of several cars and wasteful engine operations, unless some systematic scheme is followed to assign the tracks to various groups of cars.

In the present hump yard operations, the yard master does not receive information about the composition of the incoming trains sufficiently in advance to be able to plan a specific strategy of sorting cars and forming batches that would best match the particular requirements for each day. He has to rely on his experience and intuition to assign sorting tracks to various cars of an incoming train during the humping process. The typical yard master has developed a keen sense of yard operations and is generally capable of performing the sorting and train formation processes in a satisfactory manner. However, with the expected introduction of modern data-processing and communication equipment in various yards, information about incoming trains will be available sufficiently

---

References p. 387

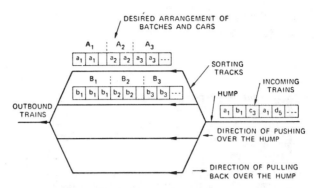

Fig. 1. Typical hump yard configuration.

in advance which would be used for planning efficient schemes of sorting. Furthermore, with the continual increase in railroad activity, resulting in an increasing number of cars to be handled per day in each hump yard, it seems desirable to investigate systematic approaches for sorting cars and forming the desired outgoing trains, to avoid delays, excessive rehumpings, and unnecessary engine operations. In some of the European hump yards, e.g., those of the Swiss Railway System, a few systematic schemes have recently been introduced. However, these schemes have been developed according to local conditions. In the United States, a need still exists to study the problem in detail and develop efficient schemes suitable for conditions in this country.

Statement of the Problem

In its basic form, the sorting and track assignment problem can be stated as follows: With the constraint that the number of sorting tracks available is significantly less than the total number of batches to be sorted and formed, find feasible schemes, most appropriate in some sense, of sorting and forming the desired batches and arranging them in a desired sequence, given advance information about the number, length, and destination of the cars in incoming trains and the number of available sorting and train formation tracks in the yard and the length of these tracks.

Four Schemes for Sorting and Train Formation with Examples

A few introductory remarks and definitions are necessary to characterize the schemes to be desired.

The process in which the cars of each incoming train are humped for the first time is referred to as "primary humping". Let the outbound trains to be formed be called Trains A, B, etc., each consisting of corresponding Batches A_1, A_2, A_3, ..., B_1, B_2, B_3, ..., etc. Let the single cars belonging to Batches A_1,

$B_1, \ldots$ be denoted by $a_1, b_1, \ldots,$ etc. When lower-case letters with subscript numbers appear in parentheses, it means that the cars are randomly mixed: e.g., $(a_1, a_2, \ldots, b_1, b_2, \ldots)$ means that in this group Cars $a_1, a_2, b_1, b_2, \ldots,$ etc., are randomly mixed. The operation when a car or group of cars is pulled back from the sorting tracks over the hump and is pushed over the hump again is called "rehumping".

With these definitions, the four schemes can be characterized by the method of grouping cars in the primary humping.

Scheme 1—Initial Grouping According to Subscript

According to this scheme, cars designated with the same subscript are grouped together during primary humping; for example, all cars with subscript 1, a_1, b_1, c_1, etc. are grouped together and assigned a common track, $a_2, b_2, \ldots$ are assigned another, and so on. Thus, after primary humping the yard may appear as in Fig. 2.

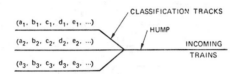

Fig. 2. Scheme 1: Arrangement after primary humping.

After the primary humping has been completed, the tracks are pulled back in such a sequence that the group having the lowest subscript precedes the group with the next lower, and so on, as shown in Fig. 3. Alternatively, the cars can be rehumped track by track in proper sequence, if enough tracks are available to receive the rehumped cars.

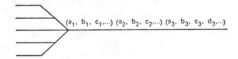

Fig. 3. Scheme 1: Arrangement of groups for second humping.

When the cars are rehumped now, a_1 can be put on one track, b_1 on another, and so on. The a_2 cars are put behind the a_1 cars on Track 1, the b_2 cars behind the b_1 cars on Track 2, and similarly with $a_3, b_3, \ldots.$ When all the cars have been rehumped, the batches are separated and are in proper sequence.

Scheme 2—Initial Grouping According to Outbound Trains

According to this scheme, the cars of outbound trains are combined in a few groups during the primary humping (e.g., cars of Trains A, B, and C in one

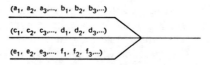

Fig. 4. Scheme 2: Arrangement after primary humping.

group, D, E, and F in another group, etc.) and the cars belonging to each group are assigned a common track during primary humping. Thus, if outbound Trains A, B, C, D, etc. are to be formed, each with Batches A_1, A_2, B_1, B_2, etc., the cars belonging to Trains A and B, i.e., all cars $(a_1, a_2, \ldots, b_1, b_2, \ldots)$ can be put on one track. Similarly, all cars $(c_1, c_2, c_3, \ldots, d_1, d_2, d_3, \ldots)$ can be put on another, etc., as shown in Fig. 4.

Each track can then be rehumped and each batch be put on a separate track, as shown in Fig. 5.

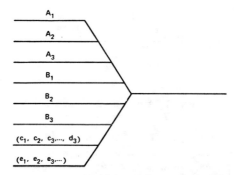

Fig. 5. Scheme 2: Arrangement after first track has been rehumped.

Figure 5 shows that either Train A or Train B can be pulled out first by coupling the respective batches. With the departure of Train A or Train B, the respective tracks become empty, so the batches of Trains C and D can easily be sorted.

Scheme 3—Triangular Scheme

In some European hump yards, the so-called "triangular" scheme of batch sorting is frequently employed. According to this scheme, the car subscript numbers are arranged as follows:

$$
\begin{array}{lllll}
1 & 3 & 5 & 8 & 12 \\
2 & 6 & 9 & 13 \\
4 & 10 & 14 \\
7 & 15 \\
11
\end{array}
$$

During primary humping, all cars having the subscripts 1, 3, 5, 8, . . . (the first horizontal set of numbers) are assigned the same track; cars with 2, 6, 9, . . . (the second horizontal set of numbers) are assigned another track, and so on. The first track is now pulled back and rehumped. The cars with subscript 1 are directed to as many separate tracks as there are batches with subscript 1, e.g., a_1 cars to one track, b_1 cars to another, c_1 to a third, etc. Cars of subscript 3 are directed on Track 2, behind the (2, 6, 9, . . .) group. Cars with subscript 5 are directed on Track 3, etc. The Track 2 is now pulled back, and cars with subscript 2 are directed to the respective tracks behind cars with subscript 1, e.g., a_2 behind a_1, b_2 behind b_1, etc.

An example of the formation of four trains is helpful in clarifying the various operations of this scheme. Let the number of batches in the four trains be 10, 9, 7, and 4, respectively. This rather special combination of batch numbers has been chosen for convenience, since the operations of the triangular scheme can be explained easily with this particular combination when four trains are to be formed.

Let the train with 10 batches be called A, with Batches $A_1, A_2, \ldots, A_{10}$.
Let the train with 9 batches be called B, with Batches $B_2, B_3, \ldots, B_{10}$.
Let the train with 7 batches be called C, with Batches $C_4, C_5, \ldots, C_{10}$.
Let the train with 4 batches be called D, with Batches $D_7, D_8, \ldots, D_{10}$.

Note that the first batch of Train B is labeled B_2 and not B_1. Similarly, in Train C, the first batch is labeled C_4 instead of C_1. The advantage of labeling the first batches in this manner is that no further sorting tracks are needed during the second and third humpings, as will become clear. Note that the numbers used for the first batches of the various trains (1, 2, 4, and 7) are the numbers that appear in the first column of the triangular scheme.

During primary humping all the cars having subscripts 1, 3, 5, and 8 are collected on one track. All the cars having subscripts 2, 6, and 9 are collected on a second track. Cars having subscript 4, 10 are collected on a third track; cars having subscript 7 are collected on a fourth track. The tracks appear as in Fig. 6.

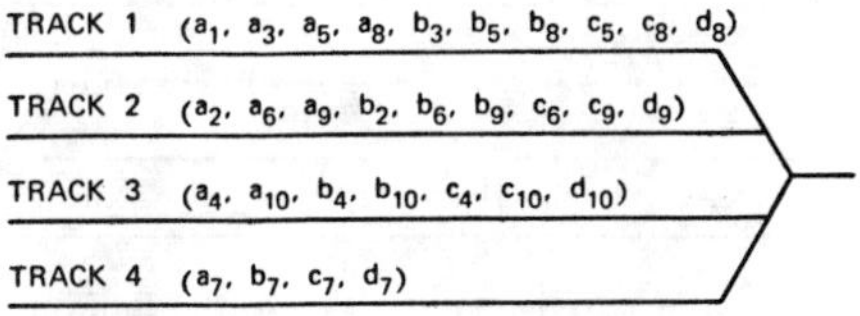

TRACK 1 $(a_1,\ a_3,\ a_5,\ a_8,\ b_3,\ b_5,\ b_8,\ c_5,\ c_8,\ d_8)$

TRACK 2 $(a_2,\ a_6,\ a_9,\ b_2,\ b_6,\ b_9,\ c_6,\ c_9,\ d_9)$

TRACK 3 $(a_4,\ a_{10},\ b_4,\ b_{10},\ c_4,\ c_{10},\ d_{10})$

TRACK 4 $(a_7,\ b_7,\ c_7,\ d_7)$

Fig. 6. Scheme 3 (Triangular Scheme): Arrangement after primary humping.

In the second phase of humping, the cars on Track 1 are pulled back and re-humped. Cars a_1 can be recollected on Track 1. The (a_3, b_3) cars are collected on Track 2 behind existing cars. The (a_5, b_5, c_5) cars are collected on Track 3 behind existing cars. The (a_8, b_8, c_8, d_8) cars are collected on Track 4 behind (a_7, b_7, c_7, d_7). The yard appears as in Fig. 7.

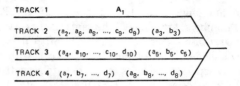

Fig. 7. Scheme 3: Arrangement after second humping of Track 1.

The Track 2 cars are now pulled back and rehumped. The a_2 cars are collected behind the A_1 batch on Track 1, the b_2 cars are collected on Track 2, and the $(a_6, b_6,$ and $c_6)$ cars are collected on Track 3 behind existing cars. The (a_9, b_9, c_9, d_9) cars are collected on Track 4 behind the $(a_8, b_8, \ldots, d_8)$ group. The a_3 cars are collected behind the a_2 cars on Track 1, and the b_3 cars behind the b_2 cars on Track 2. The yard appears as in Fig. 8.

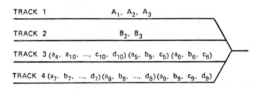

Fig. 8. Scheme 3: Arrangement after second humping of Track 2.

The Track 3 cars are now pulled back and rehumped. The a_4 cars are collected behind the a_3 cars on Track 1, the b_4 cars behind the B_3 batch on Track 2, and the c_4 cars on Track 3. The group $(a_{10}, b_{10}, c_{10},$ and $d_{10})$ are collected on Track 4 behind $(a_9, b_9, c_9,$ and $d_9)$. Then the a_5 cars are collected on Track 1, the b_5 cars on Track 2, the c_5 cars on Track 3; then the a_6 on 1, b_6 on 2, and c_6 on 3. The yard appears as shown in Fig. 9.

```
TRACK 1              A₁, A₂, ..., A₆
TRACK 2              B₂, B₃, ..., B₆
TRACK 3              C₄, C₅, C₆
TRACK 4  (a₇, b₇, ..., d₇)(a₈, ..., d₈)(a₉, ..., d₉)(a₁₀, ..., d₁₀)
```

Fig. 9. Scheme 3: Arrangement after second humping of Track 3.

The Track 4 cars are now pulled back. The d_7 cars are collected on the Track 4, a_7 on Track 1 behind the A_6 batch, b_7 on Track 2 behind the B_6 batch, and the C_7 batch behind the C_6 batch on Track 3. Then $(a_8, b_8, \ldots, d_8)$ are collected behind the respective A_7, B_7, etc. batches and similarly $(a_{10}, \ldots, d_{10})$. All four trains are now formed.

Scheme 4—Geometrical Scheme

Let the car subscript numbers be arranged in the following way. For simplicity, subscript numbers up to only 12 have been considered.

$$
\begin{array}{llllll}
1 & 3 & 5 & 7 & 9 & 11 \\
2 & 6 & 10 & & & \\
4 & 12 & & & & \\
8 & & & & &
\end{array}
$$

The numbers in each column of this scheme form a geometric series.

During primary humping, all cars having the subscripts included in the first row are assigned one common track, cars having the subscript numbers in the second row a second common track, etc. Further operations following the primary humping can conveniently be explained by using the example problem discussed in connection with the triangular scheme. Using the same batch designations as before, with the exception that the four batches of Train D are now labeled as D_8, D_9, D_{10}, D_{11}, the yard will appear as in Fig. 10, when the

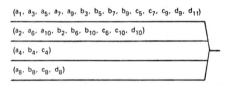

Fig. 10. Scheme 4 (Geometric Scheme): Arrangement after primary humping.

cars have been distributed according to the geometrical scheme during primary humping. The reason for labeling the first batches of Trains A, B, C, and D as 1, 2, 4, and 8, respectively, is the same as explained in connection with the triangular scheme: 1, 2, 4, and 8 are the numbers in the first column above.

The arrangements after the second humping of Track 1, Track 2, Track 3, and Track 4 are as depicted in Figs. 11 through 14, respectively, and are self-explanatory.

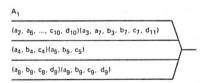

Fig. 11. Scheme 4: Arrangement after second humping of Track 1.

Characteristics and Suitability of Various Schemes

We now describe some basic characteristics of each of the four schemes and indicate the circumstances where each scheme might be suitably employed.

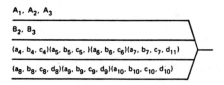

Fig. 12. Scheme 4: Arrangement after second humping of Track 2.

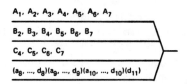

Fig. 13. Scheme 4: Arrangement after second humping of Track 3.

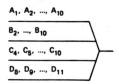

Fig. 14. Scheme 4: Arrangement after second humping of Track 4—trains formed.

Scheme 1—Initial Grouping According to Subscripts

In this scheme:

(1) Each car is humped twice.

(2) The batches are put in proper sequence after the second humping.

(3) All the outbound trains are formed simultaneously.

(4) The number of sorting tracks required during primary humping is equal to the maximum number of batches in any outbound train; for example, if three trains are to be formed, one with four, another with five, and the third with six batches, then during primary humping six sorting tracks will be required, so that cars with different subscripts may be directed to six different tracks.

(5) The length of the sorting tracks during primary humping must be such that the respective tracks can accommodate the total number of cars of the groups $(a_1, b_1, c_1,), (a_2, b_2, c_2, \ldots)$, etc.

(6) The number of empty tracks during the second humping must be equal to the number of trains to be formed; for example, considering the example above, either the cars on three of the six tracks on which cars were placed during primary humping should be pulled back and rehumped, or at least two extra tracks should be available, so that each of the six tracks can be rehumped one by one to form the three trains.

In view of the above characteristics, this scheme may be useful where the maximum number of batches in each train approximately equaled the number of trains to be formed, e.g., three trains each with three batches, or four trains each with four batches, etc. The track lengths and the capacity of the engines to pull the tracks back would also have to be taken into account before the scheme was chosen.

Scheme 2–Initial Grouping According to Outbound Trains

In this scheme:

(1) Each car is humped twice.

(2) The batches are formed separately on separate tracks, so they have to be coupled together later to form the desired trains.

(3) Trains can be formed in various desired time sequences, i.e., Train A first, then B, and then C or some other sequence.

(4) The number and length of the sorting tracks during primary humping should be adequate to accommodate respective groupings of cars, e.g., $(a_1, a_2, a_3, \ldots, b_1, b_2, b_3, \ldots), (c_1, c_2, c_3, \ldots, d_1, d_2, d_3, \ldots)$.

(5) The number of tracks during the second humping of various tracks must be equal to the number of batches in the group being rehumped; for example, if a group contains cars of Trains A and B with batches A_1, A_2, A_3 and B_1, B_2, then five tracks are needed to sort out batches A_1, A_2, A_3, B_1, and B_2 during the rehumping of this group.

This scheme is particularly useful in the following circumstances:

(1) When various outbound trains do not have to be formed simultaneously and a time priority of trains is desired.

(2) When the number of batches in various trains is not so large that more tracks than are available would be required to form batches on individual tracks.

(3) When the number of tracks available in the beginning is small but more tracks will become available later. This situation is frequently encountered in the humpyards.

Since tracks are emptied when the formed trains leave, the total track requirements for this scheme are significantly reduced.

Triangular and Geometrical Schemes

Both the triangular scheme and the geometrical scheme have the following characteristics:

(1) All cars are humped at least twice. Many are humped three, four, or five times, depending upon the maximum number of batches in the outbound trains.

(2) The batches are placed in proper sequence as the second, third, fourth, etc., humpings are executed.

(3) All outbound trains are formed simultaneously.

(4) The number of sorting tracks required is small compared to the number of batches to be sorted. In the example considered in Section III, 30 batches were sorted and four trains were formed with only four tracks being used.

(5) In general, the sorting tracks have to be fairly long. Specifically, at each stage of humping, the tracks have to be sufficiently long to accommodate groups such as $(a_1, a_3, a_5, a_8, b_3, b_5, b_8 \ldots)$ or $(a_2, a_6, a_9 \ldots c_9, d_9)(a_3, b_3)$.

Both the triangular scheme and the geometrical scheme are particularly suitable where the batches in various trains are numerous and where the available tracks are few but long. In general, triangular schemes will be found to require fewer total rehumpings than the geometrical scheme. In the four-train example considered for both schemes, it can be shown that in the triangular scheme no car is humped more than three times, whereas in the geometrical scheme some cars are humped four times. It is for this reason that the triangular scheme is generally preferred to the geometrical scheme.

The above noted characteristics have been summarized in Table I, for convenience.

TABLE I

Summary of Characteristics of Various Schemes

Schemes	Humping	Batch formation	Train formation	Remarks
1. Subscript grouping	All cars humped twice	Batches formed in desired sequence	All outbound trains formed simultaneously	Scheme likely to be useful when the number of batches in each train is small (i.e., 4 or 5) and is approximately equal to the number of outbound trains to be formed.
2. Outbound train grouping	All cars humped twice	Each batch formed on a separate track	Trains formed in various time sequences	Likely to be useful where: • trains are to be formed in some time sequence • number of batches in each train is small (2 or 3) • initially few, but later more tracks are available.
3. Triangular and Geometric schemes	All cars humped at least twice; some 3 to 4 times	Batches formed in desired sequence	All outbound trains formed simultaneously	Likely to be useful where fewer trains, each with large number of batches (10 to 15), are to be formed and tracks are fairly long. Generally, triangular scheme is better than geometric, requiring less car humps.

Concluding Remarks

In view of the large variations in hump yard design, operating policies, and size of batches, it is almost impossible to recommend one scheme as the best. However, as advance information about incoming trains and desired outgoing trains becomes available, several alternative schemes or combination of schemes can be developed, using the basic combination schemes described in this paper. Giving consideration to the existing and expected yard conditions and also to other factors (e.g., priority of some outgoing trains over others), the yard master can then choose the most appropriate scheme and plan track assignment scheduling in advance.

Simple computer programs (even hand calculations are not too difficult to perform) for various schemes can easily be written, based on typical yard conditions and covering a large set of possible variations. The output of these programs will be a listing of the track requirements during various phases of humping, the number of car rehumpings, the number of engine operations, the approximate time to execute the complete process, etc. A decision can then be made about the most appropriate sorting and train formation strategy on the basis of a quick analysis of the results of various possible approaches; if necessary and feasible, the yard master may even take suitable steps to create favorable conditions. The use of advance analysis will result in expedient and efficient operation of the yard.

Acknowledgment

The work on which this paper was based was carried out as part of a research project for the Southern Pacific Transportation Company, San Francisco. The support of the Southern Pacific Transportation Company and their permission to publish this paper are gratefully acknowledged.

References

1. M. W. Siddiqee, Sorting and Train Formation Schemes for Efficient Operation of a Hump Yard, Final Report, Phase II (Part B), SRI Project 7865 for Southern Pacific Transportation Company, San Francisco, January 1970.
2. K. J. Pentinga, Teaching simultaneous marshalling, *The Railway Gazette* (May 22, 1959).
3. H. Konig and P. Schaltegger, Optimum simultaneous marshalling of local goods trains in marshalling yards, *Monthly Bull. The International Railway Congress Association,* 4, No. 1 (January 1967).

A Strategic Model for Urban Transport Planning

J. C. Tanner
Road Research Laboratory, Department of the Environment, Crowthorne, Berkshire, England

Abstract

The paper describes the development of a strategic urban planning model. Its object is to enable rapid comparisons to be made of the traffic benefits of alternative transport investment plans, when these can be expressed in a sufficiently generalized form. Important and novel features of the model include the assumption of radially symmetric networks and trip patterns, and a representation of demand and modal split which permits straightforward economic evaluation.

Introduction

The model described in this paper arose from the need for a rapid method of assessing the economic benefits of alternative transport investment strategies. Most existing transport planning models have two limitations; in the first place, the complexity of an urban area, in terms of the numbers of links in the transport networks and the numbers of different origins and destinations of trips, is such that with current computers and techniques the evaluation of a single transport plan may occupy a considerable length of time; secondly, largely as a result of this complexity, the models do not usually contain an adequate representation of all aspects of supply, demand and the interaction between them.

With the increased understanding that is now developing of the determinants of travel behavior and of transport cost structures, it is becoming practicable to set up more satisfactory models, but computational problems remain. Because of these, it has not so far been practicable to make full use of the developing understanding of travel behavior.

It was felt that, for broad strategic planning, it would be useful to develop a model which dealt with a simplified geographical representation of a real area, but kept the fullest possible details of other aspects of travel behavior. Such a model would not be able to deal with local geographical detail, but might be able to show the effects of proposals expressed in terms of corridors, rings, etc. The particular form of geographical simplification that is being studied at present is that the area should have radial symmetry about a central point. Transport networks are assumed to consist of rings and equally spaced radials, and trip origins and destinations to lie at intersections of ring and radial roads. By this means the number of distinguishable transport links is reduced by an order of magnitude, and so is the number of distinguishable origin–destination pairs.

The development of the model, which is known by the name CRISTAL, is now complete. All the programs have been tested individually, and a set of data

389

has been assembled. The iterative behavior of the calculations is satisfactory. This paper describes the theoretical model, and opens up for constructive discussion some of the problems of strategic models, and some of the techniques employed in the present one, especially the passenger demand/modal split model, which is believed to be new, and the use of radially symmetric networks. Further papers will deal with computing problems, with the data inputs, and with applications of the model.

General Outline of the Model

The overall logical structure of the model is shown in Fig. 1. The two main components are the demand model and the supply model. The demand model specifies how many trips will occur between any pair of places, by each mode and by which route, as functions of the generalized (time plus money) costs of travel by various modes and routes. The supply model specifies how the times and costs of travel between any pair of places by any mode depend on the numbers of trips being made by various modes and routes. These two models together determine, by an iterative process, the actual numbers of trips and patterns of flows, times and costs which would arise as a result of any specified set of input parameters, including the specification of the networks. When the iterative process has reached equilibrium, the main user costs and benefits, i.e., travel times and costs, and consumer surpluses, can be evaluated, together with revenues to road and public transport authorities, and compared with similar quantities for alternative transport plans or other input parameters.

Before the model can be used, a calibration process is required. In this, certain of the parameters of the model are estimated and adjusted until the model re-

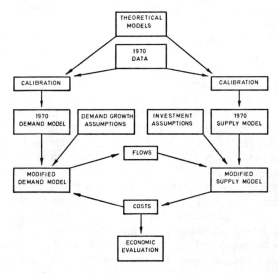

Fig. 1. Structure of the model.

produces as accurately as possible what is known about present flows, times, costs, etc.

In order to apply the model to the situation in a future year, the demand models are modified in the light of expected changes in such factors as population and car ownership, and the supply models are adjusted to take account of the network changes which are being hypothesized. The iterative process and economic evaluation indicated above can then be applied.

Separate models are being run for the morning peak period, the evening peak period and the remainder of the day.

Networks

In its present form the model contains three distinct networks: motorways, other roads, and railways. Five kinds of travel are allowed for: road goods vehicles, cars, buses, taxis, and rail. Goods vehicles, cars and taxis can use both motorways and other roads; at present buses only use other roads.

The network of ordinary roads consists of M concentric 2-way ring roads with specified radii and N equally spaced 2-way radial roads extending from the innermost ring to the outermost one. There are no roads at the center, and roads outside the outer ring are not modelled. This network has MN nodes, but because of the symmetry assumptions, the N nodes on any one ring are indistinguishable. There are MN links on ring roads; those lying on the same ring are indistinguishable from each other, and on all ring links the two directions of travel are indistinguishable. There are $(M - 1)N$ radial links; those lying at the same radius are indistinguishable, but the two directions of travel are different and separate account must be kept of their flows, costs, etc.

The motorway network may include complete rings on the lines of and in addition to selected ordinary rings. The number of motorway rings can be anything between 0 and M. If there are any radial motorways, they must exist on every radial of the network of ordinary roads; they may extend from any of the M rings to any other. In a typical application they might extend from the outermost of the M rings to the innermost ring, to the innermost motorway ring, or to some ring lying between these. As with ordinary roads, not all of the motorway links are distinguishable, nor are the two directions on ring motorways.

Access between one ordinary road and another, or between one motorway and another, can be made without penalty at any node of the respective networks. Access between an ordinary road and a motorway is possible at any node of the ordinary network at which there is also a motorway, subject to a penalty as described below.

Buses run on two kinds of routes. Ring routes operate in both directions on selected ordinary ring roads, perhaps on all of them; these routes have no termini. Radial routes operate on all N radials, travelling to and fro between one terminus and the other. These termini may, but need not, be on the inner and outer ring roads. There is in the present formulation of the model an inconsistency in that the frequency of buses on radials, insofar as this affects waiting times and operating costs, may vary along their length, implying that some buses do not traverse the full route, whereas it is also assumed that any bus at a point may be travelled on to any point on that radial to which buses run.

Train services are similar to bus services in that there are services on some but not necessarily all rings and on parts of every radial.

All trips start and finish at the nodes of the ordinary road network.

Cost Functions

The travel demand models described in the next section, as well as the economic evaluation, are based on the 'generalized costs' of travel between pairs of nodes. These generalized costs are the sum of money costs and travel times valued at various rates. In this section we describe how these costs and times are determined in the supply model, as functions of the traffic flows.

Speed-Flow Relations on Roads

The costs and travel times on roads are dependent on travel speeds, which in turn depend on the flow of traffic. On each link of the ordinary road system, it is assumed that car speeds and total flows are related as follows:

$$v_1 = a - b(q_1 + pq_2)$$

and

$$v_2 = a - b(q_2 + pq_1),$$

where

v_1, v_2 = car speeds in the two directions
q_1, q_2 = total flows in the two directions, in units of 'passenger car units' per unit time
a, b, p = constants for the link.

The passenger car unit flows are weighted sums of the flows of each kind of vehicle, the weights varying from 1 for private cars to perhaps 3 for buses. The constant p is a measure of the amount of interaction between the opposing flows; it would be near zero for a dual carriageway, but near 1 for an ordinary road.

Speeds of vehicles other than cars are derived from those for cars. For taxis, speeds are assumed to be the same as for cars. For goods vehicles, empirical relationships of the form:

$$V^{-s} = u^{-s} + v^{-s}$$

are being used, where v and V are speeds of cars and of goods vehicles, and u and s are constants, typically about 100 km/h and 4, respectively. Bus speeds are obtained from goods vehicle speeds by adding to the travel time per kilometer for goods vehicles an empirical 'stopped time' that allows for time used in stopping at bus stops.

For computing reasons, it is convenient to impose a lower limit to the car speed, of the order of 5 km/h. This applies at all flows in excess of that for which the speed-flow relations above yield this lower limit.

Speeds on motorways are defined similarly. The value of p is zero. It is also necessary to introduce a capacity, a level of flow below which the normal linear speed-flow relation applies and speeds are fairly high, but above which the speed takes the minimum value.

Vehicle Occupancies

For the most part the demand models are specified in terms of person trips, whereas the cost models are based on vehicle flows. At present, conversion from persons to vehicles is done on the basis of assumed persons per vehicle figures, specific to the type of vehicle, the distance from the center and the direction of travel. One consequence of this is that a direct effect of increasing the number of bus or train passengers on a route is to decrease their waiting time, since the latter decreases with increasing flow of buses or trains.

Car, Goods Vehicle and Taxi Times and Costs

A trip by car between two nodes involves the following elements:

Times: Initial walk to car
        Travel time on network
        Time to park and walk to destination

Costs: Cost of parking at origin
        Cost of operating car on network
        Cost of parking at destination

The terminal times have to be supplied as functions of distance from the center. The travel time on the network depends on the route used, the travel times on each link, and possibly some access time between motorways and ordinary roads. The link travel times are calculated from the speed-flow relations and the link lengths and flows. Motorway access times are calculated as an assumed access distance, divided by the average of the speeds on ordinary road links leading from the point concerned.

Costs are calculated similarly. A half of the estimated parking cost is attributed to the preceding trip and one half to the following one. Car operating costs are calculated from the speeds and distances using functions of the form,

$$\text{cost/car km} = c + d/v + ev^2 + T,$$

where T is a tax charge, discussed further below, and c, d, and e are constants.

Values of time are also required; there is one for access time and one for travel time, and these may be different for the different modes. These values are used to combine the times and costs to give generalized costs.

A conventional shortest-route algorithm is used to determine the cheapest route, in the generalized cost sense, between the origin and the destination. The cheapest route is always used. The cheapest routes on a symmetrical ring-radial

network may easily be seen to consist, in general, of a radial section at each end
of the trip and a ring section joining them. In many cases one or more of these
three sections may be absent; each of them may be either a motorway or an ordinary road. Because of the ring-radial structure, the networks do not need to
be coded or stored in the usual way; a list of input parameters defines the network sufficiently. This, together with the symmetry, leads to convenient and
rapid operation.

In the above calculations, as in all other similar ones, two kinds of costs are distinguished. Firstly there are real costs, which represent a use of real resources.
Secondly there are tax and fare costs, which represent a transfer of money from
one person to some other person or organization; these influence travellers' behavior but do not directly enter into an overall assessment of costs. In the case
of cars, two options are at present provided for: firstly a tax per car-km, which
may vary with distance from the center but not with speed, and secondly an optimum road pricing charge, equal to the marginal cost that one car-km imposes
on other vehicles, which is a function of speed.

Goods vehicle times and costs are similar in structure to car times and costs,
but no terminal times or costs are included.

Link travel times for taxis are assumed to be identical with those for cars.
Times to wait for a taxi and to pay it off, including any walking time, are constants that have to be supplied. It is assumed that, apart from a tax element,
the fares paid for taxis reflect the cost of providing the service, and are therefore
regarded as real costs, not transfer payments. The model allows for an initial
charge and a charge per kilometer.

Bus Times and Costs

The times and costs of the following elements of bus trips are included: initial
walk, initial wait, travel on bus, interchange (from one bus to another), final
walk.

The initial and final walks are assumed to vary with distance from center and,
at a given distance, to be equal; they are supplied as a set of distances, with a
value of walking speed. Walking, and also waiting and interchanging, is assumed
to have a time cost but no money cost.

The initial waiting time for the calibration stage is a supplied set of constants,
depending on distance from center and direction of travel; it is not at this stage
related to service frequency (see the section on Calibration Procedures below).
The waiting times in subsequent runs vary with service frequency, as follows:

$$w = \frac{w_0 q_0 \, W}{q_0 w_0 + q(W - w_0)},$$

where w and w_0 are waiting times in the subsequent run and the calibration stage
respectively, q and q_0 are the bus flows in the subsequent run and the calibration stage, and W is an upper limit to the waiting time, perhaps of the order of
15 min.

The bus travel times on links are obtained as described in a preceding section;

the cost to the passenger is just the fare, which is regarded as a transfer payment. The bus operator receives the fares and incurs the cost of operating the buses.

Because of the difficulties in assigning passengers to bus routes by means of a shortest route algorithm, a more straightforward alternative is being followed. For each pair of nodes the generalized cost is calculated for each of the possible alternative routes. As in the case of cars, these in general consist of a section of radial travel at each end, with a section of ring travel between. The cheapest of this set of alternatives is used.

Interchange between one bus and another is required whenever ring travel changes to radial or vice-versa. On the assumption that the average number of interchanges per trip will be about the same in the model as in real life, each interchange involves a waiting time calculated on the same basis as the initial waiting time, and also an empirically based walking time or distance.

Rail Times and Costs

These are similar to those for buses, but with a few extra complications.

Access to the stations at the beginning and end of the rail part of the journey may be on foot or by bus; access by car or taxi is not at present allowed in the model but methods of including this are being considered. The input data include the total access time, and the proportion of this spent travelling in a bus. This implies the distance travelled by bus, making use of the relevant bus speed, and hence fares and contributions to bus loadings can be calculated.

Rail travel times are in general not dependent on train flows in the same way as road travel times depend on vehicle flows. However, at peak periods, certain routes have a train capacity that cannot in the short run be exceeded, and some allowance must be made for this in the model. This is being achieved by assuming the train speeds to diminish linearly from their actual speeds to a very low value over an appropriate range of train flows.

Demand and Modal Split

The inputs to the demand models are the generalized costs of travelling from each node to each other node by each mode of transport, using the cheapest route for each mode. The outputs are the numbers of trips between each pair of nodes by each mode. A basic assumption is that the number of trips between two nodes depends only on the generalized costs between these nodes, and not directly on the costs or flows between other nodes. Thus we do not have a generation model and a distribution model in the conventional sense, since this involves a dependence of numbers of trips between two nodes on costs between other pairs of nodes. There is, however, an interaction between journeys which employ common links via the speed-flow relationships.

The rest of this section is concerned with numbers of trips and generalized costs from one particular node to another particular node. There are two demand models for a particular pair of nodes in a particular run of the model; one is for goods vehicle trips and the other is for passenger trips. The former is simpler and we shall discuss it first.

Suppose that during the calibration run the generalized cost is g_0 and the number of goods vehicle trips is t_0 (the estimation of t_0 from an observed trip matrix is discussed in the section on Data Inputs). Then for some other run of the model we postulate that the number of trips is given by

$$t = k_1 k_2 t_0 (g/g_0)^{-m},$$

where g is the new cost, k_1 and k_2 are factors that express the change in generation potential at the two ends of the trip, and m is the elasticity of demand with respect to generalized cost. The quantities k_1, k_2, t_0, g_0, and m have to be supplied, and g emerges during the iterative running of the model.

The 'consumer surplus' attributable to these trips is unambiguously defined because of the simplifying assumption that demand for trips between the two nodes is independent of costs between any other pairs of nodes. It is

$$\text{Consumer surplus} = \lim_{x \to \infty} \int_g^x k_1 k_2 t_0 (y/g_0)^{-m} \, dy.$$

If m is greater than 1, the limit converges and

$$\text{Consumer surplus} = k_1 k_2 t_0 g_0{}^m g^{1-m}/(m-1).$$

In other cases the expression becomes very large. However in applications we shall always be concerned not with this consumer surplus itself but with differences between two or more values arising from different network assumptions; these will give rise to different values of g but the same value of k_1, k_2, t_0, g_0, and m. The terms that become large will cancel out, being independent of g, and we may therefore omit them in our calculations. When m is less than unity, the above expression gives correct differences between consumer surpluses, and when $m = 1$

$$\text{Relative consumer surplus} = -k_1 k_2 t_0 g_0 \log g.$$

What we are saying here is that goods vehicles that would travel under any of the alternative costs g being considered receive a consumer surplus that is large, but that this is irrelevant to the comparison of the alternative systems.

The passenger demand model follows the same general lines, but has the added complication of several modes to distinguish. Let us suppose that for some particular run of the model the costs of the various modes are g_i, where i runs over the modes. Our first task is to set up a model which expresses the way in which the trips t_i on the individual modes depend on g_i.

We shall suppose that the number of people for whom the gross benefit from travelling between the pair of places concerned exceeds B is of the form KB^{-m}, i.e., there is an elasticity of demand m for total passenger trips. We shall also suppose that the mode used is not necessarily the cheapest one, but that for those people whose gross benefit is B, the proportion using mode i is

$$\frac{g_i^{-n}}{B^{-n} + \Sigma\, g_j^{-n}}$$

and that the proportion not making the trip is

$$\frac{B^{-n}}{B^{-n} + \Sigma\, g_j^{-n}}\,,$$

where the summations are over all modes j.

This modal split model is similar to the often-used exponential one, with an inverse power in place of the negative exponential and with 'not travelling' regarded as an additional mode. If n is large, the model tends towards one in which the cheapest mode is used in preference to the others, and in which the trip is made only if this cheapest price is less than the gross benefit.

Using the above assumptions, the number of trips by mode i is

$$t_i = \int_{B=0}^{\infty} \frac{g_i^{-n}}{B^{-n} + \Sigma\, g_j^{-n}}\; KmB^{-m-1}\; dB.$$

This integral may be evaluated with the help of the standard integral

$$\int_0^{\infty} \frac{dx}{1+x^n} = \frac{\pi/n}{\sin(\pi/n)} \qquad (n > 1)$$

and the result is

$$t_i = \frac{Kg_i^{-n}\,(\Sigma\, g_j^{-n})^{m/n-1}\; \pi m/n}{\sin(\pi m/n)}$$

provided that $n > m > 0$. The total number of trips is

$$t = \Sigma\, t_i = \frac{K(\Sigma\, g_j^{-n})^{m/n}\; \pi m/n}{\sin(\pi m/n)}. \qquad (1)$$

In the calibration stage of running the model, the values of g_j, m, n, and the total number of trips t are supplied and the value of K is determined from (1) (for each origin-destination pair) as a calibration constant. For subsequent runs this value of K is used, modified when appropriate by growth factors k_1, k_2 as in the case of goods trips, and new values of g inserted. Thus if g_{j0}, t_0 are values for calibration, the model is

$$t_i = \frac{k_1 k_2 t_0 g_i^{-n}\,(\Sigma\, g_j^{-n})^{m/n-1}}{(\Sigma\, g_{j0}^{-n})^{m/n}}$$

so that

$$t = \Sigma t_i = k_1 k_2 t_0 \left(\frac{\Sigma g_j^{-n}}{\Sigma g_{j0}^{-n}}\right)^{m/n}.$$

It may be noted that the calibration constant K is such as to make the total person trips t_0 correct; it does not ensure that the corresponding trips t_{i0} by individual modes are correctly reproduced, and some care is therefore required to ensure that the values of the g_{i0} and other parameters are such that these t_{i0} are sufficiently accurately reproduced.

The consumer surplus may be calculated by integrating, over the distribution of B, the excess of B over the cost g of the mode used for all trips actually made. Because the model allows some trips to be made for which the cost exceeds the benefit, some trips contribute a negative amount to the consumer surplus, but for reasonable numerical values this presents no problem and is compensated for by other trips with large surpluses. The consumer surplus corresponding to the t trips given by (1) is

$$\sum_i \int_{B=0}^{\infty} \frac{(B - g_i) g_i^{-n} KmB^{-m-1} \, dB}{B^{-n} + \Sigma g_j^{-n}}.$$

This sum and integral are finite provided $n > m > 1$ and the value is

$$\frac{K\pi m \, (\Sigma g_j^{-n})^{m/n-1}}{n} \left[\frac{(\Sigma g_j^{-n})^{1-1/n}}{\sin \pi(m-1)/n} - \frac{\Sigma g_j^{-n+1}}{\sin (\pi m/n)}\right]$$

which on inserting the value of K from the calibration run becomes

$$\frac{k_1 k_2 t_0}{(\Sigma g_{j0}^{-n})^{m/n}} \left[\frac{\sin (\pi m/n) (\Sigma g_j^{-n})^{(m-1)/n}}{\sin \pi(m-1)/n} - (\Sigma g_j^{-n})^{m/n-1} \Sigma g_j^{-n+1}\right].$$

When $m = 1$ or $1 > m > 0$, the consumer surplus is infinite, but as in the case of goods vehicles it is possible to derive expressions for a relative consumers' surplus, which gives correct differences between surpluses for different sets of values of the g_j; the same expression applies when m is less than unity, and when $m = 1$

Relative consumer surplus =

$$\frac{k_1 k_2 t_0}{(\Sigma g_{j0}^{-n})^{1/n}} \left[\frac{\sin (\pi/n)}{\pi} \log (\Sigma g_j^{-n}) - (\Sigma g_j^{-n})^{1/n-1} \Sigma g_j^{-n+1}\right]$$

The above calculations are carried out separately for those persons with a car available and for those without, the car mode being omitted for the latter. The proportion with a car available has to be supplied.

The main reason for adopting a demand and modal split model of this form was the desire to incorporate an explicit dependence of trip-making on cost, and to make it in such a way as to permit economic evaluation. This could have been achieved by assuming that the cheapest mode was always used, but this would not have given realistic splits between modes. The device used is a simple and not altogether satisfactory method of allowing for variations in mode choice due to such factors as convenience of access, valuation of travel time, individual preferences, and ignorance of true costs; while some of these, especially valuation of travel time, could have been explicitly modelled, no method of doing this has yet been found that did not give rise to an unacceptable increase in computing time.

Evaluation

The economic evaluation of the transport effects of a particular set of investment proposals follows straightforwardly from the above formulation of the model. The relevant economic quantities, for each relevant future year, are (i) total consumer surplus, (ii) total revenue to highway and parking authorities, and (iii) profit or loss to public transport operators.

The fare revenue to public transport operators is calculated on the lines discussed previously; this has to be set alongside the operators' costs. These latter are of up to three kinds. Firstly the cost of keeping the system open; this is fairly small for buses but substantial for rail. Secondly the cost of providing and maintaining a fleet of vehicles; this depends on peak demands and will be based on the time period giving rise to the highest vehicle-hours. Thirdly the costs that are incurred in proportion to vehicle use; these are costs on the basis of a cost per hour plus a cost per kilometer. The model makes no attempt, at this stage in its development, to adjust fares so as to enable the operator to break even: the model yields as an output the amount of subsidy required or profit achieved.

The best single figure of benefit is the sum of the items (i), (ii), and (iii) above. However various component quantities, as well as various physical rather than economic quantities, will also be of considerable interest.

So far as is practicable, the model will be run on alternative investment proposals under conditions representing each of a number of future years; benefits will then be compared on the basis of conventional discounted cash flows.

It is perhaps worth emphasizing that the model is purely a model of transport effects, costs and benefits. It does not provide a complete cost-benefit analysis of alternative strategies. This requires in addition (a) estimation of the capital costs of the alternative strategies, and (b) an assessment of the nontransport effects of the alternatives, including such factors as noise, pollution, re-housing problems, and so forth.

Calibration Procedures

A single run of the model, i.e., the evaluation of a particular transport system at a particular time of day in a particular year, requires the model to be set up with values given to a large number of parameters. Many of these are supplied as input data, but the values of certain of them have to be established by means of a calibration process.

To take an example, speed-flow relations for the network of ordinary roads are difficult to specify for the model in a way that will be consistent with what is known about present speeds and flows. For a particular kind of link, we need values of a and b in the linear relation $v = a - bq$. Now the values of b obtained from measuring the behavior of traffic on the real road network will bear little relation to the values required for the model, since the road links in the model are quite different in their levels of flow, as well as their geometry. In order to establish an appropriate value of b for some future system, it is necessary first to establish the corresponding value for current (1970) conditions, and then to modify it if necessary to take account of any assumed changes in the network or in traffic behavior. The current value is found by using current data that ought to carry over reasonably precisely from real life to the model. In this instance, the actual vehicle speeds v and the no-flow speeds a are considered to be among the appropriate items to carry over from real life to the model without modification. We therefore assign the trip matrix (in symmetric form—see the next section) to the networks on the basis of the times and costs; this gives flows on road links. The values of b are calculated as

$$b = (a - v)/q.$$

A similar procedure applies to the estimation of the demand functions. We know the current origin to destination numbers of trips, and the times and costs of travel on links and hence between origins and destinations. We have assumed a functional form for the dependence of trips on costs, and all the parameters of this relationship are estimated outside the model except for a multiplying constant. This is estimated in the calibration process by dividing the trips by the value of the remainder of the function.

A third item requiring calibration is the bus and train waiting times. As in the case of the road speed-flow relations, the vehicle flows in the model do not correspond closely with those in real life; a similar procedure is followed, as set out in the section on Bus Times and Costs.

Data Inputs

Most of the data inputs required to run the model have already been referred to or implied, and we do not need to discuss them further; they are obtained by fairly standard techniques. Two items are, however, worth mentioning.

The calibration process requires to be supplied with a current node-to-node trip matrix for goods vehicles and another one for passengers by all modes. The data from which these are to be estimated consist of zone-to-zone matrices, where the zones bear little relation to the nodes of the model. It is therefore necessary to estimate node-to-node radially symmetric matrices which reproduce the zone-to-zone matrices as accurately as possible. In the present application, zone-to-zone data were most conveniently available for a small number of very large zones, for which a substantial proportion of trips were within-zone ones.

It was convenient to express the node-to-node trip numbers as formulas, to avoid the need to handle large matrices, and the following forms were used:

$$t_{ij} = hA_iB_j f(d_{ij}) D_iD_j,$$

where

i, j represent two nodes,
t_{ij} = trips from i to j (per unit time),
h is a constant,
A_i is a function of the distance x of node i from the town center, of the form $e^{-Fx}x^{-G}$,
B_j is similarly defined for node j, has the same form, but different values of F and G,
$f(d_{ij})$ is a function of the straight line distance d_{ij} from node i to node j, and has the form $e^{-Pd}d^{-Q}$,
D_i, D_j are the areas represented by nodes i and j, and in general are geometrically determined.

This expression for t_{ij} is defined by 8 parameters. These are the constant h, the constants F and G for A_i, other values of F and G for B_j, constants P and Q for $f(d_{ij})$ and the value of D for nodes on the outer ring. The reason for fitting a constant for the outer ring is that nodes on this ring are assumed to generate not only trips with an end on this ring but also all trips from outside the ring to inside it.

If the values of the 8 parameters are given, the values of t_{ij} follow and these can be aggregated to give a zone-to-zone matrix of the same form as the data matrix. This consisted of a triangular matrix (including within-zone trips) for trips between all pairs of zones within an area smaller than that covered by the model, plus numbers of trips between each of these zones and all areas outside, these external areas being partly inside and partly out of the area covered by the model. If the actual matrix elements are denoted by O and the aggregated model estimates by E, then

$$S = \Sigma (O - E)^2/E.$$

is a measure of the goodness of fit of the model to the data. A computer optimization program was used to find the values of the parameters that minimized S.

The other item concerns travel speeds, and the relation between those in the model and those on real networks. Car speeds, for example, were measured on routes that approximated to radials and to rings and these were analyzed to give average minutes per kilometer, where the kilometers were measured along straight lines or circles. However, such results could not necessarily be used directly, because real journeys were not confined to ring and radial routes of the kind studied. Therefore a sample of real trips, taken from an original–destination survey, was timed, using test cars, and adjustments were made to the ring and radial times so that these, taken together with the shortest-route algorithm, correctly reproduced the average point-to-point travel times. A similar procedure was followed with bus and rail times.

Mode of Operation

A complete cycle of operation consists of all of the operation on Fig. 1 for each of three time periods (a.m. and p.m. peaks, off-peak) and some manual combination of the results of these, with the iterative cycle and economic evaluation being carried out for each of a number of alternative investment assumptions and possibly demand growth assumptions.

The model is being operated on the English Electric 4/70 computer at the Road Research Laboratory. The programs are stored on a disc, and data are supplied on cards. Apart from a limited use of disc storage, mainly for the calibration constants for the demand model, of which there are too many for core storage, all of the running of the model is within the fast core store. Every effort has been made to minimize storage requirements, so that rapid operation and sufficient size of model can both be achieved. Currently, the numbers of rings and radials are both 20; substantial increase of these numbers would lead to a shortage of core storage or to an increase in the use of disc storage.

The main iterative loop in Fig. 1 is shown in its simplest possible form. Careful attention to choice of starting values and to the damping of oscillations, as well as to the criteria for satisfactory convergence, will be necessary. On most runs, convergence is achieved within 5 cycles provided the flows entered into the supply model are weighted averages of those obtained in the two previous applications of the demand model, with about equal weights to each.

Discussion

The model presented in this paper has various strengths and weaknesses. Its main strength is that it is a strategic model in the sense that it is fairly quick to run on a computer and, equally important, quick to modify to test alternative proposals; no detailed coding of networks is required, for example. A second major strength is that the interaction between supply and demand for transport facilities is modelled explicitly, fairly simply, and in a way that permits relatively unambiguous economic evaluation. The model is also fairly strong on the detailed representation of travel costs and times and their dependence on flows.

A possibly significant weakness of the model is its assumption of radial symmetry. While this assumption is perhaps more useful than any equally simple alternative, it is a substantial restraint on the direct use of the model, and raises difficulties in relating the model to the real world both as regards data inputs and the interpretation of results. Other weaknesses are shared with all transport models, and arise from our present inability to represent in a few formulas and algorithms the whole complexity of human and system behavior, and to make forecasts of changes that at least in part depend on exogenous factors.

There is undoubtedly a need for strategic models that are simple and quick to operate. What is much less clear at present is what forms are the best. The present model represents one line of approach that could be developed a good deal further. Alternative approaches might retain much more geographical realism, at least as regards the major transport routes, but simplify in other directions.

In the immediate future our plans are to apply the model to a range of problems. This will show, better than any theoretical discussion, how useful the approach may be, and will indicate what further model development is required.

Acknowledgments

The work described in this paper requires the coordinated effort of a number of people, and I am grateful to all concerned for their support, especially to L. Gyenes for his work on the computing aspects and to D. A. Lynam and S. V. Magee for work on the data inputs.

This paper is contributed by permission of the Director of Road Research, Crown copyright 1971. Published by permission of the Controller of Her Britannic Majesty's Stationery Office.

Decentralization: A Mathematical Explanation

Rodney J. Vaughan*

Operational Research Group, University of Sussex, Sussex, England

Abstract

This paper assumes a theoretical city in which the homes and work places have independent bivariate normal distributions. One parameter 'a' characterizes the spread of homes and another 'b' the spread of work places. Six different routing systems are considered for this family of cities, namely rectangular, radial, outer, inner and polar or radial/arc radial. For each routing system the probability that a random journey passes through a small area about a point in the city is calculated, this probability is called the route density. The route densities give the traffic pattern for the city and the total congestion is measured by the expected number of crossings per pair of random journeys.

The average distance travelled, for this family of cities with routing systems of the kind above, has been calculated previously by Wilkins [*Trans. Sci.* **3**, 93–98 (1969)]. In this paper a linear combination of the average distance and of the expected number of crossings per pair of random paths is used to approximate the average travel time which is a measure of the overall traffic efficiency of a city. The expected number of crossings is shown to depend only on the ratio of the spread of home and work places, i.e., a/b and the average distance travelled is a linear increasing function of the spread of homes and work places, i.e., a and b.

To minimize total travel time the city is induced to decentralize its work places in all routing systems bar the radial, in which case the city is induced to centralize its activity. The radial system represents a city with strong public transport, with which a city builds a strong C.B.D. but today with the widespread use of the automobile the city is induced to disperse its C.B.D. activity. Finally it is inferred from the similarity of the results for all systems bar the radial, and from the robustness of the normal distribution that the above conclusion holds for most cities.

Introduction

Clark[1] has reported the decentralization of business and work centers in major cities. It is the purpose of this paper to develop a measure of congestion in a city and to explain in general terms one of the reasons for the above trend.

We shall use for the distribution of homes the bivariate normal distribution proposed by Sherratt[2] with probability density (in polar coordinates):

$$\frac{e^{-r^2/a^2}}{\pi a^2} \qquad 0 \leqslant r < \infty, \ 0 \leqslant \theta < 2\pi. \tag{1}$$

*Formerly Australian Road Research Board Research Fellow Mathematics Department, University of Adelaide, Adelaide, South Australia.

References p. 412

Wilkins[3] further proposed that the same form be taken for the distribution of work places, with probability density,

$$\frac{e^{-r^2/b^2}}{\pi b^2} \qquad 0 \leqslant r < \infty, \, 0 \leqslant \theta < 2\pi, \tag{2}$$

where a and b represent the spread of homes and work places throughout the city. Wilkins has summarized the advantages of this model and gives a good list of references. For more extensive references in the general area treated by this paper consult Lam and Newell.[4]

Wilkins has calculated the expected (average) distance travelled, in cities of the above type for different routing systems, and the present paper derives the expected number of crossings of any two randomly chosen vehicle paths (the route crossing probability) for these cities.

Route crossing probabilities were first introduced (for cities with uniform home and work place distributions) by Miller and Holroyd[5] and further developed by Holroyd[6]; the author uses their methods. To follow the detail of the calculations, it is necessary for the reader to be familiar with the papers of Wilkins, Miller, and Holroyd.[3,5,6] However, the readability of the paper is not affected by these details.

Mathematical Notation

The use of a route density function, to give the traffic flow in a particular direction at a particular point, is needed. Let $q(. , . , *)$ be the route density as defined by Miller and Holroyd where $. , .$ represents a pair of cartesian or polar coordinates, x, y or r, θ. The asterisk $*$ denotes the direction of traffic flow and takes the following forms;

ϕ : an angle to a fixed data line
N, S, E, W.   : the four directions of the compass.
I, O, C, CC. : directions, inward along the radials, outward along the radials, clockwise, around an arc, and counterclockwise, around an arc.

By P_i we shall mean the route crossing probability for system i and E_i is the average distance travelled in system i. These two quantities are functions of a, b and when considered as such will be denoted by $P_i(a, b)$ and $E_i(a, b)$.

The subscript 1 on either cartesian or polar coordinates $(x_1, y_1$ or $r_1, \theta_1)$ will denote a home. Whilst a work place will be represented by the subscript 2.

The densities (1) and (2) shall be represented in cartesian coordinates (x, y) by the densities;

For x:

$$\psi_1(x) = \frac{1}{\sqrt{\pi a}} \, e^{-x^2/a^2}$$

$$\psi_2(x) = \frac{1}{\sqrt{\pi b}} \, e^{-x^2/b^2}$$

the cumulative distribution,

$$\Psi_i(x) = \int_{-\infty}^{x} \psi_i(x)\,dx, \quad i = 1, 2.$$

For y: replace x by y in the above.

System 1 – Direct

Here the commuter travels as the crow flies. The route density at the point (r, θ) in the direction Φ is derived from the formula of Miller and Holroyd.

$$q(r, \theta, \phi) = \int_0^\infty \int_0^\infty f(r_1, \theta_1, r_2, \theta_2)(l_1 + l_2)\,dl_1\,dl_2, \quad (3)$$

where $f(r_1, \theta_1, r_2, \theta_2)$ is the probability per unit area that a trip starts out from the vicinity of the point (r_1, θ_1) towards the unit area about (r_2, θ_2) and l_1, l_2 are the distances from (r, θ) to (r_1, θ_1) and (r_2, θ_2). In our case,

$$f(r_1, \theta_1, r_2, \theta_2) = (\pi ab)^{-2}\,\exp - (r_1^2/a^2 + r_2^2/b^2).$$

Using the relation between r_j and l_j;

$$r_1^2 = r^2 + 2rl_1 \cos \phi + l_1^2$$
$$r_2^2 = r^2 - 2rl_2 \cos \phi + l_2^2$$

and integrating (3) we obtain

$$q(r, \theta, \phi) = [\exp\{-r^2(1/b^2 + \sin^2(\phi)/a^2)\}\,\Phi(\sqrt{2}r \cos(\phi)/a)/a$$
$$+ \exp\{-r^2(1/a^2 + \sin^2(\phi)/b^2)\}\,\Phi(\sqrt{2}r \cos(\phi)/b)/b]$$
$$\cdot \sqrt{\pi}/(2\pi^2) = q(r, \phi) \text{ say}$$

where

$$\Phi(x) = \frac{1}{\sqrt{2\pi}} \int_x^\infty e^{-z^2/2}\,dz.$$

Another formula of Miller and Holroyd[5] gives the route crossing probability as,

$$P_1 = \int_0^{2\pi} \int_0^\infty \int_0^{2\pi} \int_0^{2\pi} q(r, \phi_1)\,q(r, \phi_2)\,|\sin(\phi_1 - \phi_2)|\,d\phi_1\,d\phi_2\,rdrd\theta$$

$$= 2\pi \int_0^\infty \int_0^{2\pi} \int_0^{2\pi} q(r, \phi_1)\,q(r, \phi_2)\,|\sin(\phi_1 - \phi_2)|\,d\phi_1\,d\phi_2\,rdr$$

which the author has been able to reduce to a double integral but unable to find an explicit expression in terms of a and b. In the case $a = b = 1$ $P_1 \approx .216$.

System 2–Rectangular

In this case the commuter must take the path provided by a rectangular street grid. We follow closely a method developed by Holroyd[6] and obtain

$$q(x,y,E) = \tfrac{1}{2}\Psi_1(x)\,[1 - \Psi_2(x)]\,[\psi_1(y) + \psi_2(y)]$$
$$q(x,y,W) = \tfrac{1}{2}\Psi_2(x)\,[1 - \Psi_1(x)]\,[\psi_1(y) + \psi_2(y)]$$
$$q(x,y,N) = \tfrac{1}{2}\Psi_1(y)\,[1 - \Psi_2(y)]\,[\psi_1(x) + \psi_2(x)]$$
$$q(x,y,S) = \tfrac{1}{2}\Psi_2(y)\,[1 - \Psi_1(y)]\,[\psi_1(x) + \psi_2(x)]$$

with

$$P_2 = 2\int_{-\infty}^{\infty}\int_{-\infty}^{\infty} [q(x,y,E) + q(x,y,W)]\,[q(x,y,N) + q(x,y,S)]\,dx\,dy$$

$$= 2\left\{ \int_{-\infty}^{\infty} \tfrac{1}{2}[\psi_1(x) + \psi_2(x)]\,[\Psi_1(x) + \Psi_2(x) - 2\Psi_1(x)\,\Psi_2(x)]\,dx \right\}^2$$

$$= 2(\tfrac{1}{2}I)^2 \tag{4}$$

It can be shown that

$$I = 1 - \frac{1}{\pi}\,\text{arc sin}\,\{a/[2(a^2 + b^2)]^{1/2}\}$$

$$-\frac{1}{\pi}\,\text{arc sin}\,\{b/[2(a^2 + b^2)]^{1/2}\}$$

In the case where $a = b$ it is simple to show that $P = 2/9$ which is the value found by Holroyd[6] for uniform rectangular cities. We may note, in general, that if both the home and work place densities are equal, i.e., $\psi_1(x) = \psi_2(x) = \psi(x)$, then by the substitution $z = \psi(x)$ in Eq. (4), the result $P_2 = 2/9$ is always obtained and, by the calculus of variations, is the minimum value.

System 3–Radial

Here the commuter must travel direct to the center along a radial and then direct to his destination along another radial. The route densities are

$$q(r,\theta,C) = q(r,\theta,CC) = 0,$$
$$q(r,\theta,I) = \exp(-r^2/a^2)/2\pi r,$$
$$q(r,\theta,O) = \exp(-r^2/b^2)/2\pi r,$$

and, of course, the only place where route crossings are possible is the center. The figure of Miller and Holroyd[5] applies, i.e., $P_3 = 1/3$.

System 4—Outer (Arc then Radial)

The commuter travels along the circle $r = r_1$, till he reaches the radial upon his work place lies, and then travels on this radial to his destination. The route densities are

$$q(r, \theta, C) = q(r, \theta, CC) = \frac{1}{2} \cdot \frac{1}{4} \cdot \frac{2e^{-r^2/a^2}}{a^2} \cdot r,$$

$$q(r, \theta, I) = \gamma_2(r) \left[1 - \gamma_1(r)\right] / 2\pi r,$$

$$q(r, \theta, O) = \gamma_1(r) \left[1 - \gamma_2(r)\right] / 2\pi r,$$

where

$$\gamma_1(r) = \int_0^r \frac{2e^{-r^2/a^2}}{a^2} \, r\, dr = 1 - e^{-r^2/a^2}$$

and

$$\gamma_2(r) = \int_0^r \frac{2e^{-r^2/b^2}}{b^2} \, r\, dr = 1 - e^{-r^2/b^2}$$

The route crossing probability is

$$P_4 = 2 \int_0^\infty \int_0^{2\pi} \left[q(r, \theta, C) + q(r, \theta, CC)\right] \left[q(r, \theta, I) + q(r, \theta, O)\right] r\, dr\, d\theta$$

$$= \frac{1}{4} + \frac{1}{2} \frac{1}{1 + a^2/b^2} - \frac{1}{2 + a^2/b^2}$$

which agrees with the value, obtained by Miller and Holroyd[5] for uniform circular cities, of $P_4 = 1/6$ when $a = b$.

System 5—Inner (Radial then Arc)

The commuter travels along the radial $\theta = \theta_1$ till he reaches a point on the circle $r = r_2$ and then travels on this circle to his destination. We need only reverse the roles of home and work place in system 4 to obtain,

$$P_5 = \frac{1}{4} + \frac{1}{2} \frac{1}{1 + b^2/a^2} - \frac{1}{2 + b^2/a^2}$$

System 6—Polar

The commuter uses whichever is the shorter of systems 3, 4, 5. A judicial combination of the method of systems 3, 4, 5 and Miller and Holroyd[5] gives:

$$q(r, \theta, C) = q(r, \theta, CC) = \frac{r}{2\pi^2} \cdot \frac{a^2 + b_2}{a^2 b^2} \cdot \exp\left[r^2/a^2 + r^2/b^2\right],$$

$$q(r, \theta, O) = \frac{1}{2\pi}\left\{\frac{\pi - 2}{\pi}[1 - \gamma_2(r)] + \frac{2}{\pi}[1 - \gamma_1(r)]\,\gamma_2(r)\right\}\bigg/ r,$$

$$q(r, \theta, I) = \frac{1}{2\pi}\left\{\frac{\pi - 2}{2}[1 - \gamma_2(r)] + \frac{2}{\pi}[1 - \gamma_2(r)]\,\gamma_1(r)\right\}\bigg/ r.$$

The route crossing probability at the center is given by Miller and Holroyd[5] and is $\left(\frac{\pi - 2}{\pi}\right)^2 \cdot \frac{\pi + 4}{3\pi}.$

Hence

$$P_6 = \left(\frac{\pi - 2}{2}\right)^2 \frac{\pi + 4}{3\pi} + 2 \cdot \int_0^{2\pi} \int_0^{\infty} [q(r, \theta, O) + q(r, \theta, I)]$$

$$\cdot [q(r, \theta, C) + q(r, \theta, CC)]\, r dr d\theta$$

$$= \left(\frac{\pi - 2}{2}\right)^2 \frac{\pi + 4}{3\pi} + \left(\frac{2}{\pi}\right)^2 \frac{1}{\pi}\left(\frac{a^2 + b^2}{a^2 + 2b^2} + \frac{a^2 + b^2}{2a^2 + b^2} - 1\right)$$

$$+ 2 \cdot \frac{2}{\pi} \cdot \frac{\pi - 2}{\pi} \cdot \frac{1}{2\pi}\left(\frac{a^2 + b^2}{a^2 + 2b^2} + \frac{a^2 + b^2}{2a^2 + b^2}\right).$$

When $a = b$ a value of $P_6 = .241$ again agrees with Miller and Holroyd's result.

Note there is a mistake in the expression for E_6 in Wilkins[3] which is corrected in Wilkins[7]

General Properties of $P_i(a, b)$

We note that

$$P_i(a, b) = P_i(c \cdot a, c \cdot b),$$

where c is a constant, further,

$$P_i(a, b) = P_i(b, a), \quad \text{for } i = 1, 2, 3, 6,$$

i.e., for systems where the route taken is not affected by the interchange of home and work places. Also,

$$E_i(a, b) = E_i(b, a), \quad \text{for } i = 1, 2, 3, 6.$$

Discussion

We could use a measure of the relative efficiency of a city, the total travel time, which is a linear combination of E_i and P_i, i.e.,

$$T_i(a, b) = \alpha_i E_i(a, b) + \beta_i P_i(a, b),$$

where α_i and β_i are constants for the particular routing system i. The first term represents, the time spent actually moving while covering the distance travelled, and the second term, the time lost due to delays by cross traffic.

To obtain an understanding of the performance of this function, Fig. 1 has been plotted. The function $p_i(\sigma)$ is defined as $p_i(\sigma) = P_i(1, \sigma)$. Using the general properties of P_i described in the previous section, we may obtain $P_i(a, b)$ for all a, b in terms of $p_i(\sigma)$, σ may be thought of as the relative sprawl. The function $e_i(\sigma)$ is equal to $E_i(1, \sigma)$ and equals $E_i(\sigma, 1)$ for $i = 1, 2, 3, 6$.

For most cities we have $a > b$. Let us first hold the number of work places constant and add more homes to the city. For example take $a = 5, b = 1$, so that $\sigma = 5$ and we are operating in the region $1 < \sigma < \infty$ of Fig. 1, it is seen that both E_i and P_i are increasing functions of a. So to minimize travel time a must be kept as small as possible, i.e., the homes must cluster about the present city. Alternatively hold the number of homes constant and add more work places to the city. For example take $a = 1, b = 1/5$, so that $\sigma = 1/5$; now we operate in the region $0 < \sigma < 1$ of Fig. 1, where P_i is a decreasing function of σ, and E_i an increasing function. So that (for practical values of a, b, α_i, β_i) T_i is initially a decreasing function of b for all systems bar the radial.

We can now see one historical explanation for the development of a strong Central Business District (C.B.D.). Before the widespread use of the motor car the only transport system available was the public transport system which is essentially radial. So all other factors being equal, business concentrated in the center of the city, since the travel time is minimized as $b \to 0$. However, with the motor car, one of the other routing systems is used, and business, all

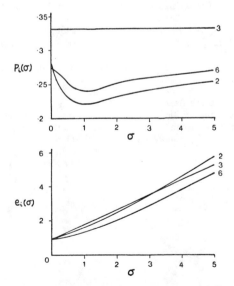

Fig. 1. Graphs of, the relative congestion $p_i(\sigma)$ and the expected distance travelled $e_i(\sigma)$, for systems, 2 rectangular, 3 radial and 6 polar.

other factors being equal, is induced to spread out into the suburbs because congestion is minimized as $\sigma \to 1$, i.e., $b \to a$, and the total travel time is minimized for a positive value of b between zero and a.

Conclusion

The above inferences depend of course, on the model used, but since the normal distribution procedures are particularly robust, and because of the general result for system 2, it is felt that given an appropriate measure of the spread of homes and work places similar results will hold for all cities. We also note that the general result holds for all routing systems bar the radial and hence would hold for the higgledy-piggledy routing systems of most cities. The author intends to apply the methods of this paper to transportation planning and in particular will treat the correlation between home and work places.

Acknowledgments

The author would like to thank the Australian Road Research Board for their financial support and the following for valuable discussion, C. A. Wilkins, R. B. Potts, C. E. M. Pearce, W. N. Venables, and D. L. Vaughan.

References

1. C. Clark, *Population Growth and Land Use,* Macmillan, London, 1967.
2. G. G. Sherratt, A Model for General Urban Growth, in *Management Sciences Models and Techniques*, Vol. 2, p. 147, Pergamon Press, N.Y., 1960.
3. C. A. Wilkins, Sherratt's model and commuter travel, *Transportation Sci.* **3**, 93–98 (1969).
4. T. N. Lam and G. F. Newell, Traffic assignment on a circular city, *Transportation Sci.* **1**, 318–362 (1967).
5. A. J. Miller and E. M. Holroyd, Route Crossings in Urban Areas, Proc. 3rd. Conf., Australian Road Research Board, Sydney (1966).
6. E. M. Holroyd, Routing Traffic in a Square Town to Minimize Route-Crossings, Proc. 4th. International Symposium of Traffic Flow, Karlsruhe (1968).
7. C. A. Wilkins, Erratum, *Transportation Sci.* **4**, 238 (1970).

Optimal Form of a Class of Collection–Distribution Networks

Ezra Hauer

Department of Civil Engineering, University of Toronto, Toronto, Ontario, Canada

Abstract

Some collection–distribution networks are characterized by a main "stem" and branching "twigs," e.g., a major interurban highway connected to intermediate towns by secondary roads, or a central sewage collector connected to houses by small diameter pipes.

The optimal form and location of two specific networks of this class is sought. One is the route of a vehicle which distributes passengers from a common origin to stops from where each passenger walks to his destination; the other is a network which consists of a central walkway and footpaths leading to separate buildings. In both cases, some travellers temporarily tied together by a common vehicle or network link will prolong their travel to accommodate other users going to intermediate destinations. Also, the length of higher cost links is increased for the same purpose with some compensation from the shortening of lower cost system components. Within the framework of such trade-offs the question of optimality is raised.

In the first stage, the properties of the optimal network are investigated. Four such properties are identified. Their exploration leads to some insight about the intrinsic properties of networks which belong to the aforementioned class. When possible, the role of parameters, mainly unit costs, is investigated.

In the second stage, the properties of the optimal network defined earlier are utilized to actually construct an optimal collection–distribution system. The graphical solution algorithm is illustrated by few numerical examples.

Introduction

Consider the following situation. Several travellers leave the same origin using a common vehicle. The vehicle follows a predetermined route. The travellers have different destinations and alight from the vehicle at that point of its route which is closest to the endpoint of their journey. The remaining part of the trip is then completed on foot.

Should the route meander in order to follow all destinations closely, walking distance would be kept short but the duration of travel for the vehicle and for most passengers would be considerable. If, on the other hand, the route is kept as straight as possible, travel times are relatively short but walking distances may be appreciable. The problem then, is to select the best route for the vehicle to follow. Such selection will be the result of a compromise between two opposing objectives: The desire to keep walking distances short and the goal of achieving short travel times for both vehicle and passengers. The situation, as described above, arises when institutions provide transportation to and from work for their employees. Problems of similar structure are encountered in a host of other

References p. 427

"distribution-collection" situations. Consider for example a complex of separate buildings interconnected by a single main walkway from which footpaths lead to each building. If people are constrained to move on the walkway and the footpaths, the determination of the shape and the location of the central walkway seems analogous to the previous problem of finding the optimal vehicle route. The vehicle route and the walkway problems will be discussed in Part I and Part II, respectively. Should the water distribution system or the sewer collection network or electricity supply to the same complex of buildings consist of a main line and secondary branches, the selection of the optimal shape and location of the main line is a problem of the same class. The basic similarity between the vehicle route problem and several larger scale situations is also marked. Such is the case when the shape and location of a major interurban highway is sought with respect to intermediate towns; or of a railroad track in relation to concentration of cargo; or of a rapid transit line in correspondence to commuter residences.

The aforementioned problematic is closely related to the task of constructing a network of minimal length; a so called Steiner Minimal Tree. An excellent exposition of the properties of the solution and its construction may be found in Gilbert and Pollak (1968)[1] or in Werner (1969).[2] Generalization of the Steiner problem in which not network length but total transport cost is minimized may be found in Beckmann (1967)[3] and Werner (1968).[4] The former mentions briefly the trunk-feeder road problem on which discussion centers in this paper.

Part I: The Vehicle Route

Assumptions and Criteria

Two basic assumptions will be made:

1. The vehicle may travel (the passenger may walk) at the same speed in all directions.

2. The order at which the vehicle approaches destinations is known in advance.

The first assumption is intended to simplify the geometrical structure of the problem; the second, its combinatorial nature. The applicability of the solution procedure to real situations will depend on the degree to which the aforementioned assumptions conform to the given circumstances. It is probably best to postpone discussion on the significance of the basic assumptions till after their role in the solution procedure becomes apparent.

The quality of the route will be judged by three attributes: walking distance for passengers, travel time for passengers and travel time of the collecting-distributing vehicle. For the time being it is not necessary to decide on quantitative indices for the three attributes. Their specification will be called for later.

Having stated the main assumptions and chosen the evaluation yardstick, it is now possible to describe several properties of the optimal route.

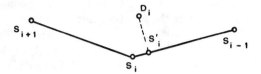

Fig. 1. A concave route.

Properties of the Optimal Route

PROPERTY I. The optimal route is a polygon the vertices of which are the points at which passengers alight—the stops.

Proof. Property I holds since the walking distances are unaffected by the shape of the route in between stops, and the travel time of both vehicle and passengers is minimized (under Assumption 1) when the route is a straight line.

PROPERTY II. The two polygon legs meeting at a stop are convex with respect to the destination served by that stop.

Proof. Let S_i in Fig. 1 be a stop at which a passenger alights and from where he walks to his destination D_i. Should the route be concave with respect to D_i as in Fig. 1, it would be better for the passenger to alight at S_i'. Also, by changing the route from $S_{i-1} \rightarrow S_i \rightarrow S_{i+1}$ into $S_{i-1} \rightarrow S_i' \rightarrow S_{i+1}$, travel time is shortened. Thus, any route which is concave with respect to the destination can be improved upon and therefore is not optimal.

PROPERTY III. The access link $\overline{D_i S_i}$ (Fig. 2) approximately bisects the angle $\angle S_{i-1} S_i S_{i+1}$.

Proof. The travel time of the vehicle between stops S_{i-1} and S_{i+1} remains fixed when S_i is located anywhere on an ellipse, the foci of which are the stops S_{i-1} and S_{i+1}. The travel time of passengers (with the exception of the passengers alighting at S_i) is also unchanged when S_i is anywhere on the ellipse. The walking distance of the passenger alighting at S_i is shortest when $\overline{D_i S_i}$ is normal to the ellipse. Therefore (neglecting changes in the travel time of the alighting passengers on the last leg of their journey), the optimal location of S_i is where

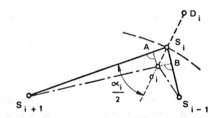

Fig. 2. The polygon angle and the access link.

$\overline{D_iS_i}$ is normal to the ellipse. Since the normal to the ellipse bisects the angle between the focal radius vectors, the proof is complete.

Property III is only approximate because of the neglect of the possible trade off between walking and travelling distances for passengers alighting at S_i. However, a barely noticeable shortening in the passenger's travel time would markedly increase his walking distance. In addition, passengers tend to attach different weights to walking and travelling the same time. For these reasons the approximation is likely to be a close one.

PROPERTY IV. Let

    C–be the cost of one passenger of prolonging the route of the vehicle by a unit of distance;

    K–the cost to the owner of the vehicle of prolonging the route by a unit of distance;

    k–the cost to the passenger of prolonging his walking distance by one unit;

    n_i–the average number of passengers aboard the vehicle between stops S_{i-1} and S_{i+1};

    a_i–the number of passengers alighting at stop i;

    α_i–the acute angle of the rouge polygon at S_i.

Then

$$\hat{\alpha}_i = \max \left\{ 2 \text{ arc cos } \frac{a_i}{2(n_i C/k + K/k)}; \quad \measuredangle S_{i-1} D_i S_{i+1} \right\}, \tag{1}$$

where $\hat{\alpha}_i$ is the value of α_i for the optimal route.

Proof. Assume that the part of the route depicted in Fig. 2 is optimal. If so, the improvement in vehicle and passenger travel distance which would result from a minute shortening of the distance $\overline{S_{i-1}S_iS_{i+1}}$ should be exactly offset by the detrimental effect of such a change on the walking distance. Let $\Delta_{\omega i}$ be the increment in walking distance associated with a shortening of the distance $\overline{S_{i-1}S_iS_{i+1}}$ by 2Δ. Then, the necessary condition for the minimization of total cost is

$$\lim_{\Delta \to 0} 2\Delta(n_i C + K) = \lim_{\Delta \to 0} k\Delta_{\omega i} a_i. \tag{2}$$

When $\Delta \to 0$,

$$\Delta = \Delta_{\omega i} \cos(\alpha_i/2). \tag{3}$$

Upon substitution into Eq. (2),

$$\cos(\bar{\alpha}_i/2) = \frac{ka_i}{2(n_i C + K)}, \tag{4}$$

where $\bar{\alpha}_i$ is a value of α_i at which the sum of travelling and walking costs is extremal. It is easy to show that the extremum is a global minimum when $\alpha_i > \measuredangle S_{i-1} D_i S_{i+1}$. If, however, $\bar{\alpha}_i$ is smaller than $\measuredangle S_{i-1} D_i S_{i+1}$, the foundations of its derivation (Eq. 2) are no longer valid. In this case, a shortening of the vehicle's route shortens also the walking distance. Consequently, the vertex angle of the optimal route must be at least as large as $\angle S_{i-1}D_iS_{i+1}$. (This follows also from Property III). On the other hand, the fact that the total cost is not extremal inside the triangle $S_{i-1}D_iS_{i+1}$ indicates that within the triangle total cost decreases when α_i decreases. Therefore, total cost in this case is minimal when $\alpha_i = \angle S_{i-1}D_iS_{i+1}$.

Intermediate Results and Discussion

It has been shown in the preceding that the optimal vehicle route is characterized by four properties. Some interpretation and discussion of these results is in order.

(a) The definition of the first three properties does not rely on quantitative measurement and valuation of the Performance Criteria. For their derivation it is sufficient to recognize only the direction of change, i.e., improvement or deterioration. Thus, irrespective of the magnitude of the costs associated with travel or walking time or distance, the general shape of the optimal vehicle route and the access links can be visualized. It is a polygon which is convex with respect to destinations such that the access links are approximate extensions of the vertex angle bisectors.

More generally, and probably more importantly, when Assumption I holds, such is the optimal form of all collection–distribution networks characterized by a central "spine" and branching links to destinations on which movement is associated with cost.

(b) The main conclusion to draw from Property I does not pertain to the shape of the route in between stops; the latter depending exclusively on the impedance to movement. Rather, the importance of the location of stops as the prime determinants of the shape of the network is emphasized. Once the position of stops (or sewer manholes, railroad stations, highway junctions, etc., in other problems) is known, the network determination task is reduced to finding the connecting minimum time (cost) paths.

(c) Only the two ratios C/k and K/k are required for the determination of $\hat{\alpha}_i$. To assign empirically based value estimates to these ratios is beside the subject matter of this paper. Nevertheless, to gain some familiarity with the concepts and feeling for the orders of magnitude, a short digression may be justified.

If the passenger cares only about the overall duration of his journey, $C/k =$ speed of walking/speed of travel on vehicle. If so, should the speed of walking and the speed of travel be 4 mph and 40 mph, respectively, $C/k = 1/10$. Should, however, 1 min of walking be regarded as bothersome as 2 min spent travelling, and with speeds as before, $C/k = 1/20$.

Similarly, a value x of the ratio K/k means, that if each of x passengers pays k in order to avoid walking one unit of distance, the amount xk suffices to bribe the vehicle owner into providing travel by vehicle for one unit of distance.

(d) Equation (4) defines the value of $\bar{\alpha}_i$ only when

$$ka_i \leqslant 2(n_iC + K). \tag{5}$$

If Ineq. (5) does not hold, the high cost of walking, small vehicle occupancy etc. make the attainment of equality in (2) impossible and the optimal route must pass directly through the destination.

(e) As a corollary of the preceding paragraph, let

$$v_i = \frac{a_i/2 - K/k}{C/k} .$$

(6)

Then, for all i at which $n_i \leqslant v_i$, the destinations and the corresponding stops of the optimal route coincide *irrespective of the mutual location of the destinations*.

(f) Speaking pictorially, one can imagine the route as a flexible string which is being pulled by forces anchored at the destinations. The larger the number of passengers aboard the stronger the resistance of the route to the pull exerted on it. Since the number of passengers on the distributing vehicle is a decreasing function of i, the resistance to the destination forces weakens as the vehicle pro-ceeds along the route. In general, then, the optimal route will be characterized by small changes in direction near the origin and by closely following the desti-nations near the end of the route.

The mode of variation of $\bar{\alpha}_i$ as a function of n_i for selected values of the pa-rameters a_i, C/k and K/k is depicted in Fig. 3.

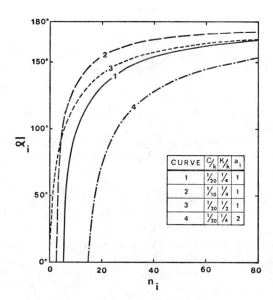

Fig. 3. The variation of $\bar{\alpha}_i$ with n_i.

The Solution Procedure

As originally posed, the problem is to find the best route for a passenger distributing vehicle to follow. So far, only the *properties* of the optimal route have been described. The time has come to deploy this knowledge in an attempt to actually find the optimal route.

A graphical solution procedure is advanced. In this form the essence of the algorithm is probably most transparent. It consists of a series of steps which are best illustrated by an example.

Example I. Let the three destinations in Fig. 4 be labelled in the order in which they will be approached by the vehicle originating from O with 8 passengers aboard. Let $a_1 = 1; a_2 = 1; a_3 = 6$, and after the last passenger alights the vehicle returns to O. Furthermore, let $C/k = 1/10$ and $K/k = 1/4$ passengers.

Then, using Eq. (4), $\bar{\alpha}_1 = 120°$ and $\bar{\alpha}_2 = 112° 30'$. For the last destination Ineq. (5) is violated and consequently the route passes directly through D_3

Step 1. Select a tentative direction for the initial link OS_1 (Fig. 4).

Step 2. Using Property III and the known value of $\bar{\alpha}_1$, determine the location of S_1 on the direction OS_1 so that $\angle OS_1D_1 = 180° - \bar{\alpha}_1/2$.

Step 3. At S_1 construct angle $\bar{\alpha}_1$ thereby fixing the direction S_1S_2.

Step 2'. Determine the location of S_2 on the direction S_1S_2 so that $\angle S_1S_2D_2 = 180° - \bar{\alpha}_2/2$.

Step 3'. Construct at S_2 the angle $\bar{\alpha}_2$ thereby fixing direction S_2S_3.

Step 4. Since the route must pass through D_3, the route is optimal if the direction S_2S_3 hits D_3. If not, the tentative direction of the initial link has to be altered to achieve closure.

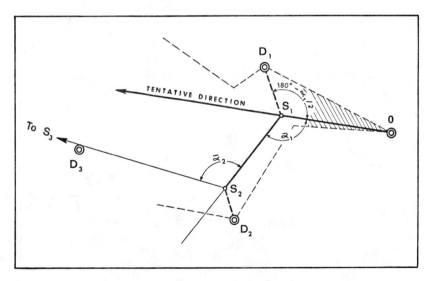

Fig. 4. Graphical solution—Example I.

The salient features of the solution procedure are apparent. Having chosen a trial direction for the initial link, the tentative location of all stops is uniquely determined through the knowledge of the angles $\bar{\alpha}_i$, $(i = 1, 2, \ldots n)$ and of Properties II and III. In other words, a certain direction of the initial link determines automatically an entire route polygon. Out of all sensible directions of the initial link (which are within the shaded area in Fig. 4) only one such direction is associated with a polygon the last leg of which passes through the last destination. This is then the only vehicle route conforming to all four properties—the optimal vehicle route.

Since Example I has been constructed to illustrate the main features of the algorithm, it may be misleading in its simplicity. A more elaborate case will therefore be used to present some of the complications and shortcuts which call for intelligent and explicit use of all properties of the optimal route.

Example II. Let the eight destinations in Fig. 5 be labelled in the order in which they will be approached by a vehicle originating at O with 8 passengers aboard. As before, let $a_i = 1$ for all stops. The values of $\bar{\alpha}_i$, $(i = 1, 2, \ldots 6)$ calculated by Eq. (4) are listed in Table I.

TABLE I

Stop label (i)	$\bar{\alpha}_i$
1	120°00′
2	112°30′
3	102°40′
4	88°50′
5	67°00′
6	0°00′

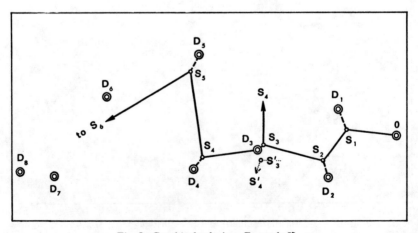

Fig. 5. Graphical solution—Example II.

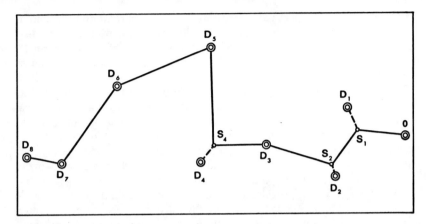

Fig. 6. Optimal vehicle route–Example II.

Already $\bar{\alpha}_6 = 0$, consequently for $i \geqslant 6$, $S_i = D_i$. Thus, for the optimal route determination task D_6 is the last destination to consider.

Assuming, as before, a tentative direction OS_1 (Fig. 5) the construction proceeds without problems till S_3. At that point, however, plotting of $\bar{\alpha}_3$ will cause direction $S_3 S_4$ to shoot off into outer space. The situation cannot be rectified by a counter-clockwise rotation of the initial link which would bring the third stop into a location such as S_3' because in this case direction $S_3' S_4'$ goes out of bounds on the opposite side. In fact, any reasonable location of S_4 will result in $\measuredangle S_2 S_3 S_4 > \bar{\alpha}_3$. If so, α_3 has to be kept as small as possible. For all sensible locations of S_4, $\measuredangle S_2 S_3 S_4$ is the smallest when S_3 coincides with D_3. Consequently, the optimal route passes through D_3.

Having established that, the original route determination task can be divided into two parts. In the first, the optimal vehicle route between O and D_3 has to be found. This, however, is exactly the problem tackled in Example I and its solution appears on the right side of Fig. 6. In the second part, the optimal vehicle route has to be found between the origin D_3 and the final destination D_6.

Once more a tentative direction is selected for the initial link $D_3 S_4$ and using the by now routine construction method a tentative route is depicted in Fig. 5. To achieve closure a clockwise rotation of $D_3 S_4$ is necessary. However, even before the direction $S_5 S_6$ hits D_6, S_5 coincides with D_5. The resulting optimal route is depicted in Fig. 6.

Some Observations

The algorithm presented in the preceding section relies more on heuristic argumentation than on precise proofs. Lack of experience in its application and the seemingly endless variety of destination constellations seem to preclude at present the possibility to formulate clear cut rules and the proof of the uniqueness of the result.

Yet, so it seems, the relative looseness of the graphical solution procedure is not devoid of advantages. It certainly provides for close contact between the analyst and the problem at hand. Also, since the amount of calculations involved is small, the need to formalize the logic in a set of rules to facilitate computer programming is not pressing.

The immediate applicability of the solution procedure to the specific task of determining vehicular routes in urban areas is of course very limited. The main obstacle being the incompatibility of Assumption I with the movement on urban streets. For one, speeds are different at different locations. For two, the planned hierarchy of carriers means radically different speeds on freeways, arterials and local streets. For three, the grid geometry of urban street networks negates the radial symmetry implied by Assumption I.

Assumption II, which provides for advance knowledge on the ordering of destinations, does not appear to be unrealistic in most instances. Should, however, uncertainty as to the best sequence of destinations arise, it is feasible to construct optimal routes for few alternative orderings to resolve such questions.

In some situations, the solution algorithm may be used in its present form at least as an indication with respect to the optimal form of the network. In other cases, it may be feasible to impart a greater degree of realism to the solution through an appropriate transformation of the physical space into one in which Assumption I is closely fulfilled.

Part II: The Walkway

Problem Description, Assumptions and Criteria

Consider a complex of buildings as in Fig. 7 which are to be interconnected by a central walkway and branching footpaths. The problem at hand is the determination of the optimal form and location of the walkway–footpaths network. The network performance will be judged by the total distance walked by the users of the system and by the cost of construction and upkeep of the network. Analysis and solution procedure rest on the same basic assumptions which have been stated in the first part of this paper.

Properties of the Optimal Walkway

The first two properties of the optimal vehicle route can be carried over from Part I without modification. The difference between the vehicle route and the

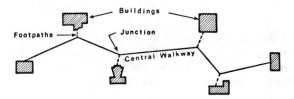

Fig. 7. The walkway–footpaths network.

walkway problems becomes manifest with respect to the third property. The considerations through which bisection of the vertex angle by the access link direction was justified in the former case are no longer valid.

Let $f_{i,i-1}$ and $f_{i,i+1}$ be the flows of walkers who either originate or are destined to D_i, and who pass on their way junctions S_{i-1} and S_{i+1}, respectively (Fig. 8a). Let f_i be the flow of all other walkers passing S_i. Also, let

k—be the cost of one walker of prolonging his walking distance by one unit,

K_f—the cost per unit of time (e.g., annual cost) of prolonging the footpath by one unit of distance, and

K_w—the cost per unit of time of prolonging the walkway by one unit of distance.

The total costs of transport per unit of time on the links $(S_{i,i+1} S_i)$, $(S_i D_i)$ and $(S_i S_{i,i-1})$ are $C_{i+1} = [K_w + k(f_{i,i+1} + f_i)]$, $C_i = [K_f + k(f_{i,i+1} + f_{i,i-1})]$ and $C_{i-1} = [K_w + k(f_{i,i-1} + f_i)]$, respectively. The cost minimizing location of S_i is where the "forces" represented by the link costs C_{i+1}, C_i and C_{i-1} are in equilibrium as in Fig. 8b (Werner 1968).[4]

Consequently,

$$\cos \alpha = \frac{C_i^2 - C_{i+1}^2 - C_{i-1}^2}{2 C_{i+1} C_{i-1}} \tag{7}$$

and

$$\cos (\sphericalangle D_i S_i S_{i-1}) = \frac{C_{i+1}^2 - C_i^2 - C_{i-1}^2}{2 C_{i-1} C_i}.$$

Knowledge of the angles at junctions facilitates the construction of an optimal walkway–footpath network in a manner which is similar to that used for the vehicle route. The solution procedure will be illustrated by a numerical example in the next section. Before doing so, the role of the cost and flow parameters in the equilibrium of the junction will be briefly explored.

Utilizing again the concept of S_i moving on an ellipse the foci of which are at S_{i+1} and S_{i-1}, it can be shown that the angle between the footpath direction and

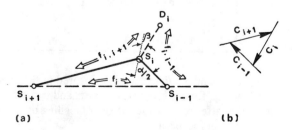

(a)   (b)

Fig. 8. Junction equilibrium.

the normal to the ellipse, β in Fig. 8a, is given by

$$\sin \beta_i = \frac{f_{i,i+1} - f_{i,i-1}}{K_f/k + f_{i,i+1} + f_{i,i-1}} \sin (\alpha_i/2) \qquad (8)$$

Inspection of Eq. (8) leads to the following conclusions:

(a) The angle of shift is (approximately) proportional to the difference between the flows of walkers approaching and leaving the destination D_i from the left and from the right. The footpath should be rotated from the direction of the normal towards the larger flow.

(b) Even if $|f_{i,i+1} - f_{i,i-1}|$ is large, β will not be significant when the walkway angle α is close to π.

By reasoning along lines similar to the proof of Property IV it can be shown that where β is small,

$$\bar{\alpha} \simeq 2 \text{ arc cos } \frac{f_{i,i+1} + f_{i,i-1} + K_f/k}{2f_i + f_{i,i+1} + f_{i,i-1} + 2K_w/k} \ . \qquad (9)$$

In this form the role of the various parameters is somewhat more transparent than in the exact expression given in Eq. (7). Particularly interesting is the double weight of the flow on the trunk line which does not turn onto the feeder (f_i) as compared to the flow on the feeder ($f_{i,i-1} + f_{i,i+1}$). A similar relationship may be observed between the construction and maintenance costs of the trunk and the feeder.

The Solution Procedure—A Numerical Example

Let the annual flow of walkers between the six buildings depicted in Fig. 9 be as given in Table II.

TABLE II

Annual Flow of Walkers (in thousands)

From building	To building					
	1	2	3	4	5	6
1	–	3	3	2	1	1
2	3	–	2	2	1	0
3	3	2	–	1	4	3
4	2	2	1	–	2	3
5	1	1	4	2	–	4
6	1	0	3	3	4	–

Furthermore, let $K_f/k = 3$ and $K_w/k = 8$, both measured in thousands of walkers per annum. The mission is, for the given flows, costs and constellation of buildings to interconnect the latter in the order of their labelling by an optimal central walkway and branching footpaths.

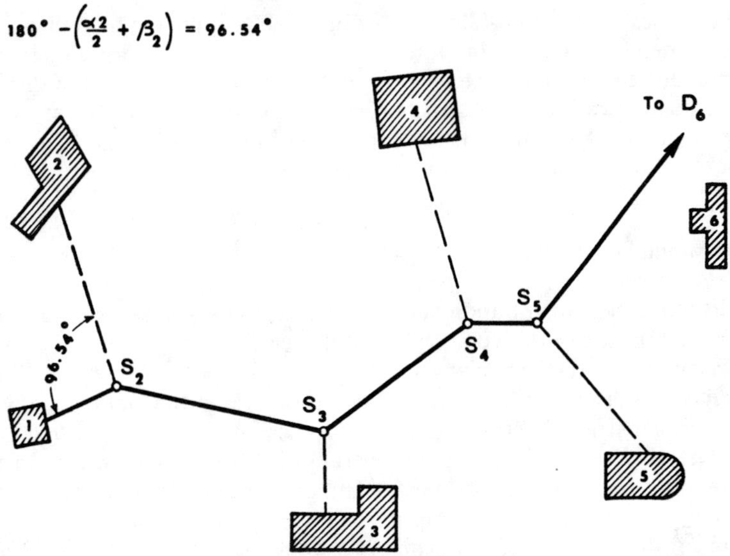

$$180° - \left(\frac{\alpha_2}{2} + \beta_2\right) = 96.54°$$

Fig. 9. Tentative solution for numerical example.

Knowing the order of the junctions, the flows $f_{i,i-1}$, f_i and $f_{i,i+1}$ are easily deduced from Table II and listed in Table III.

TABLE III

Flows Near Junctions (in thousands of walkers per annum)

Junction number	$f_{i,i-1}$	f_i	$f_{i,i+1}$
2	6	14	10
3	10	14	16
4	10	20	10
5	16	14	8

The resulting values of $\bar{\alpha}$ and β are listed in Table IV.

TABLE IV

Network Angles

Junction number	$\bar{\alpha}°$	$\beta°$
2	143.85	11.54
3	131.99	10.89
4	144.77	0.00
5	127.66	15.42

The graphical solution procedure is analogous to that used in the vehicle route case, and therefore does not merit repetition. The only variation being that junctions are located by using angle $(\bar{\alpha}/2 - \beta)$ or $(\bar{\alpha}/2 + \beta)$ instead of $\bar{\alpha}/2$ in the former case. A trial network thus constructed is depicted in Fig. 9. To convert it into an optimal network a slight clockwise rotation of the initial link $D_1 S_2$ is required so as to hit destination 6 by the direction $S_5 D_6$.

Summary and Discussion

Several collection–distribution networks are characterized by a central "spine" and minor branches. For two networks of this class, a vehicle route and a walkway system, the optimal form was sought.

Solution was reached in two stages. First, properties of the optimal network were investigated and described. Then, a graphical method was used to construct a network conforming to the properties of the optimal network. In simple cases only one such network exists. This guarantees its optimality.

There is some merit to the description of the properties of the optimal network in addition to being instrumental to the finding of the solution proper. It leads to insights and generalizations which may provide guidance in design. Also, the description of the properties of the optimal network aids in pinpointing the impact of parameters and renders the workings of the system easy to comprehend.

The iterative nature of the graphical solution procedure provides for clarity and simplicity of the construction. It might therefore be attractive to engineers and planners. Its major deficiency in the present state (or perhaps its strength) lies in the lack of formalized rules to rely on in complex situations. Thus, when a deviation from the basic steps of the solution procedure is called for, the analyst has to rely on intimate knowledge of the properties of the optimal network and on logical reasoning.

Analysis and solution are facilitated by two assumptions. As always, simplifying assumptions raise questions about the applicability of theoretical solutions to real situations. In the case of the vehicle route, the homogeneous travel speed assumption is certainly remote from reality in most instances. Although ways of relaxing this constraint may be found, direct application of the solution in a grid network seems to be inappropriate. The theoretical solution could be used to indicate the vicinity through which the optimal vehicle route should pass. For the walkway–footpath network, the homogeneity assumption is probably less restrictive. In this case, walking and construction costs are likely to be insensitive to the location on the plane. Any variation can be easily accounted for by selecting location specific cost parameters.

The second basic assumption, that of given network topology, is most likely unimportant in many cases of practical interest. Should few alternative topologies appear reasonable, it is possible to construct an optimal network for each and compare their performance.

References

1. E. N. Gilbert and H. O. Pollak, Steiner minimal trees, *SIAM J. Appl. Math.* **16** (1) (1968).
2. C. Werner, Networks of Minimum Length, *Canadian Geographer* **13**, 1 (1969).
3. M. Beckmann, Principles of optimum location for transportation networks, in *Quantitative Geography.* (Garrison and Marble, Eds.), Northwestern University, Studies in Geography No. 13, 1967.
4. C. Werner, The Role of Topology and Geometry in Optimal Network Design, *Papers of the Regional Sci. Assoc.* **21** (1968).

Minimum Cost Paths When the Cost Per Unit Length Depends on Location and Direction

John G. Wardrop

Research Group in Traffic Studies, University College London, London, England

Abstract

This paper considers the problem of finding paths of minimum cost in an area in which the cost per unit length μ depends in a known way on the location and on the direction of movement.

The special case when the cost per unit length depends only on direction, not on location, is considered first. Here $\mu = \mu(\psi)$, where ψ is the angle between the direction of movement and the x axis. The polar diagram corresponding to the distance moved directly at cost c in any direction is given by $r = c/\mu(\psi)$. This may not be convex, in which case it can be extended by drawing in common tangents, which can be reached at cost c by combinations of direct movements. The method of constructing the isocost curves connecting points which can be reached at a given cost from a given origin, and the minimum cost paths where these are unique, is described.

The case where the polar diagram changes at a boundary is considered, and the method of continuing isocost curves and, where appropriate, minimum cost paths beyond the boundary is indicated.

Finally the general case in which μ varies with location as well as direction is discussed. Equations are derived from which the isocost curves from a given origin can be obtained. From these the minimum cost paths, if they are unique, can be derived.

Introduction

A previous paper (Wardrop 1969)[1] considered the problem of finding paths of least cost if the cost associated with a given transport line depended in a known way on location, but was independent of direction.

It was shown that where a minimum cost path crosses a boundary at which the cost per unit length changes discontinuously from μ to μ', that the minimum cost path changes direction in such a way that

$$\mu \sin \phi = \mu' \sin \phi',$$

where ϕ, ϕ' are the angles between the path and the normal to the boundary at the crossing point. In the case where the cost per unit distance μ varies continuously it was shown that the curvature κ of the path at any point is given by

$$\kappa = \frac{1}{\mu} \frac{d\mu}{dn} \sin \phi,$$

References p. 437

where ϕ is the angle between the path and the normal to the curve μ = constant passing through the point, and d/dn denotes differentiation along the normal to the curve μ = constant. Problems of this type were also considered by Angel and Hyman (1970).[2]

In the present paper the condition that μ is independent of the direction of the path is relaxed. The method used is the same as that used by Huygens (1690)[3] to trace the paths of light rays through anisotropic media.

The Value of μ, Cost Per Unit Length

Suppose that μ, the cost per unit length of a path at a point (x, y) in Cartesian coordinates in a direction making an angle ψ with the x axis, is given by the function

$$\mu = \mu(x, y, \psi).$$

We shall begin by considering the special case when μ depends only on ψ.

μ Independent of Location, but Dependent on Direction

In this case we have $\mu = \mu(\psi)$. Starting from an arbitrary point P, let us consider the distance r which we can cover by a direct movement in the direction ψ at a cost of c. Clearly this is given by

$$r = \frac{c}{\mu} = \frac{c}{\mu(\psi)}.$$

We can now construct a polar diagram of r, as shown by the continuous curve in Fig. 1. This will have the property that there is only one value of r corresponding to each value of ψ. The curve will not necessarily be convex, and r may be zero for some values of ψ (corresponding to an infinite cost per unit length, e.g., an insurmountable obstacle of some kind).

We are looking for minimum cost paths. Each path PQ in Fig. 1 has the same cost c and is a straight line. We need to know whether one can reach any point outside the curve $r = c/\mu(\psi)$ at a cost no greater than c.

Consider the effect of combining movements in two directions, one parallel to PQ and of length $p(PQ)$, the other parallel to PR and of length $(1 - p)PR$ (see Fig. 2.). It is easily seen that however the complete movement is made up, whether it follows the path PAS, the path PBS, or a complex path such as $PCDEFS$, the movement will finish at the point S, and that S lies on QR. Since cost is proportional to distance in a particular direction, the cost of reaching S by any of these routes is c. This means that by combining movements parallel to PQ and PR any point on QR between Q and R can be reached at a cost of c. From this result it follows that if the original curve is convex, then alternative paths will not allow additional points to be reached at cost c. On the other hand if the original curve is nonconvex, then some points on a line joining two points

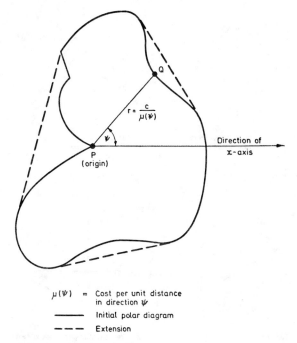

Fig. 1   Polar diagram of distance covered at a given cost *c*.

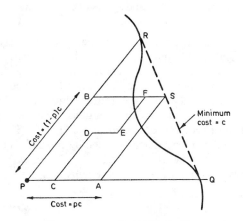

Fig. 2. Extension of initial polar diagram.

of the curve and lying between them will be outside the curve, but can be
reached at cost *c*. Thus the curve of all points which can be reached at cost *c* is
that obtained by extending the original curve by drawing in common tangents as
illustrated by the dashed lines of Fig. 1.

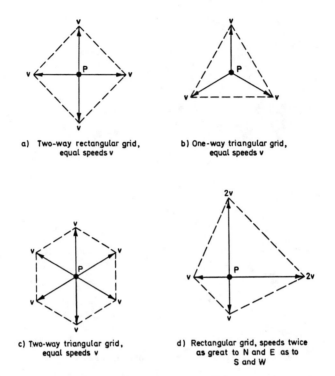

a) Two-way rectangular grid,
equal speeds v

b) One-way triangular grid,
equal speeds v

c) Two-way triangular grid,
equal speeds v

d) Rectangular grid, speeds twice
as great to N and E as to
S and W

Fig. 3. Examples of extended polar diagrams of equal journey time from *P*.

Particular cases of the polar diagram occur when paths are only possible in certain directions. The most familiar example is that of travel costs on a road network. Some cases which might occur are illustrated in Fig. 3, where cost is assumed to be equal to journey time. In Fig. 3(a), for instance, it is assumed that there is a rectangular grid of two-way roads of infinitesimal mesh and that the speed is v in each of the four possible directions. The initial polar diagram then consists of the four vectors from P shown in Fig. 3(a). The extension of the polar diagram is the diamond shape outlined by the dashed curve. Figures 3(b), (c), and (d) illustrate other similar cases of road grids with infinitesimal networks. In the case of finite networks these polar diagrams apply at the intersections only.

Minimum Cost Paths When μ is Independent of Location, but Dependent on Direction

Consider the derivation of minimum cost paths when the cost depends only on direction. Let us start from some given origin at which the cost is zero. There will in general be a variety of minimum cost paths from this origin, and on each one there will be a point which can be reached at cost c from the origin. These

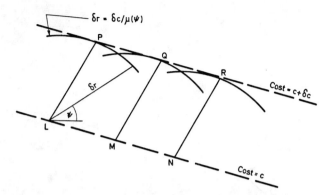

Fig. 4. Construction of isocost curves when μ is independent of location but varies with direction.

points can be joined into a curve of equal cost c from the origin (an isocost curve).

Suppose that *LMN* is such an isocost curve (see Fig. 4), and that the distance *LN* is sufficiently small for *LMN* to be regarded as a straight line. We wish to find the corresponding isocost curve for cost $c + \delta c$. At *L*, *M*, and *N* we construct the polar diagrams given by

$$\delta r = \delta c / \mu(\psi).$$

It is easy to see that points which are at the maximum distance from *LMN* at the given extra cost δc lie on the envelope of the polar diagrams. Since δr is independent of location this will be a straight line parallel to *PQR*, and tangential to the polar diagrams. Let *P* be the point at which this line touches the polar diagram whose center is *L*; we will suppose *P* to be unique. Then the most economical route from *L* is *LP*, and in this case all the most economical routes from *LMN* are parallel to *LP*. Thus for each element of the isocost curve for cost c we can construct the corresponding element of the isocost curve for cost $c + \delta c$.

Minimum Cost Path at a Boundary Where the Polar Diagram Changes

Suppose that we have a straight boundary *AB* (see Fig. 5) separating a region I in which $\mu = \mu_1(\psi)$ from a region II in which $\mu = \mu_2(\psi)$. Let *PQ* be the isocost curve at cost c from some origin. *PQ* is assumed not to be parallel to *AB*, and let *P* be the point at which it meets *AB*. Let *RP* be the corresponding minimum cost path through *P*, again assumed to be unique. Then we can find the isocost curves and minimum cost paths in Region II as follows.

Let *ST* be the isocost curve corresponding to cost $c + \delta c$ and let *QS* be the minimum cost path through *S*. (This defines *Q* as the intersection of the isocost curve "c" and the minimum cost path through *S*.) Suppose that δc is sufficiently small for the curvature of *PQ* to be neglected. Then *RP* and *QS* are

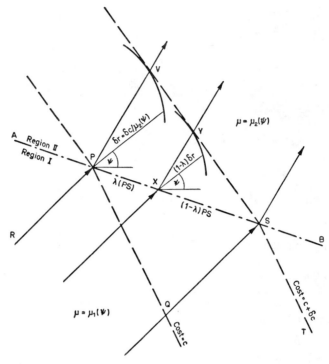

Fig. 5. Construction of isocost curves and minimum cost paths at a boundary where $\mu(\psi)$ changes.

parallel. We construct the polar diagram with center P given by

$$\delta r = \delta c / \mu_2 (\psi)$$

as shown in Fig. 5. Let SV be the tangent from S to the polar diagram, meeting it at V. Consider a parallel minimum cost path intermediate between RP and QS, meeting AB at X where $PX = \lambda(PS)$. It is evident that the cost at the point X is $c + \lambda\delta c$. If we construct a polar diagram with center X given by

$$\delta r = (1 - \lambda) \, \delta c / \mu_2(\psi)$$

this will touch SV at Y, say, the triangle XYS simply being the triangle PVS reduced by the factor $(1 - \lambda)$. It follows that the straight line SV represents the isocost curve corresponding to $c + \delta c$ in Region II, and the straight line PV represents the minimum cost path.

The Case When the Minimum Cost Path is not Unique

So far we have assumed that the minimum cost path is unique, so that, for example, the tangent from S to the polar diagram centered on P in Fig. 5 touches it at a single point. If the polar diagram contains one or more straight sections

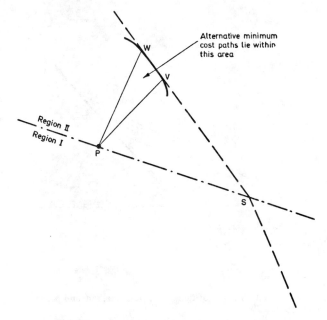

Fig. 6. The construction of isocost curves at a boundary when minimum cost paths are not unique.

(as in Fig. 1) this will not always be the case. We may then have the situation shown in Fig. 6. The isocost curve for $c + \delta c$ is still uniquely determined, but the minimum cost path is not.

The General Case

In the general case we consider an element of the isocost curve corresponding to cost c, PQ (see Fig. 7). We construct the (different) polar diagrams given by

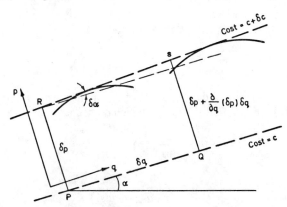

Fig. 7. Isocost curves in the general case.

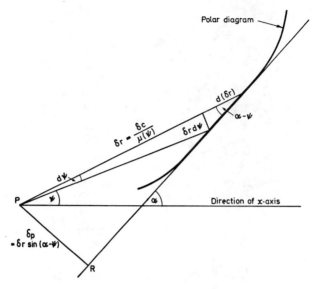

Fig. 8. Relation between direction of tangent to polar diagram and perpendicular distance between isocost curves (*PR*).

$\delta r = \delta c/\mu(\psi)$ centered at P and at Q as shown. The common tangent RS is tangential to the envelope of the polar diagrams and is determined by the perpendicular distances from P and Q, PR and QS.

Given $\mu(\psi)$ at any point, we need to find the relation between the angle between the tangent and the x axis, α (see Fig. 8) and the perpendicular distance δp. This is given by the equations

$$\cot(\alpha - \psi) = \frac{1}{\delta r}\frac{d(\delta r)}{d\psi} = -\frac{1}{\mu}\frac{\partial\mu}{\partial\psi}$$

and

$$\delta p = \delta r \sin(\alpha - \psi)$$

$$= \frac{\delta c}{\mu(\psi)}\sin(\alpha - \psi).$$

These equations give α and δp in terms of ψ, and hence δp in terms of α. It can be seen from Fig. 7 that the rate at which α changes is given by

$$\delta\alpha = \frac{\partial}{\partial q}(\delta p)$$

$$= \delta c\frac{\partial}{\partial q}\left\{\frac{\sin(\alpha - \psi)}{\mu(\psi)}\right\},$$

where q is measured along the isocost curve. Hence

$$\frac{\partial \alpha}{\partial c} = \frac{\partial}{\partial q} \left\{ \frac{\sin(\alpha - \psi)}{\mu(\psi)} \right\}.$$

Applications

The technique outlined here can clearly be used to find isocost curves, and minimum cost paths where appropriate, from any origin, if we are given a functional relation between μ and (x, y, ψ). In practice it will probably be simplest to divide the area into small zones, in each of which the polar diagram is assumed to be independent of location. We can then construct isocost curves corresponding to successive incremental values of c as the envelopes of similar polar diagrams.

References

1. J. G. Wardrop, Minimum cost paths in urban areas, 4th International Symposium on the Theory of Traffic Flow. Karlsruhe, 1968, *Strassenbau und Strassenverkehrstechnik* **86**, 184–90 (1969).
2. S. Angel, and G. M. Hyman, Urban velocity fields, *Environment and Planning* **2** (2), 211–224 (1970).
3. C. Huygens, *Traité de la lumière,* Leyden, 1690 English transl.: *Treatise on light,* by S. P. Thompson, Macmillan, London, 1912.

Urban Density Models

C. Pearce,* Pippa Simpson,[†] and W. Venables[‡]
University of Adelaide, Adelaide, South Australia

Abstract

One of the more promising mathematical techniques for describing and predicting urban traffic flow is the continuum approach of Smeed, Holroyd, and Miller, which is based on supposed known distributions of trip origins and destinations. Most relevant to this approach is a knowledge of urban population density distributions, for which a number of models is extant.

A statistical analysis based on Australian data is made of the relative adequacies of some of these distribution models. Two general empirical laws recently found by Smeed are also examined.

Introduction

In recent years many problems of urban planning and traffic flow in towns have been based on mathematical models linking trip origins and destinations and the routes whereby vehicles flow between them. Most relevant is perhaps the highly congestion-producing commuter traffic in which homes constitute the trip origins. The original gravity and opportunities models involve difficulties in satisfying associated conservation constraints (see Fairthorne[9] and Heggie[10] and an increasing number of investigations have adopted a continuum approach pioneered by Smeed,[19] Holroyd and Miller.[11] In this, unlike the gravity and opportunities models, it is usually takèn that the origins and destinations can be regarded as independent samples from relatively simple prescribed distributions. Models allowing for correlated trip ends have, however, been developed by Vaughan.[24] Ultimately studies employing origin and destination distributions must relate to empirical data obtained for the city of interest. We pursue this point in the following discussion. Elsewhere one of the authors[17] considers the optimal siting of ring roads in a city using some of the ideas developed here.

Though the idea that trip ends in a real city might follow simple mathematical distributions dates to Bleicher (1892),[2] its current widespread use is attributable to Clark[5,6] who found from data in many cities that population falls off roughly exponentially with increasing distance from the city center (usually taken as the center of the central business district).

An alternative model due to Sherratt[18] and Tanner[23] involves a normal den-

*†Mathematics Department.
‡Statistics Department.

sity law

$$f(r) \propto e^{-cr^2}$$

for the population density $f(r)$ a distance r from the city center. The normal law
has been found by Sherratt to give a better fit for Sydney and by Tan[22] to give
a better fit for Adelaide than Clark's

$$f(r) \propto e^{-cr}.$$

More recently, Newling[16] has suggested a composite law

$$f(r) \propto e^{br-cr^2}.$$

Unlike the two previous models which have a peak at the center and a monotonically decreasing density with r increasing, Newling's model can have either a
peak at the center (for $b \leqslant 0$) or a 'central crater' (for $b > 0$). The crater formation can be attributed to the development of a strong business district which
crowds out central residential space as the city grows. The growth of a city is
thus associated with a change from negative to positive values of b.

Also current in the literature are the uniform model

$$f(r) = \text{const.}$$

(Smeed[19]), the inverse model

$$f(r) \propto \frac{1}{r^n}$$

(Horwood,[12] and Smeed[20]), and some other less common models (see
Casetti[4]).

In his recent visit to Australia Professor Reuben Smeed made two interesting
observations on population data for 5 Australian cities (see Table I). He found
that to a close approximation the relation

$$\frac{\text{density at center}}{\text{average density}} = 2.4 \tag{1}$$

TABLE I

	Central density A	$\dfrac{A}{\text{Average density}}$
Sydney	12,000	2.44
Melbourne	11,900	2.48
Brisbane	7,200	2.37
Hobart	7,140	2.48
Toowoomba	6,180	2.25
		average 2.40

held for each city, and also that a plot of log (area) against total population suggested a law

$$\text{area} = aP^\gamma \tag{2}$$

where P is the total urban population. We shall term Eqs. (1) and (2) Smeed's first and second laws, respectively.

The graphical values of the parameter obtained gave a right-hand side for formula Eq. (2) close to

$$\left(\frac{P}{500}\right)^{0.7} \text{(square miles)},$$

through $\gamma = 0.8$ also provided a reasonable fit. An interesting feature of these results is that earlier attempts to find similar results with British data have been unsuccessful. The reason for this appears to be that Australian demographic practice is essentially to define a city's boundaries as occurring where the population density falls to 500 persons per square mile rather than according to somewhat arbitrary administrative areas as in the United Kingdom.

As these results are not in themselves sufficient to fix the form of the population density law in a city (at least without some strong additional assumptions) we do not attempt a proper justification of them here. In the first part of this paper we shall instead consider some of their consequences vis à vis the form of density law.

In the second part of the paper we present some results of a statistical investigation on the relative merits of the various density distributions current in the subject. The districts for which the population data was available were districts used in transportation studies, constituting a fairly detailed partitioning of the cities concerned. The data pertains to the four Australian cities Melbourne, Adelaide, Perth, and Launceston. Some typical results for Adelaide and Melbourne are shown in Tables II–VIII. The fits for Launceston and Perth are rather better, which we associate with the smaller number of districts (79 and 53, respectively) in those cities, as commented on later.

Dr. A. J. Miller has recently drawn the authors' attention to papers by Ajo[1] and Casetti.[4] Ajo suggests that different density models may be appropriate at varying distances from the city center. This hypothesis and that of the existence of a central crater are considered by Casetti, who gives strong support to the former and weak support to the latter concept.

Consequences of Smeed's Empirical Laws

We represent a real city by a fictitious circular city of radius R miles. The natural value to take for R is that given by

$$\text{actual city area} = \pi R^2.$$

For convenience we normalize f through

$$f(r) = AF(r),$$

where $F(0) = 1$, and write

$$G(x) = F(x^{\frac{1}{2}}).$$

With any particular city the normalized $F(G)$ is then determined by a parameter c (possibly vector valued). A can then be interpreted as the central density.

We adopt the convention of not counting the normalizing constant (A) when referring to the number of independent scalar parameters in a density model.

Relation (1) gives

$$\frac{\displaystyle\int_0^{R^2} G_c(s)\,ds}{R^2} = \frac{1}{2.4}. \tag{3}$$

This equation will usually be satisfied by only a finite number of values of R for a given function G_c. In particular there will be exactly one solution if G_c is a monotone decreasing function of its argument. Thus for a given form of G, R will normally be determined by c.

The Australian demographic convention for defining a city suggests the side condition

$$AG_c(R^2) = 500 \tag{4}$$

to express that $r = R$ corresponds to the periphery of the city. With Eq. (4) we have that A is also fixed by c and hence by Eq. (1) the total population of the city is also fixed by c. If c is a scalar parameter, relation (1) can be regarded as specifying (for a given class G) the city area and total population once the central density is known.

Failure of this to hold for real cities would suggest that c involves at least two independent scalar components. Whether or not the central density does actually determine the total population and area seems somewhat unclear.

We observe incidentally that if Eq. (2) is taken into account, we can demonstrate that A determines P and the city area without use of Eq. (4) or comment on the character of c. For by simple elimination between Eqs. (1) and (2) we have

$$P^{1-\gamma} = \frac{a}{2.4} A$$

and

$$\text{urban area} = \left(\frac{A}{2.4}\right)^{\gamma/(1-\gamma)} a^{1/(1-\gamma)}.$$

We shall see in the subsequent statistical analysis that best fit values of γ do vary somewhat from city to city, so that this determination of P and the city area by A will be rough.

Equations (3) and (4) also provide for some interesting degeneracies. If the left side of Eq. (3) is a function of $G_c(R^2)$ alone, then by Eq. (4) A has a fixed value independent of c and R. This situation arises with both Sherratt's model and Clark's model.

Thus for Sherratt's model, taking the density at distance r from the center as $\propto e^{-cr^2}$, we have from Eq. (3) that

$$\frac{1 - e^{-cR^2}}{cR^2} = \frac{1}{2.4}.$$

This can be solved for cR^2, and so A is now fixed, as Eq. (4) reads

$$Ae^{-cR^2} = 500.$$

Similar working applies to Clark's model. To foreshadow our subsequent discussion, this difficulty provides some additional evidence in favor of the two-parameter exponential models. It is also perhaps worthy of note that when c is a 2 component vector, knowledge of the central density A strongly delimits the possible density laws, effectively reducing them to a set with a single scalar parameter.

We now turn our attention to Smeed's second empirical law, which we express in the form

$$\pi R^2 = a\left[\int_0^{R^2} G(s)A\pi ds\right]^\gamma \tag{5}$$

This can be rewritten as

$$\frac{\int_0^{R^2} G(s)ds}{R^2} = \frac{R^2(1/\gamma - 1)}{(\pi a)^{1/\gamma}A}$$

so that on comparison with Eq. (3)

$$AR^{2(1-1/\gamma)} = \frac{2.4}{(\pi a)^{1/\gamma}}.$$

From Eq. (4) we deduce that

$$F(R) = G(R^2) = \frac{500(\pi a)^{1/\gamma}}{2.4} R^{2(1-1/\gamma)}.$$

This suggests the density law

$$f(r) \propto r^{2(1-1/\gamma)}.$$

For γ of the order 0.7, 0.8, the exponent is negative and of order $-0.5, -0.8$. Such a law would formally give an infinite density at the city center. However the population in any finite area containing the center is finite provided $\gamma \geqslant \frac{2}{3}$, so that there is no necessary conflict with real demographic data. The power law $f(r) \propto r^{-1}$, i.e., $\gamma = \frac{2}{3}$, has been found by Smeed[21] to give a reasonable fit to

London data. The form of f would, of course, need to be modified near the city center to justify our use of (1) and the normalization of f.

Suppose that the power law is perturbed only very near the city center, so that the total city population can still be calculated by integrating the unperturbed power density over the entire area of the city. If we write

$$f(r) = Br^{2(1-1/\gamma)}$$

for finite r, Eq. (2) still holds and analogously to Eq. (4) we have

$$BR^{2(1-1/\gamma)} = 500.$$

Thus two relations subsist between γ, R, B, and P. Knowledge of P and the city area then prescribes the density function.

If 500 persons per square mile represents an arbitrary level at which the city periphery is demarcated rather than a level at which an abrupt physical discontinuity in population occurs, the same power law can be derived rigorously from the boundary condition and Eq. (2) under the assumption only that a and γ do not depend on the cutoff density (so far taken as 500 persons per square mile).

We vary the cutoff value ρ and consider the induced variation in R.

We write Eq. (2) generally as

$$\pi R^2 = aP^\gamma \tag{6}$$

and the side condition as

$$f_c(R) = \rho. \tag{7}$$

The population P is given as

$$P = \int_0^R 2\pi r f_c(r)dr. \tag{8}$$

Equations (7) and (8) can be regarded as defining

$$P = P(\rho, c)$$
$$R = R(\rho, c)$$

as functions of ρ and c.

These equations implicitly also define P as $P(r, c)$ and we have

$$\frac{\partial P}{\partial \rho} = -\left(\frac{\partial P}{\partial R}\right)_c \frac{\partial R}{\partial \rho}.$$

Logarithmic differentiation of Eq. (6) with respect to ρ now yields

$$\frac{2}{R}\frac{\partial R}{\partial r} = -\frac{\gamma}{P} 2\pi R f_c(R) \frac{\partial R}{\partial \rho}$$

so that

$$f_c(R) = -\frac{P}{\gamma}\frac{1}{\pi R^2}$$

and

$$\frac{\partial}{\partial R} f_c(R) = -\frac{2}{\gamma} f_c(R)(1-\gamma)/R.$$

Hence

$$f_c(R) = \text{function of } c \times R^{2(1-1/\gamma)}$$

or, since R is being regarded as a variable,

$$f_c(r) = \text{function of } c \times r^{2(1-1/\gamma)}$$

as desired.

Independently of the above laws, we expect the density law to be expressible in the form

$$f(r) = \frac{P}{\text{area}} \cdot h\left[\frac{r}{(\text{area})^{1/2}}\right],$$

for some function h.

This is easily derived from dimensional arguments as follows:

The density $f(r)$ at a point a distance r from the city center can quite generally be expected to depend on P, the town area, and r. It would therefore be expected to depend on a number of terms of the form

$$P^i(\text{area})^j r^k$$

Since f has dimensions P^{+1} length^{-2}, we have

$$i = 1$$

$$2j + k = -2.$$

Hence

$$f = \frac{P}{\text{area}} \sum_{\text{terms}} \left[\frac{r}{(\text{area})^{1/2}}\right]^j$$

or

$$f = \frac{P}{\text{area}} h\left(\frac{r}{(\text{area})^{1/2}}\right).$$

Statistical Analysis

The basic data for each city consisted of the populations and areas of a large number of districts and the locations of their centroids. The distance of each

centroid from the center of the central business district was derived, and that
distance taken as representing the distance of the district from the city center.
It can be seen that only second order effects are introduced in the model by
neglecting the variation of distance from the center over the district. The best-
fit parameters for several density models were then calculated on an unweighted
least squares basis and a statistical t-test taken of their significance.

The 95% level of significance was adopted though many of the parameter estimates were significant to the 99.99% level or better. Consequently standard
errors of the significant parameters tend to be small, frequently being only a
quarter of the parameter value or less.

In Tables II–VIII significant estimates are marked with an asterisk. The sub-
model considered in each line is given by the b labellings on the column for
which that line has entries.

The problem of discriminating between hypotheses which relate to different
functional forms rather than merely to different values of a parameter with a
common functional form is quite a difficult one. The most complete discussion
is given by Cox.[7,8] The simplest approach for the present problem, that of find-

TABLE II

Melbourne

Model: $\log Y = b_0 + b_1 r + b_2 r^2 + b_3 r^3 + \epsilon.$[a]

	b_0	b_1	b_2	b_3	ρ^2
Original	1.93*	0.23	−0.042*	0.0013*	0.22
(125 districts)	(0.46)	(0.18)	(0.019)	(0.0006)	
Grouped	2.42*	0.14	−0.42*	0.0014*	0.44
(70 districts)	(0.49)	(0.18)	(0.019)	(0.0006)	
Center shifted	2.56*	0.07	−0.036	0.0014**	0.27
(125 districts)	(0.56)	(0.22)	(0.024)	(0.0008)	

[a] Figures in brackets denote standard errors.
** Significant at 90% level.

TABLE III

Model: $\log Y = b_0 + b_1 r + b_2 r^2 + b_3 r^3 + \epsilon.$

	No. districts	ρ^2
Melbourne	125	0.22
Launceston	79	0.37
Adelaide	75	0.44
Melbourne grouped	70	0.44
Perth	53	0.64
Adelaide grouped	38	0.64

TABLE IV

Melbourne (125 Districts)
Full Model: $\log Y = b_0 + b_1 r + b_2 r^2 + b_3 r^3 + \epsilon$.

b_0	b_1	b_2	b_3	ρ^2
2.59*	-0.12			0.18
2.10*		-0.0053*		0.16
1.90*			0.00022*	0.11
1.93*	0.23	-0.042*	0.0013*	0.22
2.69*	-0.15*	-0.0015		0.18
2.47*		-0.019	0.00066*	0.21
2.74*	-0.16*		0.00009	0.19

TABLE V

Adelaide (75 Districts)
Full Model: $\log Y = b_0 + b_1 r + b_2 r^2 + b_3 r^3 + \epsilon$.

b_0	b_1	b_2	b_3	ρ^2
2.73*	-0.25*			0.41
1.81*		-0.012*		0.35
1.50*			-0.0006*	0.27
2.35*	0.047	-0.05	0.002	0.44
3.22*	-0.40*	0.008		0.42
2.45*		-0.044*	0.0019*	0.44
3.16*	-0.35*		0.00035	0.42

TABLE VI

Melbourne (125 Districts)
Full Model: $Y = b_0 + b_1 r + b_2 r^2 + b_3 r^3 + \epsilon$.

b_0	b_1	b_2	b_3	ρ^2
16.1*	-0.86*			0.35
12.4*		-0.034*		0.25
11.1*			-0.001*	0.17
17.4*	-0.85	-0.06	0.0030	0.39
19.1*	-1.7*	0.04*		0.38
15.4*		-0.14*	0.005*	0.38
18.5*	-1.4*		0.001*	0.38

TABLE VII

Adelaide (75 Districts)

Full Model: $Y = b_0 + b_1 r + b_2 r^2 + b_3 r^3 + \epsilon.$

b_0	b_1	b_2	b_3	ρ^2
10.4*	−0.67*			0.44
7.9*		−0.03*		0.34
7.1*			−0.0016*	0.25
11.2*	−0.56	−0.069	0.0039	0.48
12.8*	−1.40	0.040		0.47
9.96*		−0.14*	0.006*	0.47
12.3*	−1.11*		−0.0015*	0.47

TABLE VIII

Melbourne (125 Districts)

Full Model: $\log Y = b_0 + b_1 \log r + b_2 (\log r)^2 + b_3 (\log r)^3 + \epsilon.$

b_0	b_1	b_2	b_3	ρ^2
2.89*	−0.70*			0.13
2.59*		−0.25*		0.18
2.38*			−0.084*	0.19
1.80*	1.48	−0.92	0.080	0.21
1.87*	1.06*	−0.56*		0.20
2.25*		0.14	−0.13	0.19
2.04*	0.34		−0.11*	0.20

ing which of two urban density models gives a better fit to actual data, is to work in terms of a composite model containing the two models of interest as special cases. Thus for the Clark and Sherratt models we consider the composite model

$$f(r) = e^{c_0 + c_1 r + c_2 r^2},$$

i.e., the form proposed by Newling. The c's are estimated by least squares.

Suppose we have N regions with centroids at distances $r_i, i = 1 \ldots N$, from the city center and corresponding average densities Y_i.

On the assumption that

$$\log Y_i = c_0 + c_1 r_i + c_2 r_i^2 + \epsilon_i,$$

where ϵ_i is a normally distributed error term, we can ascertain which of c_0, c_1, c_2 are significant. A plot of points will indicate whether the assumption regarding ϵ is reasonable. If c_1, say, were significant and c_2 nonsignificant, we would conclude that the Clark model gives the better fit.

The same method could be used to compare the Clark model and the power law. The Clark model would regress $\log Y$ on r and the power law $\log Y$ on log *r*. For moderate values of *r*, log *r* is much less sensitive than is *r* to changes Δr, so that *ceteris paribus* the power law could be expected to supply a bette₁ fit by virtue of this variation damping effect. Such a gain would, however, be compensated for by a corresponding loss whenever the power model was used for prediction purposes, so that the superior fit would be somewhat spurious. For this reason, this and some other direct comparisons were not made.

In each table in the analysis the proportion of variation between districts accounted for by the model as measured by ρ^2 is given. The extended power law appears comparable with the exponential models on this basis, but in view of the accuracy loss involved in prediction and a heteroscedasticity which appeared in the computer plot, evidence would seem to favor the exponential model.

Summary of Results

The complete models examined were:
(a) the generalized power law,

$$\log Y = \sum_{j=0}^{3} b_j (\log r)^j + \text{.}$$

(b) the polynominal law,

$$Y = \sum_{j=0}^{3} b_j r^j + \epsilon;$$

(c) the exponential law,

$$\log Y = \sum_{j=0}^{3} b_j r^j + \epsilon.$$

The data plots displayed nonheteroscedasticity for all except the exponential model.

The basic complete exponential model was checked by (a) lumping adjacent districts, and (b) (in the case of Melbourne) shifting the point taken on the city center to minimize the error sum of squares. The shift was 2.8 miles. As can be seen from Table II these changes did not affect the significance patterns or parameter values, although grouping did improve the ρ^2 value appreciably, as might be expected. It was also found that in a comparison of the same model between different cities the ρ^2 value increased with a decreasing number of districts (see Table III).

In a study of the goodness of fit of the Clark model for American cities, Muth[15] noted a very large range in ρ^2 values. Our results suggest this was probably in large part a reflection of the variability of the number of districts between one city and another rather than of differences in the adequacy of the model.

Likewise the relatively high values of ρ^2 given by Casetti[4] in his statistical analyses seems attributable to his use, as basic data, of population densities in a comparatively small number of concentric annuli about city centers.

Exponential Law Regression

An analysis was carried out with the model

$$f(r) = \exp\left[\sum_{j=0}^{6} b_j r^j + \epsilon\right]$$

and it was found that the quartic and higher coefficients were not significant. In all cases the linear (Clark) model gave the best one-parameter fit, followed by the quadratic (Sherratt) and the cubic. The two-parameter models were all better than the Clark model, the best being clearly the quadratic-cubic

$$f(r) \propto e^{b_2 r^2 + b_3 r^3}.$$

The quadratic-cubic model is in fact as good as the full three-parameter model. We find b_2 is negative and b_3 positive. Neither Newling (quadratic) nor cubic additions individually gave any significant improvement on the linear model. The values of the regression coefficients obtained gave no evidence of a "central crater" effect.

Polynomial Regression

An unexpected result was that a cubic polynomial fitted the data rather better than the exponential model for Melbourne and Adelaide although less well for Perth and Launceston.

Significance patterns consistently found the linear polynomial to give the best one parameter fit, and a significant contribution arising from an added quadratic or cubic form. The quadratic plus cubic form was generally less good. The three parameter model gives no significant improvement over the two parameter model.

A closer look at the possible use of quadratic models in practical calculations is warranted.

Extended Power Law Regression

The full model was

$$\log Y = b_0 + b_1 \log r + b_2 (\log r)^2 + b_3 (\log r)^3 + \epsilon.$$

The significance patterns were somewhat unclear. All three one parameter models compared reasonably with the exponential models, although as suggested earlier this result is meretricious. There seems some evidence that the linear-quadratic models, i.e.,

$$f(r) \propto r^{b_1 + b_2 \log r},$$

is worthwhile. The regression coefficients indicate that if a power law is used the exponent will vary from one city to another.

General Remarks

In view of the great complexity and diversity of cities it is perhaps surprising that any simple law can account in large part for the variation in density within a city. The consistency of the signs and magnitudes of the regression coefficients for different cities, and especially the significance patterns for the exponential laws, gives evidence of meaningful fits. With the exponential models we find such features as an r^2 term being highly significant on its own but not when compounded with an r term. This arises from the nonorthogonality of the variates r, r^2, r^3 against which we are regressing the density data and tends to obscure the structure of the density laws. The use of orthogonal polynomials is unlikely to resolve the issue as the coefficients in such polynomials are not fixed but dependent on the particular data. Further heuristic models along the lines of those of Miller[14] and Casetti[3] would be of great value.

The reason for the multiplicity of candidates to describe population densities is that over the range of distances from which most of the data derives, these density functions are sensibly much the same function—at least in the neighborhood of the best-fit values of the parameters involved. Divergences occur only near the city center and in the tail of the distribution. This notion can be sharpened through the concept of invariants of distribution functions (see Jeffreys[13]). Jeffreys introduces the invariants

$$I_m(F_1, F_2) = \int \left| (dF_1)^{1/m} - (dF_2)^{1/m} \right|^m,$$

and

$$J(F_1, F_2) = \int \log \frac{dF_1}{dF_2} d(F_1 - F_2)$$

(defined over suitable domains) as measures of the discrepancies between two probability distribution functions F_1, F_2.

The models discussed in this paper and prevalent in the literature have the common feature of radial symmetry. Vaughan[24] has proposed a generalization of the Sherratt model which has instead of reflection symmetry about a pair of suitably chosen axes, OX, OY, viz.,

$$f(x, y) \propto e^{-x^2/a^2 - y^2/b^2}.$$

It seems likely that this three parameter model may give better fits than the standard models simply by virtue of the extra parameters accommodating the departures from circularity of a real city.

Acknowledgments

The authors wish to thank Professor Reuben Smeed for making his empirical results available to them and Dr. A. J. Miller for bringing their attention to two of the references. Thanks are due also for helpful referee's comments.

References

1. R. Ajo, On the structure of population in London's field, *Acta Geographica* 18, 1–17 (1965).
2. H. Bleicher, *Statistische Beschreiburg der Stadt Frankfurt am Main und ihrer Bevolkerung,* Frankfurt am main, 1892.
3. E. Casetti, Urban population density patterns: An alternate explanation, *Canadian Geographer* 11, 96–100 (1967).
4. E. Casetti, Urban population density models, in *Studies in Regional Science* (A. J. Scott, Ed.), pp. 105–116, Pion, London, 1970.
5. C. Clark, Urban population densities, *J. Roy Statist. Soc. A.* 114, 490–496 (1951).
6. C. Clark, *Population Growth and Land Use,* Macmillan, London, 1968.
7. D. R. Cox, Tests of separate families of hypotheses, *Fourth Berkeley Symposium on Mathematical Statistics and Probability* (J. Neyman, Ed.), pp. 105–123, (University of California), 1961.
8. D. R. Cox, Further results on tests of separate families of hypotheses, *J. Roy. Statist. Soc. B* 24, 406–424 (1962).
9. D. B. Fairthorne, Description and shortcomings of some urban road traffic models, *Op. Res. Quart.* 15, 17–28 (1964).
10. I. G. Heggie, Are gravity and interactance models a valid technique for planning regional transport facilities? *Op. Res. Quart.* 20, 93–110 (1969).
11. E. M. Holroyd and A. J. Miller, Route crossings in urban areas, *Proc. 3rd Conference A.R.R.B.,* Vol. 3 (1), pp. 394–419, Aust. Road Res. Board, Melbourne, 1966.
12. E. M. Horwood, A three-dimensional calculus model of urban settlement, *Highway Res. Board Bull.* No. 347, 143–146 (1962).
13. H. Jeffreys, *Theory of Probability,* Oxford, 1948.
14. A. J. Miller, The intervening-opportunities model applied to land use in a uniform city, Transportation Research 4, 145–149 (1970).
15. R. F. Muth, The spatial structure of the housing market, *Papers and Proc. of the Regional Sci. Assoc.* 7, 207–220 (1961).
16. B. E. Newling, The spatial variation of urban population densities, *Geographical Rev.* 59, 242–252 (1959).
17. C. Pearce, On the optimal positioning of concentric ring roads in a city (in preparation).
18. G. C. Sherratt, A model for general urban growth, in Vol. 2 of *Management Sciences, Models and Techniques:* Proc. of 6th Int. Meeting of the Inst. (of Management Sciences), (C. W. Churchman and M. Verhulst, Eds), pp., 147–159, New York, 1960.
19. R. J. Smeed, The traffic problem in towns, Manchester Statistical Society, 8th Feb. 1961, Manchester.
20. R. J. Smeed, Road development in urban areas, *J. Inst. High Eng.* 10(1), 5–30 (1963).
21. R. J. Smeed, Traffic intensity in cities, Proc. Tewksbury Symposium 1970, University of Melbourne 4.1–4.75.

22. T. Tan, *Mathematical Model for Commuter Traffic in Cities,* Ph.D. Thesis, University of Adelaide, 1967.
23. J. C. Tanner, Factors affecting the amount of travel, Road Res. Tech. Paper no. 51, Dept. of Scientific and Industrial Research, London, 1961.
24. R. J. Vaughan, *Urban Road Traffic Patterns,* Ph.D. Thesis, University of Adelaide, 1970.

Sources of Data

Basic Data for Transport Planning, Metropolitan Adelaide Transportation Study; Launceston Area Transportation Study; Melbourne Metropolitan Transportation Study; Perth Transportation Study.